THIRTY-FIRST EDITION

Alan J. Wright

Introduction

The 'G' prefixed four letter registration system was adopted in 1919 after a short-lived spell with serial numbers beginning at K–100. Until 1928 the UK allocations were in the G–Exxx range, but as a result of further International agreements, this series was ended at G–EBZZ, the replacement becoming G–Axxx. From this point the registrations were issued in a reasonably orderly manner through to G–AZZZ. To avoid possible confusion with signal codes the G–AQxx sequence was omitted, while G–AUxx was reserved for Australian use originally. However, individual requests for marks in the latter batch have been granted by the Authorities. In July 1972 the next logical sequence was commenced at G–BAAA.

Unfortunately in recent years the strictly applied rules relating to aircraft registration have largely been relaxed, allowing widespread use of personalised marks upon application. Re-registrations have also become quite a common feature, a practice virtually unheard of in the past. Some aircraft have also been allowed to wear military markings without displaying their civil identity. In this volume the serial number actually carried in such cases is shown in parenthesis after the type's name. For example Gladiator G–AMRK flies as L8032 in RAF colours. As a further aid a military conversion list is provided.

Where re-registration has taken place at some time, the previous UK civil identity appears in parenthesis after the owner's/operator's name. An example of this practice is One-Eleven G–BBMG of British Airways which originally carried G–AWEJ.

Preserved aircraft which either fly infrequently or are purely static exhibits are shown with an asterisk in front of the type.

Any new registrations issued by the Civil Aviation Authority after this publication went to press will inevitably not be included until the next edition. To aid the recording of later in-sequence marks logged during the year, a grid has been provided at the end of the book.

Acknowledgements

Once again thanks are extended to the Registration Department of the Civil Aviation Authority for continuing to allow regular researches into their files. The useful suggestions and information supplied by H. W. Gandy, D. Lewis, P. R. March, Stansted Aviation Society and the PR staffs of numerous airlines was gratefully received, while the valuable assistance of A. S. Wright and C. P. Wright must also be recorded.

AJW

Cover: Boeing 747-227B of Braniff International./*Boeing*

This edition published 1981

ISBN 0 7110 1104 4

Published by Ian Allan Ltd, Shepperton, Surrey, and printed in the United Kingdom by Jarrold and Sons Ltd, Norwich

International Civil Aircraft Markings

A2– Botswana
A6– United Arab Emirates
A7– Qatar
A9C– Bahrain
A40– Oman
AN– Nicaragua
AP– Pakistan
B– China/Taiwan
C– Canada
C2– Nauru
C5– Gambia
C6– Bahamas
C9– Mozambique
CC– Chile
CCCP–* Soviet Union
CF– Canada
CN– Morocco
CP– Bolivia
CR– Portuguese Overseas Provinces
CS– Portugal
CU– Cuba
CX– Uruguay
D– German Federal Republic (West)
D2– Angola
D6 Comores Islands
DM– German Democratic Republic (East)
DQ– Fiji
EC– Spain
EI, EJ– Republic of Ireland
EL– Liberia
EP– Iran
ET– Ethiopia
F– France, Colonies and Protectorates
G– United Kingdom
H4– Solomon Islands
HA– Hungarian People's Republic
HB– Switzerland and Liechtenstein
HC– Ecuador
HH– Haiti
HI– Dominican Republic
HK– Colombia
HL– Korea (South)
HMAY– Mongolia
HP– Panama
HR– Honduras
HS– Thailand
HZ– Saudi Arabia
I– Italy
J5– Guinea Bissau
JA– Japan
JY– Jordan
LN– Norway
LQ–, LV– Argentine Republic
LX– Luxemburg
LZ– Bulgaria
N– United States of America
OB– Peru
OD– Lebanon
OE– Austria
OH– Finland
OK– Czechoslovakia
OO– Belgium
OY– Denmark
P– Korea (North)
P2– Papua New Guinea
PH– Netherlands
PJ– Netherlands Antilles
PK– Indonesia and West Irian
PP–, PT– Brazil
PZ– Surinam
RDPL– Laos
RP– Philippine Republic
S2– Bangladesh
S7– Seychelles
S9– São Tomé
SE– Sweden
SP– Poland
ST– Sudan
SU– Egypt
SX– Greece
TC– Turkey
TF– Iceland
TG– Guatemala
TI– Costa Rica
TJ– United Republic of Cameroon
TL– Central African Republic
TN– Republic of Congo (Brazzaville)
TR– Gabon
TS– Tunisia
TT– Chad
TU– Ivory Coast
TY– Benin
TZ– Mali
VH– Australia
VP–F Falkland Islands
VP–H Belize
VP–LAA/LJZ Antigua
VP–LKA/LLZ St. Kitts-Nevis
VP–LMA/LUZ Montserrat
VP–LVA/LZZ Virgin Islands
VP–P Kiribati
VP–V St. Vincent
VP–W, VP–Y Zimbabwe/Rhodesia
VQ–G Grenada
VQ–H St. Helena
VQ–L St. Lucia
VR–B Bermuda
VR–C Cayman Islands
VR–G Gibraltar (not used: present Gibraltar Airways aircraft registered G–)
VR–H Hong Kong
VR–U Brunei
VT– India
XA–, XB–, XC– Mexico
XT– Upper Volta
XU– Kampuchea (formerly Khmer Republic)
XV– Vietnam
XY–, XZ– Burma
YA– Afghanistan
YI– Iraq
YK– Syria
YR– Romania
YS– El Salvador
YU– Yugoslavia
YV– Venezuela
ZA– Albania
ZK–, ZL–, ZM– New Zealand
ZP– Paraguay
ZS–, ZT–, ZU– South Africa
3A– Monaco
3B– Mauritius
3C– Equatorial Guinea
3D– Swaziland
3X– Guinea
4R– Sri Lanka
4W– Yemen Arab Republic

* Cyrillic letters for SSSR.

4X– Israel
5A– Libya
5B– Cyprus
5H– Tanzania
5N– Nigeria
5R– Malagasy Republic (Madagascar)
5T– Mauritania
5U– Niger
5V– Togo
5W– Western Samoa (Polynesia)
5X– Uganda
5Y– Kenya
6O– Somalia
6V–, 6W– Senegal
6Y– Jamaica
7O– Democratic Yemen
7P– Lesotho
7Q– Malawi
7T– Algeria
8P– Barbados
8Q– Maldives
8R– Guyana
9G– Ghana
9H– Malta
9J Zambia
9K– Kuwait
9L– Sierra Leone
9M– Malaysia
9N– Nepal
9Q– Zaïre
9U– Burundi
9V– Singapore
9XR– Ruanda
9Y– Trinidad and Tobago

Aircraft Type Designations

(e.g. PA–28 Piper Type 28)

A. Beagle, Auster
AA– American Aviation, Grumman American
AB Agusta-Bell
AS Aerospatiale
A.S. Airspeed
A.W. Armstrong Whitworth
B. Blackburn, Bristol, Boeing, Beagle
BAC British Aircraft Corporation
BN Britten-Norman
Bo Bolkow
Bu Bucker
C.H. Chrislea
CLA Comper
CP. Piel
D Druine
DC– Douglas Commercial
D.H. de Havilland
D.H.C. de Havilland Canada
DR. Jodel (Robin-built)
EP Edgar Percival
F. Fairchild, Fokker
G. Grumman
GA Gulfstream American
G.A.L. General Aircraft
G.C. Globe
GY Gardan
H Helio
HM. Henri Mignet
HP. Handley Page
HR. Robin
H.S. Hawker Siddeley
IL Ilyushin
J. Auster
L. Lockheed
L.A. Luton
M. Miles, Mooney
MBB Messerschmitt-Bölkow-Blohm
M.S. Morane-Saulnier
P. Hunting (formerly Percival), Piaggio
PA– Piper
PC. Pilatus
R. Rockwell
S. Short, Sikorsky
SA., SE, SO. Sud-Aviation, Aérospatiale, Scottish Aviation
S.R. Saunders-Roe, Stinson
ST SOCATA
T. Tipsy
Tu Tupolev
UH. United Helicopters (Hiller)
V. Vickers- Armstrongs, BAC
V.S. Vickers-Supermarine
W. S. Westland

BRITISH CIVIL AIRCRAFT REGISTRATIONS

Reg.	*Type*	*Owner or Operator*	*Notes*
G–EACN	*BAT BK23 Bantam (K123)	Shuttleworth Trust	
G–EBHX	*D.H.53 Humming Bird	Shuttleworth Trust	
G–EBIA	*S.E.5A (F904)	Shuttleworth Trust	
G–EBIB	*S.E.5A (F939)	Science Museum	
G–EBIC	*S.E.5A (F938)	RAF Museum	
G–EBIR	*D.H.51	Shuttleworth Trust *Miss Kenya*	
G–EBJG	*Parnall Pixie III	Midland Aircraft Preservation Soc.	
G–EBJO	*ANEC II	Shuttleworth Trust	
G–EBKN	*Avro 504K (E449)	RAF Museum	
G–EBKY	*Sopwith Pup (N5180)	Shuttleworth Trust	
G–EBLV	D.H.60 Cirrus Moth	Hawker Siddeley Aviation Ltd.	
G–EBMB	*Hawker Cygnet I	RAF Museum	
G–EBQP	*D.H.53 Humming Bird	Russavia Collection	
G–EBWD	D.H.60X Hermes Moth	Shuttleworth Trust	
G–EBYY	*Cierva C.8L	Musée de l'Air, Paris	
G–EBZM	*Avro 594 Avian IIIA	Torbay Aircraft Museum	
G–AAAH	*D.H.60G Gipsy Moth	Science Museum	
G–AACN	*H.P.39 Gugnunc	Science Museum	
G–AAHW	Klemm L.25-1A	Anvil Aviation Ltd.	
G–AAIN	*Parnall Elf II	Shuttleworth Trust	
G–AAMY	D.H.60M Moth	C. C. & Mrs. J. M. Lovell	
G–AAPZ	*Desoutter I (mod.)	Shuttleworth Trust	
G–AARO	Arrow A2-60 Sport	P. A. Mann	
G–AAUP	Klemm L.25-1A	R. S. Russell	
G–AAWO	D.H.60G Gipsy Moth	J. F. W. Reid	
G–AAYX	*Southern Martlet	Shuttleworth Trust	
G–AAZP	D.H.80A Puss Moth	G. R. Mollison	
G–ABAA	*Avro 504K (H2311)	RAF Museum	
G–ABAG	D.H.60G Moth	Shuttleworth Trust	
G–ABDW	D.H.80A Puss Moth	Strathallan Aircraft Collection	
G–ABEE	*Avro 594 Avian IVM (Sports)	Aeroplane Collection Ltd.	
G–ABEV	D.H.60G Moth	R. I. & Mrs. J. O. Souch	
G–ABEW	D.H.60G-111 Moth Major	B. P. Gardner	
G–ABLM	*Cierva C.24	Mosquito Aircraft Museum	
G–ABLS	D.H.80A Puss Moth	R. C. F. Bailey	
G–ABMR	*Hart 2 (J9941)	Hawker Siddeley Aviation Ltd.	
G–ABNT	Civilian Coupe	Shipping & Airlines Ltd.	
G–ABNX	Redwing 2	J. Pothecary	
G–ABOI	Wheeler Slymph	A. H. Wheeler	
G–ABTC	CLA.7 Swift	P. Channon	
G–ABUS	CLA.7 Swift	R. C. F. Bailey	
G–ABUU	CLA.7 Swift	J. Pothecary	
G–ABVE	Arrow Active 2	L. J. Benjamin	
G–ABWP	Spartan Arrow	R. E. Blain	
G–ABXL	*Granger Archaeopteryx	Shuttleworth Trust	
G–ABYA	D.H.60G Gipsy Moth	Dr. I. D. C. Hay & J. F. Moore	
G–ABZB	D.H.60G-111 Moth	R. E. & B. A. Ogden	
G–ACCB	*D.H.83 Fox Moth	Midland Aircraft Preservation Soc.	
G–ACDC	D.H.82A Tiger Moth	N. H. Jones	
G–ACDJ	D.H.82A Tiger Moth	F. J. Terry	
G–ACEJ	D.H.83 Fox Moth	A. Haig-Thomas	
G–ACGT	*Avro 594 Avian IIIA	K. Smith	
G–ACIT	D.H.84 Dragon	J. Beaty (Historic Aircraft Museum)	
G–ACLL	D.H.85 Leopard Moth	A. Haig-Thomas	
G–ACMA	D.H.85 Leopard Moth	S. J. Filhol	
G–ACMN	D.H.85 Leopard Moth	H. D. Labouchere	
G–ACSP	*D.H.88 Comet	Veteran & Vintage Aircraft (Engineering) Ltd.	
G–ACSS	*D.H.88 Comet	Shuttleworth Trust	
G–ACTF	CLA.7 Swift	A. J. Chalkley	
G–ACUS	D.H.85 Leopard Moth	C. W. Annis	
G–ACUU	Cierva C.30A	G. S. Baker (Skyfame Collection)	
G–ACUX	*S.16 Scion	Strathallan Aircraft Collection	
G–ACVA	*Kay Gyroplane	Glasgow Museum of Transport	
G–ACWP	*Cierva C.30A (AP507)	Science Museum	

Notes	Reg.	Type	Owner or Operator
	G–ACXE	B.K. L-25C Swallow	D. G. Ellis
	G–ADAH	*D.H.89A Dragon Rapide	Museum of Flight, E. Fortune
	G–ADEV	*Avro 504K (E3404)	Shuttleworth Trust
	G–ADFV	*Blackburn B-2	The Aeroplane Collection
	G–ADGP	M.2L Hawk Speed Six	A. M. Stow
	G–ADGT	D.H.82A Tiger Moth	D. R. & Mrs. M. Wood
	G–ADGV	D.H.82A Tiger Moth	Air Cdre. A. H. Wheeler
	G–ADIA	D.H.82A Tiger Moth	J. Beaty
	G–ADJJ	D.H.82A Tiger Moth	J. M. Preston
	G–ADKC	D.H.87B Hornet Moth	E. J. Roe
	G–ADKK	D.H.87B Hornet Moth	C. W. Annis
	G–ADKM	D.H.87B Hornet Moth	F. R. E. Hayter
	G–ADLY	D.H.87B Hornet Moth	A. Haig-Thomas
	G–ADMT	D.H.87B Hornet Moth	Strathallan Aircraft Collection
	G–ADMW	*M.2H Hawk Major (DG590)	RAF Museum
	G–ADND	D.H.87B Hornet Moth	Shuttleworth Trust
	G–ADNE	D.H.87B Hornet Moth	P. A. Mann
	G–ADNZ	D.H.82A Tiger Moth	R. W. & Mrs. S. Pullan
	G–ADOT	*D.H.87B Hornet Moth	Mosquito Aircraft Museum
	G–ADPJ	B.A.C. Drone	G. Eastell
	G–ADPR	*P.3 Gull	Shuttleworth Trust *Jean*
	G–ADPS	Swallow 2	Strathallan Aircraft Collection
	G–ADRA	Pietenpol Aircamper	A. J. Mason & R. J. Barrett
	G–ADRC	K. & S. Jungster J-1	J. J. Penney & L. R. Williams
	G–ADRY	*Pou-du-Ciel (Replica) (BAPC 29)	P. Roberts
	G–ADUR	D.H.87B Hornet Moth	R. C. Lenton
	G–ADXS	*Pou-du-Ciel	Historic Aircraft Museum, Southend
	G–ADXT	D.H.82A Tiger Moth	J. & J. M. Pothecary
	G–ADYS	Aeronca C.3	J. Willmot
	G–AEBB	*Pou-du-Ciel	Shuttleworth Trust
		Blackburn B-2	British Aerospace
	G–AEDB	Kronfield Drone	M. C. Russell
	G–AEDU	D.H.90 Dragonfly	A. Haig-Thomas & M. Barraclough
	G–AEEG	M.3A Falcon	Shipping & Airlines Ltd.
	G–AEEH	*Pou-du-Ciel	RAF St. Athan
	G–AEFT	Aeronca C-3	C. E. Humphreys & ptnrs.
	G–AEGV	*HM.14 Pou-du-Ciel	Midland Aircraft Preservation Soc.
	G–AEHM	*Pou-du-Ciel	Science Museum
	G–AEKR	*Flying Flea (Replica) (BAPC 121)	Nostell Aviation Museum
	G–AEKV	Kronfield Drone	Wg. Cdr. J. E. McDonald
	G–AELO	D.H.87B Hornet Moth	S. N. Bostock
	G–AEML	D.H.89 Dragon Rapide	J. P. Filhol Ltd.
	G–AEOA	D.H.80A Puss Moth	A. Haig-Thomas (Shuttleworth)
	G–AEOE	*HM.14 Pou-du-Ciel (BAPC 22)	Newark Air Museum
	G–AEOH	*HM.14 Pou-du-Ciel	Gordon Riley
	G–AEPH	*Bristol F.2B (D8096)	Shuttleworth Trust
	G–AERD	P.3 Gull Six	C. C. & Mrs. J. M. Lovell
	G–AERV	M.11A Whitney Straight	R. T. Boyes
	G–AESE	D.H.87B Hornet Moth	H. J. Shaw
	G–AETA	*Caudron G.3 (3066)	RAF Museum
	G–AEUJ	M.11A Whitney Straight	R. E. Mitchell
	G–AEVS	*Aeronca 100	Historic Aircraft Preservation Soc.
	G–AEVZ	*Swallow 2	Museum of Flight, E. Fortune
	G–AEXD	Aeronca 100	Mrs. M. A. & R. W. Mills
	G–AEXF	P.6 Mew Gull	M. C. Barraclough & T. M. Storey
	G–AEXT	Dart Kitten II	M. W. Rice
	G–AEXZ	Piper J-2 Cub	Mrs. M. & J. R. Dowson
	G–AEYY	*Martin Monoplane	Martin Monoplane Syndicate
	G–AFBS	*M14A Hawk Trainer	Skyfame Collection (G–AKKU)
	G–AFCL	B. A. Swallow 2	A. M. Dowson
	G–AFFD	*Percival Q-6	Midland Aircraft Preservation Soc.
	G–AFFI	*Pou-du-Ciel (BAPC 76)	Nostell Aviation Museum
	G–AFGC	B. A. Swallow 2	H. Plain
	G–AFGD	B. A. Swallow 2	B. Arden & ptnrs.
	G–AFGE	B. A. Swallow 2	Donald G. Ellis
	G–AFGH	Chilton D.W.I.	J. T. Hayes & J. West
	G–AFGI	*Chilton D.W.I.	J. E. McDonald

G–AAMY (Top) D.H.60M Moth/*A. S. Wright*

G–ABNX (Centre) Robinson Redwing 2

G–AERD (Bottom) Percival P.3 Gull Six

Reg.	Type	Owner or Operator	Notes
G–AFHA	*Mosscraft M.A.I.	C. V. Butler	
G–AFIN	*Chrislea Airguard	Aeroplane Collection Ltd.	
G–AFIU	*Luton Minor (Parker C.A.4)	S. P. Connatty & I. V. Jones (N.A.P.S.)	
G–AFJA	Watkinson Dingbat	K. Woolley	
G–AFJB	Foster-Wickner G.M.I. Wicko	K. Woolley	
G–AFJR	Tipsy Trainer I	M. E. Vaisey	
G–AFJV	Mosscraft MA.2	C. V. Butler	
G–AFLW	M.17 Monarch	Strathallan Aircraft Collection	
G–AFNG	D.H.94 Moth Minor	A. Haig-Thomas (Shuttleworth)	
G–AFNI	D.H.94 Moth Minor	B. M. Welford	
G–AFOB	D.H.94 Moth Minor	R. E. Ogden	
G–AFOJ	D.H.94 Moth Minor	R. M. Long	
G–AFPN	D.H.94 Moth Minor	Strathallan Aircraft Collection	
G–AFRV	Tipsy Trainer I	Capt. R. C. F. Bailey	
G–AFSC	Tipsy Trainer I	G. P. Hermer & ptnrs.	
G–AFSV	Chilton D.W.IA	R. Nerou	
G–AFTA	*Hawker Tomtit (K1786)	Shuttleworth Trust	
G–AFTN	*Taylorcraft Plus C2	Leicestershire County Council Museums	
G–AFVE	D.H.82 Tiger Moth	G. C. Masterson & ptnrs.	
G–AFVN	Tipsy Trainer I	W. Callow & ptnrs.	
G–AFWI	D.H.82A Tiger Moth (BB814)	Portsmouth Naval Gliding Club	
G–AFWT	Tipsy Trainer I	H. Scanlon	
G–AFYD	Luscombe 8E Silvaire	J. D. Iliffe	
G–AFYO	Stinson H.W.75	B. A. Dunlop & P. Elliott	
G–AFZE	Heath Parasol	K. C. D. St. Cyrien	
G–AGBN	G.A.L.42 Cygnet 2	Strathallan Aircraft Collection	
G–AGHB	Hawker Sea Fury XI (WH589)	—	
G–AGJG	D.H.89A Dragon Rapide	Aerial Enterprises Ltd.	
G–AGLK	Auster 5D	W. C. E. Tazewell	
G–AGNV	*Avro 685 York I (MW100)	Aerospace Museum, Cosford	
G–AGOH	J/I Autocrat	Leicester Museum of Technology	
G–AGOS	R.S.3 Desford I (VZ728)	Strathallan Aircraft Collection	
G–AGOY	M.48 Messenger 3	T. Clark	
G–AGPG	*Avro 19 Srs. I	Historic Aircraft Museum, Southend	
G–AGRU	*V.498 Viking IA	Aerospace Museum, Cosford	
G–AGSH	D.H.89A Dragon Rapide 6	B. Haddican and ptnrs.	
G–AGTM	D.H.89A Dragon Rapide 6 (NF875)	Sealion Shipping Ltd.	
G–AGTO	J/I Autocrat	M. J. Barnett & D. J. T. Miller	
G–AGTT	J/I Autocrat	D. W. Prince & R. K. J. Hadlow	
G–AGVG	J/I Autocrat	T. F. McDonald	
G–AGVN	J/I Autocrat	P. J. Ellitt	
G–AGWE	Avro 19 Srs. 2	Strathallan Aircraft Collection	
G–AGXN	J/IN Alpha	P. A. Davey	
G–AGXT	*J/IN Alpha	Stockport Aircraft Preservation Soc.	
G–AGXU	J/IN Alpha	Mrs. J. Lewis	
G–AGXV	J/I Autocrat	Vintage Aircraft Flying Club	
G–AGYD	J/IN Alpha	F. K. Smith & P. Herring	
G–AGYH	J/IN Alpha	G. E. Twyman & P. J. Rae	
G–AGYK	J/I Autocrat	R. W. & Mrs. N. M. Biggs	
G–AGYT	J/IN Alpha	Portsmouth Naval Gliding Club	
G–AGYU	D.H.82A Tiger Moth (DE208)	P. & A. Wood	
G–AHAL	J/IN Alpha	Skegness Air Taxi Services Ltd.	
G–AHAM	J/I Autocrat	D. W. Philp	
G–AHAU	J/I Autocrat	B. J. W. Foley	
G–AHAV	J/I Autocrat	R. H. G. Kingsmill	
G–AHAY	J/I Autocrat	Aviation Advisory Services Ltd.	
G–AHBL	D.H.87B Hornet Moth	Dr. Ursula H. Hamilton	
G–AHBM	D.H.87B Hornet Moth	P. Franklin	
G–AHCK	J/IN Alpha	P. A. Woodman	
G–AHCL	J/IN Alpha	R. Willis & ptnrs.	
G–AHCN	J/IN Alpha	FTS Flying Group	
G–AHCR	Gould-Taylorcraft Plus D Special	D. E. H. Balmforth & D. R. Shepherd	
G–AHED	*D.H.89A Dragon Rapide (RL962)	RAF Museum (Cardington)	
G–AHGD	D.H.89A Dragon Rapide (Z7258)	M. R. L. Astor *Women of the Empire*	
G–AHGW	Taylorcraft Plus D	C. V. Butler	
G–AHGZ	Taylorcraft Plus D	S. J. Ball	

Notes	Reg.	Type	Owner or Operator
	G-AHHH	J/I Autocrat	D. Bridge
	G-AHHK	J/I Autocrat	W. J. Ogle
	G-AHHN	J/I Autocrat	KK Aviation
	G-AHHP	J/IN Alpha	J. T. Sime
	G-AHHT	J/IN Alpha	P. M. A. Parrett
	G-AHIC	Avro 19 Srs. 2	Strathallan Aircraft Collection
	G-AHIZ	D.H.82A Tiger Moth	C.F.G. Flying Ltd.
	G-AHKX	Avro 19 Srs. 2	Strathallan Aircraft Collection
	G-AHKY	*Miles M.18 Series 2	Strathallan Aircraft Collection
	G-AHLI	Auster 3	G. A. Leathers
	G-AHLK	Auster 3	A. A. Bucknell
	G-AHLT	D.H.82A Tiger Moth	R. C. F. Bailey
	G-AHMJ	*Cierva C.30A (K4235)	Shuttleworth Trust
	G-AHMN	D.H.82A Tiger Moth (N6985)	George House (Holdings) Ltd.
	G-AHNG	Taylorcraft Plus D	D. J. Pratt
	G-AHRI	*D.H.104 Dove I	Lincolnshire Aviation Museum
	G-AHSA	*Avro 621 Tutor (K3215)	Shuttleworth Trust
	G-AHSD	Taylorcraft Plus D	A. Tucker
	G-AHSO	J/IN Alpha	Skegness Air Taxi Services Ltd.
	G-AHSP	J/I Autocrat	J. E. & J. M. Wicks
	G-AHSS	J/IN Alpha	D. B. Connelly
	G-AHST	J/IN Alpha	P. H. Lewis
	G-AHSW	J/I Autocrat	K. W. Brown
	G-AHTE	*P.44 Proctor V	S. Wales Historical Aircraft Preservation Soc.
	G-AHTW	*A.S.40 Oxford (V3388)	Skyfame Collection
	G-AHUI	*M.38 Messenger 2A	Humberside Aircraft Preservation Soc.
	G-AHUJ	M.14A Hawk Trainer 3 (R1914)	Strathallan Aircraft Collection
	G-AHUV	D.H.82A Tiger Moth	W. G. Gordon
	G-AHVV	D.H.82A Tiger Moth	R. Jones
	G-AHWJ	Taylorcraft Plus D	A. Tucker
	G-AHXE	Taylorcraft Plus D (LB312)	Mrs. J. M. C. Pothecary
	G-AHZT	M.38 Messenger 2A	E. Pratt
	G-AIBE	*Fulmar II (N1854)	F.A.A. Museum
	G-AIBH	J/IN Alpha	Skegness Air Taxi Services Ltd.
	G-AIBM	J/I Autocrat	J. L. Goodley
	G-AIBW	J/IN Alpha	J. L. Way
	G-AIBX	J/I Autocrat	Wasp Flying Group
	G-AIBY	J/I Autocrat	D. Morris
	G-AIDK	*M.38 Messenger 2A	J. L. Rees & B. A. Llewelln
	G-AIDL	D.H.89A Dragon Rapide 6	Southern Joyrides Ltd.
	G-AIDN	V.S.502 Spitfire T.8	G. F. Miller
	G-AIEK	M.38 Messenger 2A (RG333)	J. Buckingham
	G-AIFZ	J/IN Alpha	J. J. Young
	G-AIGD	J/I Autocrat	A. G. Batchelor
	G-AIGF	J/IN Alpha	A. R. C. Mathie
	G-AIGM	J/IN Alpha	S. T. Raby
	G-AIGR	J/IN Alpha	Burton & Derby Gliding Club
	G-AIGU	J/IN Alpha	T. Pate
	G-AIIH	Piper L-4H Cub	J. W. Benson
	G-AIIZ	D.H.82A Tiger Moth	D. E. Baker
	G-AIJI	*J/IN Alpha	Humberside Aircraft Preservation Soc.
	G-AIJR	Auster J/4	B. A. Harris
	G-AIJS	Auster J/4	A. R. Weston
	G-AIJT	Auster J/4	Aberdeen Flying Group
	G-AILL	M. 38 Messenger 2A	H. Best-Devereux
	G-AIPR	Auster J/4	R. W. & M. A. Mills
	G-AIPV	J/I Autocrat	J. Linegar
	G-AIPW	J/I Autocrat	E. M. P. Cadman
	G-AIRC	J/I Autocrat	A. G. Martlew
	G-AIRI	D.H.82A Tiger Moth	M. E. Vaisey
	G-AIRK	D.H.82A Tiger Moth	R. C. Teverson & ptnrs.
	G-AISA	Tipsy B Srs. I	B. T. Morgan & A. Liddiard
	G-AISB	Tipsy B Srs. I	R. E. Barker
	G-AISC	Tipsy B Srs. I	Wagtail Flying Group
	G-AIST	V.S.300 Spitfire I (AR213)	The Hon. P. Lindsay
	G-AISU	V.S. Spitfire 5B (AB910)	RAF Battle of Britain Historical Aircraft Flight
	G-AITB	*A.S.10 Oxford (MP425)	RAF Museum

Reg.	Type	Owner or Operator	Notes
G–AIUA	M.14A Hawk Trainer 3	Shuttleworth Trust	
G–AIUL	D.H.89A Dragon Rapide 6	A. F. Ward	
G–AIVW	D.H.82A Tiger Moth Seaplane	N. H. Jones	
G–AIWA	P.28B Proctor I (R7524)	N. C. Jenson	
G–AIXA	Taylorcraft Plus D	P. J. Anderson	
G–AIXD	D.H.82A Tiger Moth	D. L. Lloyd	
G–AIXN	Benes-Mraz M.lc Sokol	J. F. Evetts & D. Patel	
G–AIYR	D.H.89A Dragon Rapide	C. D. Cyster & ptnrs.	
G–AIYS	D.H.85 Leopard Moth	Strathallan Aircraft Collection	
G–AIZE	*F.24W Argus 2	RAF Museum	
G–AIZG	*V.S. Walrus (L2301)	Historic Aircraft Preservation Soc./ Royal Navy	
G–AIZU	J/1 Autocrat	T. A. Collins & A. H. R. Stansfield	
G–AIZY	J/1 Autocrat	B. J. Richards	
G–AIZZ	J/1 Autocrat	Peter Long International Ltd.	
G–AJAB	J/1N Alpha	A. E. Poulson	
G–AJAC	J/1N Alpha	R. C. Hibberd	
G–AJAE	J/1N Alpha	M. G. Stops	
G–AJAJ	J/1N Alpha	A. R. Milne & T. F. L. Bessant	
G–AJAM	J/2 Arrow	D. A. Porter	
G–AJAS	J/1N Alpha	C. J. Baker	
G–AJCP	D.31 Turbulent	H. J. Shaw	
G–AJDR	*M.14A Hawk Trainer 3 (P6382)	Shuttleworth Trust	
G–AJDW	J/1 Autocrat	D. R. Hunt	
G–AJDY	J/1 Autocrat	General Aviation Services Ltd.	
G–AJEB	*J/1N Alpha	Aeroplane Collection Ltd.	
G–AJEE	J/1 Autocrat	F. A. Arnold	
G–AJEH	J/1N Alpha	Astral Surveys Ltd.	
G–AJEI	J/1 Autocrat	Skegness Air Taxi Services Ltd.	
G–AJEM	J/1 Autocrat	H. A. Nind	
G–AJGJ	Auster 5	J. J. McLaughlin & ptnrs.	
G–AJHO	D.H.89A Dragon Rapide	East Anglian Aviation Soc. Ltd.	
G–AJHS	D.H.82A Tiger Moth	W. G. Fisher	
G–AJHU	D.H.82A Tiger Moth	F. P. Le Coyte	
G–AJID	J/1 Autocrat	L. W. Usherwood	
G–AJIH	J/1 Autocrat	D. F. Campbell & ptnrs.	
G–AJIS	J/1N Alpha	A. Tucker	
G–AJIU	J/1 Autocrat	G. H. Mathews	
G–AJIW	J/1N Alpha	B. M. Baker (Hatfield) Ltd.	
G–AJJP	*Jet Gyrodyne (XJ389)	Aerospace Museum, Cosford	
G–AJKK	M.38 Messenger 2A	A. M. Lambourne & T. C. Evans	
G–AJOA	D.H.82A Tiger Moth (T5424)	F. P. Le Coyte	
G–AJOC	M.38 Messenger 2A	Strathallan Aircraft Collection	
G–AJOE	M.38 Messenger 2A (RH378)	Aircraft Mart	
G–AJOV	*Sikorsky S-51	Aerospace Museum, Cosford	
G–AJOZ	*F.24W Argus 2	Aeroplane Collection Ltd.	
G–AJPI	F.24W-41a Argus 2 (EV851)	Leisure Sport Ltd.	
G–AJPZ	J/1 Autocrat	K. Pyle	
G–AJRB	J/1 Autocrat	T. Booth	
G–AJRC	J/1 Autocrat	Severn Valley Aviation Group	
G–AJRE	J/1 Autocrat	D. J. Ronayne	
G–AJRH	J/1N Alpha	N. H. Ponsford	
G–AJTH	M.65 Gemini 1A	K. A. Learmonth	
G–AJUD	J/1 Autocrat	J. A. Hughes & K. M. Bowen	
G–AJUE	J/1 Autocrat	M. A. G. Westman	
G–AJUL	J/1N Alpha	L. A. Brome	
G–AJVH	Swordfish (LS326)	F.A.A. Museum	
G–AJVT	Auster 5	I. N. M. Cameron	
G–AJWB	M.38 Messenger 2A	W. F. & M. R. Higgins	
G–AJWC	*M.65 Gemini	Stockport Aircraft Preservation Soc.	
G–AJXC	Auster 5 (TJ343)	J. E. Graves	
G–AJXV	Auster 4 (NJ695)	P. C. J. Farries	
G–AJYO	J/5B Autocar	A. H. Chaplin & ptnrs.	
G–AKAA	Piper L-4H Cub	J. W. Benson	
G–AKAT	*M.14A Magister (T9738)	Newark Air Museum	
G–AKBM	*M.38 Messenger 2A	Humberside Aircraft Preservation Soc.	
G–AKBO	M.38 Messenger 2A	J. R. A. Ramshaw	
G–AKDN	D.H.C. 1A Chipmunk 10	K. R. Nunn	

Notes	Reg.	Type	Owner or Operator
	G–AKEL	M.65 Gemini 1A	J. M. Bisco
	G–AKER	*M.65 Gemini 1A	Humberside Aircraft Preservation Soc.
	G–AKEZ	*M.38 Messenger 2A (RG333)	Torbay Aircraft Museum
	G–AKGD	*M.65 Gemini 1A	D. G. Addicott
	G–AKGE	M.65 Gemini 3C	R. E. Winn
	G–AKHP	M.65 Gemini 1A	C. Spencer-Thomas
	G–AKHW	M.65 Gemini 1A	Bee Gee Aviation
	G–AKHZ	*M.65 Gemini 7	Humberside Aircraft Preservation Soc.
	G–AKIF	D.H.89A Dragon Rapide	Airborne Taxi Services Ltd.
	G–AKIN	M.38 Messenger 2A	A. J. Spiller
	G–AKJU	J/1N Alpha	Arrow Air Services (Engineering) Ltd.
	G–AKKB	M.65 Gemini 1A	S.A.C. Bristol Ltd.
	G–AKKH	M.65 Gemini 1A	M. C. Russell
	G–AKKR	*M.14A Magister (T9707)	RAF Museum
	G–AKOE	D.H.89A Dragon Rapide 4	J. E. Pierce
	G–AKOW	Auster 5	A. M. L. McLean
	G–AKPF	*M.14A Hawk Trainer (N3788)	L. N. D. Taylor
	G–AKPI	Auster 5 (NJ703)	B. H. Hargrave
	G–AKSZ	Auster 5	A. R. C. Mathie
	G–AKUW	C.H.3 Super Ace	C. V. Butler
	G–AKVF	C.H.3 Super Ace	P. V. B. Longthorp
	G–AKVZ	M.38 Messenger 4B	P. H. Louks
	G–AKWS	Auster 5	J. E. Homewood
	G–AKWT	*Auster 5 (MT360)	Humberside Aircraft Preservation Soc.
	G–AKXP	Auster 5	F. E. Telling
	G–AKXS	D.H.82A Tiger Moth	P. A. Colman
	G–AKZN	*P.30 Proctor 2E (Z7197)	RAF Museum
	G–ALAH	*M.38 Messenger 4A (RH377)	Aeroplane Collection Ltd.
	G–ALAX	*D.H.89A Dragon Rapide	Durney Aeronautical Collection
	G–ALBD	D.H.82A Tiger Moth	C. H. Schoonbeck
	G–ALBJ	Auster 5	R. H. Elkington
	G–ALBK	Auster 5	S. J. Wright & Co. (Farmers) Ltd.
	G–ALBN	*Bristol 173 (XF785)	RAF Museum
	G–ALCK	*Proctor (LZ766)	Skyfame Collection
	G–ALCS	M.65 Gemini 3C	R. E. Winn
	G–ALCU	*D.H.104 Dove 2	Midland Air Museum
	G–ALFA	Auster 5	Alpha Flying Group
	G–ALFT	*D.H.104 Dove 6	Torbay Aircraft Museum
	G–ALGT	V.S.379 Spitfire 14 (RM619)	Rolls-Royce (1971) Ltd.
	G–ALJF	P.34A Proctor 3	J. F. Moore & ptnrs.
	G–ALMU	*M.65 Gemini	Stockport Aircraft Preservation Soc.
	G–ALNA	D.H.82A Tiger Moth	Wessex Flying Group
	G–ALND	D.H.82A Tiger Moth (N9191)	Buccaneer Aviation Ltd.
	G–ALSP	*Bristol 171 (WV783)	RAF Museum
	G–ALSS	*Bristol 171 (WA576)	Strathallan Aircraft Collection
	G–ALSX	*Bristol 171 (G-48-1)	Rotorcraft Museum (Duxford)
	G–ALTW	D.H.82A Tiger Moth	C. J. Musk
	G–ALUC	D.H.82A Tiger Moth	D. R. & Mollie Wood
	G–ALUG	*M.65 Gemini	Stockport Aircraft Preservation Soc.
	G–ALWB	D.H.C.1. Chipmunk 22A	K. C. Keegan & ptnrs.
	G–ALWC	Dakota 4	Fairey Surveys Ltd.
	G–ALWF	*V.701 Viscount	Viscount Preservation Trust
	G–ALWW	D.H.82A Tiger Moth	F. W. Fay & ptnrs.
	G–ALXZ	Auster 5-150	T. Bessart & R. Witheridge
	G–ALYB	Auster 5	G. F. Kilsby
	G–ALYG	Auster 5D	A. L. Young
	G–ALZE	*BN-1F	Aerospace Museum, Cosford
	G–ALZO	*A.S.57 Ambassador	Dan-Air Preservation Group
	G–AMAU	Hurricane 2C (PZ865)	RAF Battle of Britain Historic Aircraft Flight
	G–AMAW	Luton L.A.4 Minor	J. R. Coates
	G–AMCA	Dakota 4	Air Atlantique Ltd.
	G–AMDA	*Avro 652A Anson 1 (N4877)	Skyfame Collection
	G–AMDN	Hiller UH-12A	Bristow Helicopters Ltd.
	G–AMHJ	Dakota 6	Jersey European Airways
	G–AMIU	D.H.82A Tiger Moth	R. & Mrs. J. L. Jones
	G–AMKU	J/1B Aiglet	Southdown Flying Group

Reg.	Type	Owner or Operator	Notes
G-AMLZ	P.50 Prince 6E	J. F. Coggins	
G-AMME	*M.65 Gemini 3A	Stockport Aircraft Preservation Soc.	
G-AMMS	J/5K Aiglet Trainer.	G. White & Sons	
G-AMOG	*V.701 Viscount	Aerospace Museum, Cosford	
G-AMPO	Dakota 4	Eastern Airways Ltd.	
G-AMPP	*Dakota 3 (G-AMSU)	Dan-Air Preservation Group	
G-AMPW	J/5B Autocar	J. V. Inglis	
G-AMPY	Dakota 4	Jersey European Airways	
G-AMRA	Dakota 6	Eastern Airways Ltd.	
G-AMRF	J/5F Aiglet Trainer	A. I. Topps	
G-AMRK	G.37 Gladiator (L8032)	Shuttleworth Trust	
G-AMSV	Dakota 4	Skyways Cargo Airline	
G-AMSZ	Auster 5	G. & D. Knight & D. G. Pridham	
G-AMTA	J/5F Aiglet Trainer	H. J. Jauncey	
G-AMTM	J/I Autocrat	R. Stobo & D. Clewley	
G-AMUF	D.H.C.I Chipmunk 21	Chipmunk Flying Group	
G-AMUH	D.H.C.I Chipmunk 21	D. H. Whalley	
G-AMVD	Auster 5	C. G. Clarke & I. M. Callier	
G-AMVP	Tipsy Junior	A. R. Wershat	
G-AMWW	Dakota 4	Skyways Cargo Airline	
G-AMXA	*D.H.106 Comet C.2 (XK655)	Strathallan Aircraft Collection	
G-AMYD	J/5L Aiglet Trainer	G. H. Maskell	
G-AMYJ	Dakota 6	Eastern Airways Ltd.	
G-AMZI	J/5F Aiglet Trainer	Surrey Aviation	
G-AMZT	J/5F Aiglet Trainer	A. H. Roscoe	
G-AMZU	J/5F Aiglet Trainer	R. N. Goode & ptnrs.	
G-ANAF	Dakota 4	Air Atlantique Ltd.	
G-ANAP	*D.H.104 Dove 6	Brunel Technical College, Lulsgate	
G-ANCF	B.175 Britannia 308F	Invicta International Airlines Ltd.	
G-ANCS	D.H.82A Tiger Moth (R4907)	R. H. Reeves	
G-ANCX	D.H.82A Tiger Moth	D. R. Wood	
G-ANDE	D.H.82A Tiger Moth	Stapleford Tiger Group	
G-ANDP	D.H.82A Tiger Moth	A. H. Diver	
G-ANEF	D.H.82A Tiger Moth (T5493)	RAF College Flying Club Co. Ltd.	
G-ANEL	D.H.82A Tiger Moth (N9238)	W. P. Maynall	
G-ANEW	D.H.82A Tiger Moth	A. L. Young	
G-ANEZ	D.H.82A Tiger Moth	J. W. Benson	
G-ANFC	*D.H.82A Tiger Moth	Mosquito Aircraft Museum	
G-ANFH	*Westland S.55	British Rotorcraft Museum	
G-ANFI	D.H.82A Tiger Moth	K. L. Backus & D. S. Kirkham	
G-ANFM	D.H.82A Tiger Moth	S. A. Brook & ptnrs.	
G-ANFP	*D.H.82A Tiger Moth	Mosquito Aircraft Museum	
G-ANFV	D.H.82A Tiger Moth (DF155)	Strathallan Aircraft Collection	
G-ANFW	D.H.82A Tiger Moth	G. M. Fraser	
G-ANHK	D.H.82A Tiger Moth	J. D. Iliffe	
G-ANHR	Auster 5	J. O. C. Warner	
G-ANHS	Auster 4	P. S. Sturdgess	
G-ANHX	Auster 5D	D. J. Baker	
G-ANHZ	Auster 5	J. H. D. Newman	
G-ANIE	Auster 5	H. O. Stibbard	
G-ANIJ	Auster 5D	Army Aviation Museum	
G-ANIS	Auster 5	J. Clark-Cockburn	
G-ANJA	D.H.82A Tiger Moth	J. J. Young	
G-ANJD	D.H.82A Tiger Moth	H. J. Jauncey	
G-ANJK	D.H.82A Tiger Moth	Montgomery Ultra Light Flying Club	
G-ANKK	D.H.82A Tiger Moth (T5854)	Birmingham Tiger Group	
G-ANKT	D.H.82A Tiger Moth (T6818)	The Shuttleworth Collection	
G-ANLS	D.H.82A Tiger Moth	P. A. Gliddon	
G-ANLW	W.B.I. Widgeon	Helicopter Hire Ltd.	
G-ANMV	D.H.82A Tiger Moth (T7404)	George House (Holdings) Ltd.	
G-ANMZ	D.H.82A Tiger Moth	The Tiger Club	
G-ANNK	D.H.82A Tiger Moth	Mrs. P. J. Wilcox	
G-ANOD	D.H.82A Tiger Moth	D. R. & Mrs. M. Wood	
G-ANOH	D.H.82A Tiger Moth	S. J. Carr	
G-ANOK	S-91C Safir	Museum of Flight, E. Fortune	
G-ANON	D.H.82A Tiger Moth	A. C. Mercer	
G-ANOO	D.H.82A Tiger Moth	T. J. Hartwell & ptnrs.	
G-ANOR	D.H.82A Tiger Moth	A. J. Cheshire	
G-ANOV	*D.H.104 Dove 6	Museum of Flight, E. Fortune	
G-ANPE	D.H.82A Tiger Moth	R. W. & P. R. Budge	
G-ANPK	D.H.82A Tiger Moth	J. W. Benson	
G-ANPP	P.34A Proctor 3	C. P. A. & Mrs. J. Jeffery	

Notes	Reg.	Type	Owner or Operator
	G–ANRF	D.H.82A Tiger Moth	C. D. Cyster
	G–ANRN	D.H.82A Tiger Moth	J. J. V. Elwes
	G–ANRP	Auster 5	J. W. Weeks
	G–ANSM	D.H.82A Tiger Moth	Torbay Aviation Ltd.
	G–ANTE	D.H.82A Tiger Moth	T. I. Sutton & B. J. Champion
	G–ANTK	*Avro 685 York	Dan Air Preservation Group
	G–ANTS	*D.H.82A Tiger Moth	Strathallan Aircraft Collection
	G–ANUO	D.H.114 Heron 2D	The GEC Co. Ltd.
	G–ANUW	D.H.104 Dove 6	Civil Aviation Authority
	G–ANVU	D.H.104 Dove 1B	T. D. Keegan
	G–ANWB	D.H.C.1 Chipmunk 21	G. Briggs
	G–ANWX	J/5L Aiglet Trainer	Newcastle & Tees-side Gliding Club
	G–ANXB	D.H.114 Heron 1B	Pan Universal Aircraft Services (UK) Ltd.
	G–ANXR	P.31C Proctor 4 (RM221)	J. R. Batt
	G–ANYP	*P.31C Proctor 4 (NP184)	Torbay Aircraft Museum
	G–ANZJ	*P.31C Proctor 4 (NP303)	Historic Aircraft Museum, Southend
	G–ANZR	D.H.82A Tiger Moth	D. R. & Mrs. M. Wood
	G–ANZU	D.H.82A Tiger Moth	P. A. Jackson
	G–ANZZ	D.H.82A Tiger Moth	The Tiger Club
	G–AOAA	D.H.82A Tiger Moth	The Tiger Club
	G–AOAR	*P.31C Proctor 4 (NP181)	Historic Aircraft Preservation Soc.
	G–AOBO	D.H.82A Tiger Moth	T. J. Bolt & J. N. Moore
	G–AOBU	*P.84 Jet Provost	Shuttleworth Trust
	G–AOBV	J/5P Autocar	P. E. Champney
	G–AOBX	D.H.82A Tiger Moth (T7187)	Leisure Sport Ltd.
	G–AOCR	Auster 5D	D. T. Cheetham
	G–AOCU	Auster 5	S. J. Ball
	G–AODA	Westland S.55 Srs.3	Bristow Helicopters Ltd.
	G–AODR	D.H.82A Tiger Moth	D. R. & Mrs. M. Wood
	G–AODT	D.H.82A Tiger Moth	N. A. Brett & A. H. Warminger
	G–AOEG	D.H.82A Tiger Moth	Midland Tiger Flying Group
	G–AOEH	Aeronca 7AC Champion	M. Weeks & ptnrs.
	G–AOEI	D.H.82A Tiger Moth	C.F.G. Flying Ltd.
	G–AOEL	D.H.82A Tiger Moth	Strathallan Aircraft Collection
	G–AOES	D.H.82A Tiger Moth	Mrs. J. M. Goodger & G. A. Cordery
	G–AOET	D.H.82A Tiger Moth	Jack W. Benson
	G–AOFE	D.H.C.1 Chipmunk 22A	B. Webster
	G–AOFJ	Auster 5	Miss M. R. Innocent
	G–AOFM	J/5P Autocar	C. M. Barnes
	G–AOFS	J/5L Aiglet Trainer	G. W. Howard
	G–AOFW	*ATL.98 Carvair	British Air Ferries Ltd. *Big John*
	G–AOGA	M.75 Aries	R. E. Winn
	G–AOGE	P.34A Proctor 3	R. J. Sewell
	G–AOGR	D.H.82A Tiger Moth	J. O. Hodgson & H. C. Adkins
	G–AOGV	J/5R Alpine	ABH Aviation
	G–AOHZ	J/5P Autocar	M. R. Gibbons & G. W. Brown
	G–AOIL	D.H.82A Tiger Moth	The Shuttleworth Collection
	G–AOIM	D.H.82A Tiger Moth	R. M. Wade & J. F. Terry
	G–AOIR	Thruxton Jackaroo	Stevenage Flying Club
	G–AOIS	D.H.82A Tiger Moth	V. B. & R. G. Wheele
	G–AOIY	J/5G Autocar	L. N. Garner
	G–AOJC	*V.802 Viscount	Wales Aircraft Museum
	G–AOJH	D.H.83C Fox Moth	J. S. Lewery
	G–AOJJ	D.H.82A Tiger Moth	E. Lay
	G–AOKH	P.40 Prentice 1	J. F. Moore
	G–AOKL	P.40 Prentice 1 (VS610)	J. R. Batt
	G–AOKO	*P.40 Prentice 1	J. F. Coggins
	G–AOLK	P.40 Prentice 1	Hilton Aviation Ltd.
	G–AOLP	P.40 Prentice 1	D. N. Downton & ptnrs.
	G–AOLU	P.40 Prentice 1 (VS356)	Strathallan Aircraft Collection
	G–AORL	D.H.C.1 Chipmunk 22	D. Gardner
	G–AORR	D.H.C.1 Chipmunk 22A	T. L. P. Delaney
	G–AORW	D.H.C.1 Chipmunk 22A	A. R. Lockyer
	G–AOSK	D.H.C.1 Chipmunk 22	J. G. Cullen
	G–AOSO	D.H.C.1 Chipmunk 22	D. Blackburn
	G–AOSU	D.H.C.1 Chipmunk 22 (Lycoming)	RAFGSA
	G–AOSY	D.H.C.1 Chipmunk 22	J. A. W. Clowes
	G–AOSZ	D.H.C.1 Chipmunk 22A	J. Pothecary & P. Villa
	G–AOTD	D.H.C.1 Chipmunk 22	Shuttleworth Trust
	G–AOTF	D.H.C.1 Chipmunk 23	RAFGSA

Reg.	Type	Owner or Operator	Notes
G–AOTI	D.H.114 Heron 2D	Rolls-Royce Ltd.	
G–AOTK	D.53 Turbi	The T. K. Flying Group	
G–AOTR	D.H.C.1 Chipmunk 22	London Gliding Club (Pty) Ltd.	
G–AOTY	D.H.C.1 Chipmunk 22A	West London Aero Services Ltd.	
G–AOUG	D.H.104 Dove 6	Farihan Dajami	
G–AOUH	D.H.104 Dove 5	Farihan Dajami	
G–AOUJ	*Fairey Ultra-light	British Rotorcraft Museum	
G–AOUO	D.H.C.1 Chipmunk 22 (Lycoming)	RAFGSA	
G–AOUP	D.H.C.1 Chipmunk 22	Wessex Flying Group	
G–AOVF	B.175 Britannia 312F	Invicta International Airways Ltd.	
G–AOVS	B.175 Britannia 312F	Redcoat Air Cargo *Amy*	
G–AOVT	*B.175 Britannia 312	Duxford Aviation Soc.	
G–AOVW	Auster 5	B. Marriott	
G–AOXG	D.H.82A Tiger Moth (XL717)	F.A.A. Museum	
G–AOXN	D.H.82A Tiger Moth	D. W. Mickleburgh	
G–AOYG	V.806 Viscount	British Airways	
G–AOYH	V.806 Viscount	British Airways	
G–AOYI	V.806 Viscount	British Airways (Stored)	
G–AOYJ	V.806 Viscount	British Airways	
G–AOYL	V.806 Viscount	British Airways	
G–AOYM	V.806 Viscount	British Airways	
G–AOYN	V.806 Viscount	British Airways (Stored)	
G–AOYO	V.806 Viscount	British Airways	
G–AOYP	V.806 Viscount	British Airways	
G–AOYR	V.806 Viscount	British Airways	
G–AOYS	V.806 Viscount	British Airways (Stored)	
G–AOZB	D.H.82A Tiger Moth	H. D. Labouchere	
G–AOZH	*D.H.82A Tiger Moth	Brighton Transport Museum	
G–AOZL	J/5Q Alpine	L. C. Cole	
G–AOZP	D.H.C.1 Chipmunk 22	M. E. Darlington	
G–APAF	Auster 5	D. G. Pridham & ptnrs.	
G–APAH	Auster 5	Executive Flying Services Ltd.	
G–APAM	Thruxton Jackaroo	R. P. Williams	
G–APAO	Thruxton Jackaroo	C. K. Irvine	
G–APAP	Thruxton Jackaroo	H. D. V. Leech	
G-APAS	*D.H.106 Comet 1XB	Aerospace Museum, Cosford	
G–APBC	Dakota 4	Skyways Cargo Airline	
G–APBD	PA-23 Apache 160	E. A. Clack & T. Pritchard	
G–APBE	Auster 5	G. W. Clarke	
G–APBI	D.H.82A Tiger Moth (EM903)	R. Devaney & ptnrs.	
G–APBO	D.53 Turbi	R. Johnson	
G–APBW	Auster 5	F. J. Wiseman	
G–APCB	J/5Q Alpine	G. Jones	
G–APCC	D.H.82A Tiger Moth	R. J. Moody	
G–APCU	D.H.82A Tiger Moth	K. C. K. Virtue	
G–APCY	J/1N Alpha	J. R. Pearson	
G–APDB	*D.H.106 Comet 4	Duxford Aviation Soc.	
G–APDK	*D.H.106 Comet 4	Dan-Air Preservation Group	
G–APDT	*D.H.106 Comet 4	British Airways (used by Apprentices at Heathrow)	
G–APDV	Hiller UH-12C	S. E. Davidson	
G–APEG	V.953C Merchantman	Air Bridge Carriers Ltd.	
G–APEJ	V.953C Merchantman	Air Bridge Carriers Ltd.	
G–APEK	V.953C Merchantman	Air Bridge Carriers Ltd.	
G–APEP	V.953C Merchantman	Air Bridge Carriers Ltd.	
G–APES	V.953C Merchantman	Air Bridge Carriers Ltd.	
G–APET	V.953C Merchantman	Air Bridge Carriers Ltd.	
G–APEX	V.806 Viscount	British Airways (Stored)	
G–APEY	V.806 Viscount	British Airways	
G–APFA	D.54 Turbi	W. J. Evans	
G–APFD	Boeing 707-436	British Airtours	
G–APFF	Boeing 707-436	British Airtours	
G–APFJ	Boeing 707-436	British Airtours	
G–APFO	Boeing 707-436	British Airtours	
G–APFU	D.H.82A Tiger Moth	M. R. Coward & D. M. White	
G–APGL	*D.H.82A Tiger Moth (NM140)	Strathallan Aircraft Collection	
G–APHV	Avro 19 Srs. 2 (VM360)	Strathallan Aircraft Collection	
G–APIE	Tipsy Belfair B	J. J. Penney & ptnrs.	
G–APIG	D.H.82A Tiger Moth (T6553)	Leisure Sport Ltd.	
G-APIH	D.H.82A Tiger Moth	A. J. Detheridge	

Notes	Reg.	Type	Owner or Operator
	G–APIK	J/IN Alpha	T. D. Howe
	G–APIM	V.806 Viscount	British Airways
	G–APIT	P.40 Prentice I	J. R. Batt
	G–APIY	*P.40 Prentice I (VR249)	Newark Air Museum
	G–APJB	P.40 Prentice I	City Airways
	G–APJJ	*Fairey Ultra-light	Midland Aircraft Preservation Soc.
	G–APJN	Hiller UH-12B	Bristow Helicopters Ltd.
	G–APJO	D.H.82A Tiger Moth	D. R. & Mrs. M. Wood
	G–APJZ	J/IN Alpha	L. Goddard & E. Amey
	G–APKH	D.H.85 Leopard Moth	P. Franklin (G–ACGS)
	G–APKM	J/IN Alpha	K. C. Bachmann
	G–APKN	J/IN Alpha	N. Simpson
	G–APKY	Hiller UH-12B	Shackleton Aviation Ltd.
	G–APLG	J/5L Aiglet Trainer	T. Dann & K. Wales
	G–APLK	M-100 Student	Discovery (R & D) Ltd.
	G–APLO	D.H.C.I Chipmunk 22A	Channel Islands Aero Holdings Ltd.
	G–APMH	J/IU Workmaster	R. E. Neal and S. R. Stevens
	G–APML	Dakota 6	Martin-Baker (Engineering) Ltd.
	G–APMM	D.H.82A Tiger Moth (DE 419)	M. C. Russell
	G–APMP	Hiller UH-12C	Morland Beazley Helicopters Ltd.
	G–APMR	Hiller UH-12C	Bristow Helicopters Ltd.
	G–APMS	Hiller UH-12C	Bristow Helicopters Ltd.
	G–APMX	D.H.82A Tiger Moth	K. B. Palmer
	G–APMY	PA-23 Apache 160	Mooney Aviation Ltd.
	G–APNJ	Cessna 310	Shackleton Aviation Ltd.
	G–APNR	Hiller UH-12C	Bristow Helicopters Ltd.
	G–APNS	Garland-Bianchi Linnet	Mrs. P. B. Voice
	G–APNT	Currie Wot	L. W. Richardson & ptnrs.
	G–APNZ	D.31 Turbulent	The Tiger Club
	G–APOA	J/IN Alpha	Bristow Helicopters Ltd.
	G–APOD	Tipsy Belfair	R. J. Miller
	G–APOI	Saro Skeeter Srs. 8	B. G. Heron
	G–APOL	D.36 Turbulent	A. R. Leith
	G–APPA	D.H.C.I Chipmunk 22	D. H. Parkhouse & R. Turner-Warwick
	G–APPL	P.40 Prentice I	Miss S. J. Saggers
	G–APPM	D.H.C.I Chipmunk 22	L. T. Mersh
	G–APRF	Auster 5	P. Elliott & ptnrs.
	G–APRJ	*Avro 694 Lincoln B.2 (G-29-1)	Historic Aircraft Museum, Southend
	G–APRL	AW650 Argosy 101	Air Bridge Carriers Ltd.
	G–APRM	AW650 Argosy 102	Rolls-Royce Ltd.
	G–APRN	AW650 Argosy 101	Air Bridge Carriers Ltd.
	G–APRO	Auster 6A	A. H. Wheeler
	G–APRR	Super Aero 45	C. J. Reed
	G–APRT	Taylor JT-1 Monoplane	C. J. Chambers & ptnrs.
	G–APRU	M.S.760 Paris	Cranfield Institute of Technology
	G–APSH	Hiller UH-12B	Bristow Helicopters Ltd.
	G–APSO	D.H.104 Dove 5	Miss J. D. Baker
	G–APSZ	Cessna 172	R. M. Stross & D. P. Hood
	G–APTH	Agusta-Bell 47J	W. R. Finance Ltd.
	G–APTM	Hiller UH-12B	Bristow Helicopters Ltd.
	G–APTP	PA-22 Tri-Pacer 150	J. R. Williams
	G–APTR	J/IN Alpha	C. J. & D. J. Baker
	G–APTS	D.H.C.I Chipmunk 22A	B. R. Pickard
	G-APTU	Auster 5	P. Bowers
	G–APTW	*W.B.I Widgeon	Cornwall Air Park
	G–APTY	Beech G.35 Bonanza	G. E. Brennard & J. M. Fish
	G–APTZ	D.31 Turbulent	G. Edmiston
	G–APUB	Beech 95 Travel Air	Gp. Capt. Sir Douglas Bader
	G–APUD	*Bensen B.7M (modified)	Aeroplane Collection Ltd.
	G–APUE	L-40 Meta Sokol	P. Phipps
	G–APUK	J/I Autocrat	P. L. Morley
	G–APUP	Sopwith Pup (N5182)	D. W. Arnold
	G–APUR	PA-22 Tri-Pacer 160	G. A. Allen & ptnrs.
	G–APUW	Auster J-5V-160	J. P. Webster
	G–APUY	D.31 Turbulent	C. Jones & ptnrs.
	G–APUZ	PA-24 Comanche 250	P. N. Martin & Co. Ltd.
	G–APVA	PA-22 Tri-Pacer 160	D. M. Witham
	G–APVG	J/5L Aiglet Trainer	CIBA-Geigy (UK) Ltd.
	G–APVN	D.31 Turbulent	R. Sherwin
	G–APVS	Cessna 170B	P. E. L. Lamyman
	G–APVU	L-40 Meta-Sokol	D. Kirk
	G–APVV	Mooney M-20A	Telecon Associates

Reg.	Type	Owner or Operator	Notes
G–APVY	PA-25 Pawnee 150	A.D.S. (Aerial) Ltd.	
G–APVZ	D.31 Turbulent	A. F. Bullock	
G–APWA	HPR-7 Herald 100	British Air Ferries *Mandy Keegan*	
G–APWE	HPR-7 Herald 201	Air UK	
G–APWF	HPR-7 Herald 201	Air UK	
G–APWG	HPR-7 Herald 201	Air UK	
G–APWH	HPR-7 Herald 201	Air UK	
G–APWJ	HPR-7 Herald 201	Air UK	
G–APWM	W.S.55 Whirlwind 1	Farihan Dajami	
G–APWN	W.S.55 Whirlwind 3	Bristow Helicopters Ltd.	
G–APWR	PA-22 Tri-Pacer 160	Bencray Ltd.	
G–APWY	Piaggio P. 166	The Marconi Co. Ltd.	
G–APWZ	EP.9 Prospector	Sussex Agricultural Services	
G–APXF	W.S.55 Whirlwind 1	Farihan Dajami	
G–APXJ	PA-24 Comanche 250	J. W. Quine & G. F. Lindsay	
G–APXM	PA-22 Tri-Pacer 160	E. J. Skinner	
G–APXR	PA-22 Tri-Pacer 160	D. H. Rees & E. Hooton	
G–APXT	PA-22 Tri-Pacer 150	McKenzie & Tapp Ltd.	
G–APXU	PA-22 Tri-Pacer 125	D. F. S. Aviation	
G–APXW	EP.9 Prospector	Sussex Agricultural Services	
G–APXY	Cessna 150	Merlin Flying Club Ltd.	
G–APYB	T.66 Nipper 2	B. O. Smith	
G–APYD	*D.H.106 Comet 4B	Science Museum Store, Wroughton	
G–APYG	D.H.C.1 Chipmunk 22	E. J. I. Musty & P. A. Colman	
G–APYI	PA-22 Tri-Pacer 135	Air Farm Ltd.	
G–APYN	PA-22 Tri-Pacer 160	W. D. Stephens	
G–APYT	7FC Tri-Traveller	C. H. Morris & R. W. Brown	
G–APYU	7FC Tri-Traveller	K. Collins	
G–APYW	PA-22 Tri-Pacer 150	B. A. Wallace	
G–APYX	PA-23 Aztec 250	Tamavia Ltd.	
G–APYZ	D.31 Turbulent	A. R. Lee	
G–APZE	PA-23 Apache 160	J. P. Dodd	
G–APZG	PA-24 Comanche 250	Steve Stephens Ltd.	
G–APZJ	PA-18 Super Cub 150	Southern Sailplanes	
G–APZK	PA-18 Super Cub 95	W. T. Knapton	
G–APZL	PA-22 Tri-Pacer 160	F. J. Holdcroft Ltd.	
G–APZR	Cessna 150	C. R. E. S. Taylor	
G–APZS	Cessna 175A	A. J. House	
G–APZU	D.H.104 Dove 6	Acraman Holdings Ltd.	
G–APZX	PA-22 Tri-Pacer 150	M. G. Montgomerie & ptnrs.	
G–ARAB	Cessna 150	E. G. Whelan	
G–ARAI	PA-22 Tri-Pacer 160	J. E. Fox	
G–ARAJ	PA-22 Tri-Pacer 160	Wearside Flying Group	
G–ARAM	PA-18 Super Cub 150	B. A. Eastwell & J. P. H. Gresham	
G–ARAN	PA-18 Super Cub 150	Yorkshire Gliding Club (Pty) Ltd.	
G–ARAO	PA-18 Super Cub 95	Alidair Ltd.	
G–ARAP	7FC Tri-Traveller	R. W. Brown	
G–ARAS	7FC Tri-Traveller	A. Bruniges	
G–ARAT	Cessna 180C	Old Warden Flying Group	
G–ARAU	Cessna 150	R. N. R. Bellamy	
G–ARAW	Cessna 182C Skylane	Eco Systems Ltd.	
G–ARAX	PA-22 Tri-Pacer 150	Megacirc Ltd.	
G–ARAY	H.S.748 Srs. 2	Dan-Air Services Ltd.	
G–ARAZ	D.H.82A Tiger Moth	M. V. Gauntlett	
G–ARBE	D.H.104 Dove 8	British Aerospace	
G–ARBG	T.66 Nipper 2	Felthorpe Tipsy Group	
G–ARBH	D.H.104 Dove 1	International Travel Ltd.	
G–ARBL	D.31 Turbulent	C. C. Taylor	
G–ARBN	PA-23 Apache 160	S. Brunt	
G–ARBO	PA-24 Comanche 250	R. K. Blewett	
G–ARBP	T.66 Nipper 2	A. Cambridge & D. B. Winstanley	
G–ARBS	PA-22 Tri-Pacer 160	Garb Enterprises	
G–ARBV	PA-22 Tri-Pacer 150	C. R. Turner	
G–ARBX	PA-18 Super Cub 95	A. A. Alderdice	
G–ARBY	V.708 Viscount	Alidair Ltd.	
G–ARBZ	D.31 Turbulent	D. G. H. Hilliard	
G–ARCC	PA-22 Tri-Pacer 150	Fainville Ltd.	
G–ARCF	PA-22 Tri-Pacer 150	A. L. Scadding	
G–ARCI	Cessna 310D	R. H. B. & M. C. le Brocq	
G–ARCL	Cessna 175A	C. E. Sharp Plant Hire and Sales Ltd.	
G–ARCM	Cessna 172B Skyhawk	D. H. van Stoweren	
G–ARCS	Auster D6/180	E. A. Matty	

Notes	Reg.	Type	Owner or Operator
	G–ARCT	PA-18 Super Cub 95	M. Kirk
	G–ARCV	Cessna 175A	P. J. O. Hanlon
	G–ARCW	PA-23 Apache 160	E. M. Brain & R. Chew
	G–ARCX	*AW Meteor 14 (WM261)	Museum of Flight, E. Fortune
	G–ARCZ	D.31 Turbulent	Stapleford Turbulent Group
	G–ARDB	PA-24 Comanche 250	G. K. Hare
	G–ARDC	Cessna 210 Centurion	De Ville Investment Holdings Ltd.
	G–ARDD	CP.301C1 Emeraude	A. Mackintosh
	G–ARDE	D.H.104 Dove 6	Hunting Surveys & Consultants Ltd.
	G–ARDG	EP.9 Prospector	Sussex Agricultural Services
	G–ARDJ	Auster D.6/180	J. D. H. Radford
	G–ARDO	Jodel D.112	P. J. H. McCraig
	G–ARDP	PA-22 Tri-Pacer 150	G. M. Jones
	G–ARDS	PA-22 Caribbean 150	D. V. Asher
	G–ARDT	PA-22 Tri-Pacer 160	A. A. Whiter
	G–ARDV	PA-22 Tri-Pacer 160	M. R. Angood
	G–ARDY	T.66 Nipper 2	J. K. Davies
	G–ARDZ	Jodel D.140A	W. R. Dryden
	G–AREA	D.H.104 Dove 8	British Aerospace
	G–AREB	Cessna 175B Skylark	R. J. Postlethwaite & ptnrs.
	G–ARED	PA-23 Apache 160	T & S Associates
	G–AREE	PA-23 Aztec 250	Montague Travel Ltd.
	G–AREF	PA-23 Aztec 250	Tricolore Trading Co. Ltd.
	G–AREH	D.H.82A Tiger Moth	T. Pate
	G–AREI	Auster 3	R. Alliker & ptnrs.
	G–AREJ	Beech 95 Travel Air	D. Huggett
	G–AREL	PA-22 Caribbean 150	H. H. Cousins
	G–AREO	PA-18 Super Cub 150	Lasham Gliding Soc. Ltd.
	G–ARET	PA-22 Tri-Pacer 160	T. Haigh
	G–AREV	PA-22 Tri-Pacer 160	Echo Victor Group
	G–AREX	Aeronca 15AC Sedan	A. E. Poulson
	G–AREZ	D.31 Turbulent	C. M. D. Roberts
	G–ARFB	PA-22 Caribbean 150	Borrowash Estates Ltd.
	G–ARFD	PA-22 Tri-Pacer 160	C. Fergusson & ptnrs.
	G–ARFG	Cessna 175A Skylark	C. S. & Mrs. B. A. Frost
	G–ARFH	PA-24 Comanche 250	L. M. Walton
	G–ARFL	Cessna 175B Skylark	P. M. & R. G. Rees
	G–ARFM	Cessna 175B Skylark	Foyle Aviation Ltd.
	G–ARFO	Cessna 150A	R. J. Penwardes & T. Thornton
	G–ARFS	PA-22 Caribbean 150	M. H. Armstrong & C. McFadden
	G–ARFT	Jodel D.R. 1050	S. R. White
	G–ARFV	T.66 Nipper 2	D. J. Hockings
	G–ARGB	Auster 6A	A. M. Witt
	G–ARGG	D.H.C.1 Chipmunk 22	Air Navigation & Trading Co. Ltd.
	G–ARGK	Cessna 210	G. H. K. Rogers
	G–ARGO	PA-22 Colt 108	J. E. Backhouse & ptnrs.
	G–ARGR	V.708 Viscount	Alidair Ltd.
	G–ARGV	PA-18 Super Cub 150	Deeside Gliding Club (Aberdeenshire) Ltd.
	G–ARGY	PA-22 Tri-Pacer 160	D. H. Tanner & I. J. Enoch
	G–ARGZ	D.31 Turbulent	J. P. Luck
	G–ARHB	Forney F-1A Aircoupe	C. M. Robertson
	G–ARHC	Forney F-1A Aircoupe	W. F. Barnes
	G–ARHF	Forney F-1A Aircoupe	West London Aero Services Ltd.
	G–ARHI	PA-24 Comanche 180	B.C.F.G. Ltd.
	G–ARHL	PA-23 Aztec 250	J. J. Freeman & Co. Ltd.
	G–ARHM	Auster 6A	D. Hollowell & ptnrs.
	G–ARHN	PA-22 Caribbean 150	P. H. Pickford
	G–ARHP	PA-22 Tri-Pacer 160	W. Wardle
	G–ARHR	PA-22 Caribbean 150	J. A. Hargraves
	G–ARHT	PA-22 Caribbean 150	J. S. Lewery
	G–ARHU	PA-22 Tri-Pacer 160	G. K. Hare & G. W. Worley
	G–ARHW	D.H.104 Dove 8	British Aerospace
	G–ARHZ	D.62 Condor	D. H. Wilson-Spratt
	G–ARIA	Bell 47G	Decca Navigator Co. Ltd.
	G–ARID	Cessna 172B	N. Law
	G–ARIE	PA-24 Comanche 250	R. Rackwanski
	G–ARIF	O-H7 Minor Coupe	A. W. J. G. Ord-Hume
	G–ARIH	Auster 6A	B. D. Husband
	G–ARIK	PA-22 Caribbean 150	C. J. Berry
	G–ARIL	PA-22 Caribbean 150	G. N. Richardson Motors
	G–ARIN	PA-24 Comanche 250	Fisher Douglas Aviation Ltd.
	G–ARIR	V.708 Viscount	Alidair Ltd.

Reg.	Type	Owner or Operator	Notes
G-ARIU	Cessna 172B Skylark	P. A. Howell	
G-ARIV	Cessna 172B	J. A. Hood	
G-ARIW	CP.301B Emeraude	P. A. Brook	
G-ARJB	D.H.104 Dove 8	J. C. Bamford (Excavators) Ltd.	
G-ARJC	PA-22 Colt 108	C. E. Holland	
G-ARJE	PA-22 Colt 108	G. W. Doran	
G-ARJF	PA-22 Colt 108	E. G. Gilbert	
G-ARJG	PA-22 Colt 108	Sqd. Ldr. G. R. Sharp	
G-ARJH	PA-22 Colt 108	A. Walmsley	
G-ARJR	PA-23 Apache 160	R. F. Wanbon	
G-ARJS	PA-23 Apache 160	Bencray Ltd.	
G-ARJT	PA-23 Apache 160	R. D. Dickson	
G-ARJU	PA-23 Apache 160	Sega Aviation Ltd.	
G-ARJV	PA-23 Apache 160	Gordon King (Aviation) Ltd.	
G-ARJW	PA-23 Apache 160	Gordon King Aviation Ltd.	
G-ARJZ	D.31 Turbulent	The Tiger Club	
G-ARKG	J/5G Autocar	C. Thompson	
G-ARKJ	Beech N35 Bonanza	R. J. Guise	
G-ARKK	PA-22 Colt 108	A. W. Baxter	
G-ARKM	PA-22 Colt 108	G. M. Barber	
G-ARKN	PA-22 Colt 108	J. H. Underwood & A. J. F. Tabenor	
G-ARKP	PA-22 Colt 108	C. J. & J. Freeman	
G-ARKR	PA-22 Colt 108	D. J. Beale	
G-ARKS	PA-22 Colt 108	D. W. Mickleburgh	
G-ARLB	PA-24 Comanche 250	Stakehill Engineering Ltd.	
G-ARLD	H-395 Super Courier	P. F. Hall	
G-ARLG	Auster D.4/108	W. R. Dryden	
G-ARLI	PA-23 Apache 150	K. A. Learmonth	
G-ARLK	PA-24 Comanche 250	M. Walker & C. Robinson	
G-ARLL	PA-24 Comanche 250	E. J. Spiers	
G-ARLP	A.61 Terrier	Tavair Ltd.	
G-ARLR	A.61 Terrier	G. Griffith	
G-ARLT	Cessna 172B Skyhawk	A. R. German & Sons	
G-ARLU	Cessna 172B Skyhawk	Hanoflite Ltd.	
G-ARLV	Cessna 172B Skyhawk	K. W. & W. R. Morris	
G-ARLW	Cessna 172B Skyhawk	T. J. Evason & ptnrs.	
G-ARLX	Jodel D.140B	Gladaircraft Ltd.	
G-ARLY	J/5P Autocar	J. Alva & ptnrs.	
G-ARLZ	D.31A Turbulent	R. W. Rushton	
G-ARMA	PA-23 Apache 160	Oxford Air Training School	
G-ARMB	D.H.C.1 Chipmunk 22A	College of Air Training	
G-ARMC	D.H.C.1 Chipmunk 22A	W. London Aero Services Ltd.	
G-ARMD	D.H.C.1 Chipmunk 22A	College of Air Training	
G-ARMF	D.H.C.1 Chipmunk 22A	College of Air Training	
G-ARMG	D.H.C.1 Chipmunk 22A	College of Air Training	
G-ARMH	PA-23 Aztec 250	Royle Aviation Co. Ltd.	
G-ARMI	PA-23 Apache 160	Stapleford Flying Club Ltd.	
G-ARMJ	Cessna 185 Skywagon	British Skysports	
G-ARML	Cessna 175B Skylark	Woolmer Aircraft Ltd.	
G-ARMN	Cessna 175B Skylark	H. A. Claireaux & T. D. Gray	
G-ARMO	Cessna 172B Skyhawk	Sangria Designs Ltd. & BRM Plastics Ltd.	
G-ARMP	Cessna 172B	Southport & Merseyside Aero Club (1979) Ltd.	
G-ARMR	Cessna 172B Skyhawk	J. Braithwaite	
G-ARMW	H.S.748 Srs. 1	Dan-Air Services Ltd.	
G-ARMX	H.S.748 Srs. 1	Dan-Air Services Ltd.	
G-ARMZ	D.31 Turbulent	Frederick A. Shepherd	
G-ARNA	Mooney M.20B	R. Travers	
G-ARNB	J/5G Autocar	M. T. Jeffrey	
G-ARND	PA-22 Colt 108	Richard Rimington Ltd.	
G-ARNE	PA-22 Colt 108	J. A. Clyde-Smith	
G-ARNG	PA-22 Colt 108	W. London Aero Services Ltd.	
G-ARNI	PA-22 Colt 108	T. Rundle	
G-ARNJ	PA-22 Colt 108	MKM Flying Group	
G-ARNK	PA-22 Colt 108	D. P. Golding	
G-ARNL	PA-22 Colt 108	Mr. J. A. & Miss J. A. Dodsworth	
G-ARNN	GC-1B Swift	K. E. Sword	
G-ARNO	A.61 Terrier	M. B. Hill.	
G-ARNP	A.109 Airedale	T. A. J. Luther & ptnrs.	
G-ARNY	Jodel D.117	A. D. Henderson	
G-ARNZ	D.31 Turbulent	P. L. Cox & ptnrs	
G-AROA	Cessna 172B Skyhawk	D. E. Partridge	

Notes	Reg.	Type	Owner or Operator
	G–AROC	Cessna 175B Skylark	Yorkshire Flying Services Ltd.
	G–AROD	Cessna 175B	Medical Co. Hospital Supplies Ltd.
	G–AROE	Aero 145	G. S. & Mrs. P. Galt
	G–AROF	L.40 Meta-Sokol	B. G. Barber
	G–AROJ	A.109 Airedale	D. J. Shaw
	G–AROK	Cessna 310F	S. E. Berry
	G–ARON	PA-22 Colt 108	P. Pickford
	G–AROO	Forney F-1A Aircoupe	W. I. McMeekan
	G–AROT	F.8L Falco	H. J. A. Morris
	G–AROW	Jodel D.140B	Kent Gliding Club Ltd.
	G–AROY	Stearman A.75N.1	W. A. Jordan
	G–ARPD	H.S.121 Trident 1C	British Airways
	G–ARPH	H.S.121 Trident 1C	British Airways
	G–ARPK	H.S.121 Trident 1C	British Airways
	G–ARPL	H.S.121 Trident 1C	British Airways
	G–ARPN	H.S.121 Trident 1C	British Airways
	G–ARPO	H.S.121 Trident 1C	British Airways
	G–ARPP	H.S.121 Trident 1C	British Airways
	G–ARPR	H.S.121 Trident 1C	British Airways
	G–ARPW	H.S.121 Trident 1C	British Airways
	G–ARPX	H.S.121 Trident 1C	British Airways
	G–ARPZ	H.S.121 Trident 1C	British Airways
	G–ARRD	Jodel DR. 1050	N. L. E. Dupee
	G–ARRE	Jodel DR. 1050	E. H. Ellis
	G–ARRF	Cessna 150A	Cornwall Flying Club
	G–ARRH	Cessna 175B Skylark	AJ Aviation
	G–ARRI	Cessna 175B Skylark	C. L. Thomas
	G–ARRL	J/1N Alpha	A. J. Brown
	G–ARRM	*Beagle B.206-X	Brighton Transport Museum
	G–ARRP	PA-28 Cherokee 180	Dyce Flying Group
	G–ARRS	CP-301A Emeraude	J. Y. Paxton
	G–ARRT	Wallis WA-116-1	K. H. Wallis
	G–ARRU	D.31 Turbulent	E. R. Gourd & ptnrs.
	G–ARRW	H.S.748 Srs. 1	Dan-Air Services Ltd.
	G–ARRY	Jodel D.140B	R. G. Andrews
	G–ARRZ	D.31 Turbulent	J. S. Barker
	G–ARSB	Cessna 150A	P. G. Upton & D. J. Orbell
	G–ARSC	PA-24 Comanche 180	Fenland Aviation
	G–ARSJ	CP.301-C2 Emeraude	G. G. Milton
	G–ARSL	A.61 Terrier	C. W. Thomas
	G–ARSN	D.H.104 Dove 8	Staravia Ltd.
	G–ARSP	L.40 Meta-Sokol	K. K. Johnson
	G–ARSU	PA-22 Colt 108	P. E. Palmer
	G–ARSW	PA-22 Colt 108	J. P. Smith
	G–ARSX	PA-22 Tri-Pacer 160	AF Aviation Ltd.
	G–ARTB	Mooney M.20B	R. E. Dagless
	G–ARTD	PA-23 Apache 160	Dr. D. A. Jones
	G–ARTF	D.31 Turbulent	L. J. Pennell
	G–ARTG	Hiller UH-12C	E. C. Francis
	G–ARTH	PA-12 Super Cruiser	R. Battle & D. Mallinson
	G–ARTL	D.H.82A Tiger Moth (T7281)	S. E. Marples
	G–ARTT	M.S.880B Rallye Club	J. Berry
	G–ARUE	D.H.104 Dove 7	Staravia Ltd.
	G–ARUG	J/5G Autocar	N. P. Biggs
	G–ARUH	Jodel DR. 1050	PFA Group
	G–ARUI	A.61 Terrier	D. C. Cullen
	G–ARUL	Cosmic Wind	J. Cull
	G–ARUM	D.H.104 Dove 8	National Coal Board
	G–ARUO	PA-24 Comanche 180	Uniform Oscar Group
	G–ARUR	PA-28 Cherokee 160	Falconash Ltd.
	G–ARUV	CP.301A Emeraude	J. Tanswell
	G–ARUY	J/1N Alpha	A. J. Brown
	G–ARUZ	Cessna 175C Skylark	J. E. Sansome & M. D. Faiers
	G–ARVF	V.1101 VC10	United Arab Emirates
	G–ARVJ	V.1101 VC10	British Airways
	G–ARVM	*V.1101 VC10	Aerospace Museum, Cosford
	G–ARVS	PA-28 Cherokee 160	Stapleford Flying Club Ltd.
	G–ARVT	PA-28 Cherokee 160	C. R. Knapton
	G–ARVU	PA-28 Cherokee 160	Palestar Aviation Ltd.
	G–ARVV	PA-28 Cherokee 160	R. J. Jackson
	G–ARVW	PA-28 Cherokee 160	Bolton Air Training School Ltd.
	G–ARVZ	D.62B Condor	O. G. Stuart-Lee
	G–ARWB	D.H.C.1 Chipmunk 200	Aero Bonner Co. Ltd.

Reg.	Type	Owner or Operator	Notes
G–ARWC	Cessna 150B	Worldwide Wheels Ltd.	
G–ARWD	Boeing 707-465	British Airtours	
G–ARWH	Cessna 172C Skyhawk	N. M. Meadows	
G–ARWM	Cessna 175C	Sylman Aviation	
G–ARWO	Cessna 172C Skyhawk	T. A. Cox & R. C. Jackman	
G–ARWR	Cessna 172C Skyhawk	Howe Aviation Ltd.	
G–ARWS	Cessna 175C Skylark	J. Mudd	
G–ARWW	Bensen B.8M	B. McIntyre	
G–ARWX	Luton L.A.5A Major	A. G. Cameron	
G–ARWY	Mooney M.20A	P. J. Hatcher	
G–ARXC	A.109 Airedale	E. A. Wright	
G–ARXD	A.109 Airedale	D. Howden	
G–ARXF	PA-23 Aztec 250B	Air Atlantique Ltd.	
G–ARXG	PA-24 Comanche 250	F. Stewart	
G–ARXH	Bell 47G	A.C.C. Builders & Capricorn Studios Ltd.	
G–ARXN	Tipsy Nipper 2	Griffon Flying Group	
G–ARXP	Luton L.A.4A Minor	W. C. Hymas	
G–ARXT	Jodel DR. 1050	A. W. Webster	
G–ARXU	Auster 6A	Bath & Wilts Gliding Club Ltd.	
G–ARXW	M.S.885 Super Rallye	E. W. Noakes	
G–ARXX	M.S.880B Rallye Club	M. S. Bird	
G–ARXY	M.S.880B Rallye Club	Horizon Flying Group	
G–ARYC	*H.S.125 Srs. 1	The Mosquito Aircraft Museum	
G–ARYF	PA-23 Aztec 250	Express Aviation Services Ltd.	
G–ARYH	PA-22 Tri-Pacer 160	Filtration (Water Treatment Engineers) Ltd.	
G–ARYI	Cessna 172C	J. T. Mirley & A. E. Wall	
G–ARYK	Cessna 172C	Mrs. K. M. & T. Hemsley	
G–ARYM	D.H.104 Dove 8	Canbaria (Aircraft Spares) Ltd.	
G–ARYR	PA-28 Cherokee 180	D. J. Grant	
G–ARYS	Cessna 172C Skyhawk	K. J. Squires	
G–ARYV	PA-24 Comanche 250	P. Meeson	
G–ARYZ	A.109 Airedale	J. D. Reid	
G–ARZA	Wallis WA.116 Srs. 1	N. D. Z. de Ferranti	
G–ARZB	Wallis WA.116 Srs. 1	K. H. Wallis	
G–ARZD	Cessna 172C	T. Poole & P. C. Thompson	
G–ARZF	Cesna 150B	J. E. F. Aviation Ltd.	
G–ARZM	D.31 Turbulent	The Tiger Club	
G–ARZN	Beech N35 Bonanza	Beech Aircraft Ltd.	
G–ARZP	A.109 Airedale	Airedale Aero Services	
G–ARZW	Currie Wot	D. F. Faulkener-Byrant	
G–ARZX	Cessna 150B	J. M. Clitherow	
G–ASAA	Luton L.A.4A. Minor	Four Counties Flying Syndicate	
G–ASAI	A.109 Airedale	A. C. Watt	
G–ASAJ	A.61 Terrier 2	R Skingley & P. G. Bolton	
G–ASAK	A.61 Terrier 2	Rochford Hundred Flying Group	
G–ASAL	SAL Bulldog 120	British Aerospace	
G–ASAM	D.31 Turbulent	The Tiger Club	
G–ASAN	A.61 Terrier 2	M. R. E. & M. V. Stillingfleet	
G–ASAT	M.S.880B Rallye Club	J. N. Burke	
G–ASAU	M.S.880B Rallye Club	W. J. Armstrong	
G–ASAV	M.S.880B Rallye Club	McAully Flying Group	
G–ASAX	A.61 Terrier 2	B. Rhodes	
G–ASAZ	Hiller UH-12 E4	Morland Beazley Helicopters Ltd.	
G–ASBA	Currie Wot	M. A. Kaye	
G–ASBB	Beech 23 Musketeer	D. Silver	
G–ASBD	Hughes 269A	C. R. McNeil	
G–ASBG	HPR-7 Herald 203	Air UK	
G–ASBH	A.109 Airedale	H. B. Fox & J. Fox	
G–ASBS	C.P.301A Emeraude	D. M. Upfield & A. Newton	
G–ASBU	A.61 Terrier 2	G. Strathdee	
G–ASBY	A.109 Airedale	K. P. Donnan	
G–ASCC	Beagle E.3 AOP Mk. II	A. D. Heath & M. D. N. Fisher	
G–ASCH	A.61 Terrier 2	Enstone Eagles Flying Group	
G–ASCJ	PA-24 Comanche 250	Telspec Ltd.	
G–ASCM	Isaacs Fury II (K2050)	D. Toms & ptnrs.	
G–ASCU	PA-18A-150 Super Cub	Farm Aviation Services Ltd.	
G–ASCZ	CP.310A Emeraude	Hylton Flying Group	
G–ASDA	Beech 65–80 Queen Air	Parker & Heard Ltd.	
G–ASDD	D.H.104 Dove 5	Haywards Aviation Ltd.	
G–ASDL	A.61 Terrier 2	C. P. Lockyer & C. E. Mason	
G–ASDN	PA-24 Comanche 250	J. W. Hodgson	

Notes	Reg.	Type	Owner or Operator
	G–ASDO	Beech 95A.55 Baron	Executive Aviation Ltd.
	G–ASDY	Wallis WA-116/F	K. H. Wallis
	G–ASEA	Luton L.A.4A Minor	Toon Ghose Aviation Ltd.
	G–ASEB	Luton L.A.4A Minor	R. K. Lynn
	G–ASEE	J/IN Alpha	H. C. J. & Sara L. G. Williams
	G–ASEG	A.61 Terrier	J. T. Hogben
	G–ASEO	PA-24 Comanche 250	A. van Daalen
	G–ASEP	PA-23 Apache 235	G. R. Selbert
	G–ASEU	D.62A Condor	The Tiger Club
	G–ASEV	PA-23 Aztec 250	Chad (Air Services) Ltd.
	G–ASFA	Cessna 172D	R. A. Marven
	G–ASFB	Beech 23 Musketeer	R. P. Whalley
	G–ASFD	L-200A Morava	Pochin (Wales) Ltd.
	G–ASFG	PA-23 Aztec 250	Air Conditioning (Transport) Ltd.
	G–ASFJ	Beech P.35 Bonanza	Tadavia Ltd.
	G–ASFK	J/5G Autocar	Orman (Carrolls Farm) Ltd.
	G–ASFL	PA-28 Cherokee 180	K. Winfield & ptnrs.
	G–ASFR	Bo 208 Junior	Sunderland Flying Club Ltd.
	G–ASFX	D.31 Turbulent	E. F. Clapham & W. B. S. Dobie
	G–ASGA	V.1151 Super VC10	British Airways
	G–ASGB	V.1151 Super VC10	British Airways
	G–ASGC	*V.1151 Super VC10	Imperial War Museum, Duxford
	G–ASGD	V.1151 Super VC10	British Airways (withdrawn)
	G–ASGE	V.1151 Super VC10	British Airways (withdrawn)
	G–ASGF	V.1151 Super VC10	British Airways
	G–ASGG	V.1151 Super VC10	British Airways
	G–ASGH	V.1151 Super VC10	British Airways (withdrawn)
	G–ASGI	V.1151 Super VC10	British Airways (withdrawn)
	G–ASGJ	V.1151 Super VC10	British Airways (withdrawn)
	G–ASGK	V.1151 Super VC10	British Airways
	G–ASGL	V.1151 Super VC10	British Airways (withdrawn)
	G–ASGM	V.1151 Super VC10	British Airways (withdrawn)
	G–ASGP	V.1151 Super VC10	British Airways
	G–ASGR	V.1151 Super VC10	British Airways (withdrawn)
	G–ASHA	Cessna F.172D	R. L. Fogg & Co. Ltd. & R. Soar
	G–ASHB	Cessna 182F	G. Theldrake Ltd.
	G–ASHH	PA-23 Aztec 250	G. Everington
	G–ASHJ	Brantly B.2B	A. G. Dean
	G–ASHK	Brantly B.2A	S. N. Cole
	G–ASHO	Cessna 182F Skylane	G. & H. Skydivers Ltd.
	G–ASHR	Beech B35–33 Debonair	C. M. Fraser & E. A. Perry
	G–ASHS	Stampe SV.4B	The Tiger Club
	G–ASHT	D.31 Turbulent	D. L. Riley
	G–ASHU	PA-15 Vagabond	G. J. Romanes
	G–ASHV	PA-E23 Aztec 250	Haywards Aviation Ltd.
	G–ASHW	D. H.104 Dove 8	L. de la Hay (Fishing & Marine) Salvage Ltd.
	G–ASHX	PA-28 Cherokee 180	R. Lee & D. Morris
	G–ASIB	Cessna F.172D	K. D. Horton
	G–ASII	PA-28 Cherokee 180	Worldwide Wheels Ltd.
	G–ASIJ	PA-28 Cherokee 180	B. H. Pearce
	G–ASIL	PA-28 Cherokee 180	F. W. Shaw & Sons (Worthing) Ltd.
	G–ASIP	Auster 6A	Bristol Gliding Club (Pty) Ltd.
	G–ASIT	Cessna 180	A. J. Spiller (Gt. Staughton) Ltd.
	G–ASIU	Beech B.65-80 Queen Air	Vernair Transport Services
	G–ASIY	PA-25 Pawnee 235	A.D.S. (Aerial) Ltd.
	G–ASJC	BAC One-Eleven 201	British Caledonian Airways *City of Glasgow*
	G–ASJE	BAC One-Eleven 201	British Caledonian Airways *City of Dundee*
	G–ASJF	BAC One-Eleven 201	British Caledonian Airways *Burgh of Fort William*
	G–ASJG	BAC One-Eleven 201	British Caledonian Airways *Burgh of Paisley*
	G–ASJH	BAC One-Eleven 201	British Caledonian Airways *Burgh of Hawick*
	G–ASJI	BAC One-Eleven 201	British Caledonian Airways *Royal Burgh of Nairn*
	G–ASJL	Beech H.35 Bonanza	P. M. Coulten
	G–ASJM	PA-30 Twin Comanche 160	Gainsborough Precision Engineering (Birmingham) Ltd.
	G–ASJO	Beech B.23 Musketeer	Laxtonbridge Ltd.
	G–ASJU	Aero Commander 520	Interflight Ltd.

Reg.	Type	Owner or Operator	Notes
G–ASJV	V.S.361 Spitfire IX (MH434)	Airborne Taxi Service Ltd.	
G–ASJY	GY-80 Horizon 160	A. D. Hemley	
G–ASJZ	Jodel D.117A	Wolverhampton Ultra-light Flying Group	
G–ASKB	D.H.98 Mosquito 35 (RS712)	Strathallan Aircraft Collection	
G–ASKC	*D.H.98 Mosquito 35 (TA719)	Skyfame Collection	
G–ASKH	D.H.98 Mosquito T.3 (RR299)	Hawker Siddeley Aviation	
G–ASKJ	A.61 Terrier 1	Bristol & Gloucestershire Gliding Club	
G–ASKK	HPR-7 Herald 211	Air UK	
G–ASKL	Jodel D.150A	J. M. Graty	
G–ASKM	Beech B.65–80 Queen Air	Texalon International Ltd.	
G–ASKP	D.H.82A Tiger Moth	The Tiger Club	
G–ASKS	Cessna 336 Skymaster	M. J. Godwin	
G–ASKT	PA-28 Cherokee 180	Capel & Co. (Printers) Ltd.	
G–ASKV	PA-25 Pawnee 235	Westwick Distributors Ltd.	
G–ASLA	PA-25 Pawnee 235	Westwick Distributors Ltd.	
G–ASLE	PA-30 Twin Comanche 160	Allisons Automated Machinery Ltd.	
G–ASLF	Bensen B.7	S. R. Hughes	
G–ASLH	Cessna 182F	Celahurst Ltd.	
G–ASLK	PA-25 Pawnee 235	Westwick Distributors Ltd.	
G–ASLN	Forney F.1A Aircoupe	Cornwall Flying Club	
G–ASLR	Agusta-Bell 47J-2	David Barriball Ltd.	
G–ASLV	PA-28 Cherokee 235	C.S.E. (Aircraft Services) Ltd.	
G–ASLX	CP.301A Emeraude	K. C. Green	
G–ASMA	PA-30 Twin Comanche 160	M. G. Edmunds	
G–ASMC	P.56 Provost T.1.	W. Walker	
G–ASME	Bensen B.7M	C. R. Pepper & A. J. Tabenor	
G–ASMF	Beech D.95A Travel Air	R. M. & C. T. Harlock (Thorney) Ltd.	
G–ASMG	D.H.104 Dove 8	British Aerospace	
G–ASMH	PA-30 Twin Comanche 160	Surface Craft Ltd. & MLP Aviation Ltd.	
G–ASMJ	Cessna F.172E	Conguess Aviation Ltd.	
G–ASML	Luton L.A.4A Minor	L. Davies	
G–ASMM	D.31 Turbulent	Kenneth Browne	
G–ASMN	PA-23 Apache 160	W. London Aero Services Ltd.	
G–ASMO	PA-23 Apache 160	Segadaloes Ltd.	
G–ASMS	Cessna 150A	K. R. & T. W. Davies	
G–ASMT	Fairtravel Linnet 2	J. H. Hartley	
G–ASMU	Cessna 150D	Stapleford Flying Club Ltd.	
G–ASMV	CP1310-C3 Super Emeraude	P. F. D. Waltham	
G–ASMW	Cessna 150D	D. M. Whitham	
G–ASMY	PA-23 Apache 160	Thurston Aviation Ltd.	
G–ASMZ	A.61 Terrier 2	R. B. Humphries	
G–ASNA	PA-23 Aztec 250	Margate Motors Plant & Aircraft Hire Ltd.	
G–ASNB	Auster 6A	M. Pocock & ptnrs.	
G–ASNC	Beagle D.5/180 Husky	D. B. Winstanley	
G–ASND	PA-23 Aztec 250	Survair Ltd.	
G–ASNE	PA-28 Cherokee 180	J. L. Dexter	
G–ASNF	Ercoupe 415-CD	Charles Robertson (Developments) Ltd.	
G–ASNH	PA-23 Aztec 250	Derek Crouch (Contractors) Ltd.	
G–ASNI	CP1310-C3 Super Emeraude	D. Chapman	
G–ASNK	Cessna 205	Transgap Ltd.	
G–ASNL	Sikorsky S-61N	British Airways Helicopters Ltd.	
G–ASNN	Cessna 182F Skylane	Banbury Plant Hire Ltd.	
G–ASNO	Beech B55 Baron	Can Handling Developments Ltd.	
G–ASNP	Mooney M.20C Mark 21	W. C. Smeaton	
G–ASNU	H.S.125 Srs. 1	Northern Engineering Industries Ltd.	
G–ASNV	Agusta-Bell 47J-2	S. W. Electricity Board	
G–ASNW	Cessna F.172E	Taregate Ltd.	
G–ASNY	Bensen B.8M	D. L. Wallis	
G–ASNZ	Bensen B.7M	W. H. Turner	
G–ASOB	PA-30 Twin Comanche 160	M. A. Grayburn	
G–ASOC	Auster 6A	Aquila Gliding Club	
G–ASOF	Beagle B.206 Srs. 1	Teespeed Ltd.	
G–ASOH	Beech B.55A Baron	Steer Aviation Ltd.	
G–ASOI	A.61 Terrier 2	Brooklands Flying Group	
G–ASOK	Cessna F.172E	Okay Flying Group	
G–ASON	PA-30 Twin Comanche 160	Miss P. J. Rourke	
G–ASOO	PA-30 Twin Comanche 160	John Bisco (Cheltenham) Ltd. & ptnrs.	
G–ASOV	PA-25 Pawnee 235	A.D.S. (Aerial) Ltd.	
G–ASOX	Cessna 205A	Berrard Ltd.	

Notes	Reg.	Type	Owner or Operator
	G–ASPA	D.H.104 Dove 8	Staravia Ltd.
	G–ASPF	Jodel D.120	G. H. Farr
	G–ASPI	Cessna F.172E	A. M. Castleton & ptnrs.
	G–ASPK	PA-28 Cherokee 140	R. T. Love
	G–ASPL	H.S.748 Srs. 2	Dan-Air Services Ltd. *City of Berne*
	G–ASPS	Piper J-3C-65 Cub	A. J. Chalkley
	G–ASPT	Cessna 172D	C. H. Parker
	G–ASPU	D.31 Turbulent	I. Maclennen
	G–ASPV	D.H.82A Tiger Moth	B. S. Charters
	G–ASPX	Bensen B-8S	L. D. Goldsmith
	G–ASRB	D.62B Condor	Rollason Aircraft & Engines Ltd.
	G–ASRC	D.62B Condor	N. H. Jones
	G–ASRE	PA-23 Aztec 250	S.P.T. Aircraft Ltd.
	G–ASRF	Jenny Wren	G. W. Gowland
	G–ASRH	PA-30 Twin Comanche 160	IOM & General Life Assurance Co. Ltd. & Bigland Holdings Ltd.
	G–ASRI	PA-23 Aztec 250	Meridian Airmaps Ltd.
	G–ASRK	A.109 Airedale	Miss C. I. Clarry
	G–ASRO	PA-30 Twin Comanche 160	A. G. Perkins
	G–ASRP	Jodel DR 1050	D. J. M. Edmondston
	G–ASRR	Cessna 182G	John V. White Ltd.
	G–ASRT	Jodel D.150	H. M. Kendall
	G–ASRU	PA-30 Twin Comanche 160	M. L. Sparks
	G–ASRW	PA-28 Cherokee 180	K. R. Deering
	G–ASRX	Beech 65 A80 Queen Air	Seismograph Service (England) Ltd. & Parker & Heard Ltd.
	G–ASSB	PA-30 Twin Comanche 160	E. Berks Boat Co. Ltd.
	G–ASSE	PA-22 Colt 108	J. B. King
	G–ASSF	Cessna 182G Skylane	A. Newsham
	G–ASSI	H.S.125 Srs. 1	TBF (Transport) Ltd.
	G–ASSO	Cessna 150D	W. H. & J. Rogers Group Ltd.
	G–ASSP	PA-30 Twin Comanche 160	The Mastermix Engineering Co. Ltd.
	G–ASSR	PA-30 Twin Comanche 160	Direct Air Ltd.
	G–ASSS	Cessna 172E	D. H. N. Squires
	G–ASST	Cessna 150D	F. R. H. Parker
	G–ASSU	CP 301 A Emeraude	R. W. Millward
	G–ASSW	PA-28 Cherokee 140	C. J. Plummer
	G–ASSX	Cessna 172E	Bridlington Aviation Ltd.
	G–ASSY	D.31 Turbulent	G–ASSY Group
	G–ASTA	D.31 Turbulent	D. J. R. Williams
	G–ASTD	PA-23 Aztec 250	Peregrine Air Services Ltd.
	G–ASTG	Nord 1002	L. M. Walton
	G–ASTI	Auster 6A	M. Pocock
	G–ASTJ	BAC One-Eleven 201	British Caledonian Airways *Royal Burgh of Dunfermline*
	G–ASTL	*Fairey Firefly I (Z2033)	Skyfame Collection
	G–ASTM	Hiller UH-12B	Bristow Helicopters Ltd.
	G–ASTP	Hiller UH-12C	L. Goddard
	G–ASTR	Hiller UH-12B	Bristow Helicopters Ltd.
	G–ASTZ	Hughes 269B	Twyford Moors (Helicopters) Ltd.
	G–ASUB	Mooney M.20E Super 21	T. J. Pigott
	G–ASUD	PA-28 Cherokee 180	Peterborough Aero Club Ltd.
	G–ASUE	Cessna 150D	D. Huckle
	G–ASUG	*Beech E18S	Royal Scottish Museum
	G–ASUH	Cessna F.172E	G. H. Wilson & E. Shipley
	G–ASUI	A.61 Terrier 2	Britrange Ltd.
	G–ASUL	Cessna 182G Skylane	G. H. Reeve
	G–ASUP	Cessna F.172E	W. T. Jenkins & ptnrs.
	G–ASUR	Dornier Do 28A-1	Sheffair Ltd.
	G–ASUS	Jurca MJ.2B Tempete	D. G. Jones
	G–ASVG	CP.301B Emeraude	K. R. Jackson
	G–ASVH	Hiller UH-12B	P. W. Hicks
	G–ASVI	Hiller UH-12B	Bristow Helicopters Ltd.
	G–ASVK	Hiller UH-12B	Bristow Helicopters Ltd.
	G–ASVL	Hiller UH-12B	Bristow Helicopters Ltd.
	G–ASVM	Cessna F.172E	J. White
	G–ASVN	Cessna U.206 Super Skywagon	British Skysports
	G–ASVO	HPR-7 Herald 214	British Air Ferries
	G–ASVP	PA-25 Pawnee 235	A.D.S. (Aerial) Ltd.
	G–ASVV	Cessna 310 I	A. R. J. Propafloor
	G–ASVZ	PA-28 Cherokee 140	A. D. Henley
	G–ASWA	PA-28 Cherokee 140	Southend Light Aviation Centre Ltd.
	G–ASWB	A.109 Airedale	Gainsborough Cars

Reg.	Type	Owner or Operator	Notes
G-ASWE	Bo 208A2 Junior	F. M. Wallis	
G-ASWF	A.109 Airedale	D. W. Wastell	
G-ASWG	PA-25 Pawnee 235	A.D.S. (Aerial) Ltd.	
G-ASWH	Luton L.A.5A Major	N. P. Gething	
G-ASWI	W.S.-58 Wessex 60 Srs. I	Bristow Helicopters Ltd.	
G-ASWJ	*Beagle 206 Srs. I (8449M)	RAF Halton	
G-ASWL	Cessna F.172F	C. Wilson	
G-ASWN	Bensen B.8M	D. R. Shepherd	
G-ASWO	Cessna 210D	DK Investments Ltd.	
G-ASWP	Beech A.23 Musketeer	Tenair Ltd.	
G-ASWT	Aero 145 Series 20	A. C. Frost	
G-ASWW	PA-30 Twin Comanche 160	Bristol & Wessex Flying Club Ltd.	
G-ASWX	PA-28 Cherokee 180	H. I. Jones (Whitby Garage) Ltd.	
G-ASXB	D.H.82A Tiger Moth	G. W. Bisshopp	
G-ASXC	SIPA 901	Waterside Flying Group	
G-ASXD	Brantly B.2B	Brantly Enterprises	
G-ASXF	Brantly 305	Express Aviation Services Ltd.	
G-ASXI	T.66 Nipper 2	J. Shrimpton	
G-ASXJ	Luton L.A.4A Minor	P. D. Lea & E. A. Lingard	
G-ASXR	Cessna 210	Dolphin Wetsuits Ltd.	
G-ASXS	Jodel DR.1050	J. P. I. Lloyd-Bostock	
G-ASXT	G.159 Gulfstream I	The Ford Motor Co. Ltd.	
G-ASXU	Jodel D.120A	R. W. & Mrs. J. Thompsett	
G-ASXX	*Avro 683 Lancaster 7 (NX611)	RAF Scampton Gate Guard	
G-ASXY	Jodel D.117A	G. Ward & ptnrs.	
G-ASXZ	Cessna 182G Skylane	P. M. Robertson	
G-ASYD	BAC One-Eleven 670	B.A.C. (Holdings) Ltd.	
G-ASYG	A.61 Terrier 2	B. J. Guest	
G-ASYJ	Beech D.95A Travel Air	Crosby Aviation Ltd.	
G-ASYK	PA-30 Twin Comanche 160	Heath Street Car Hirings Ltd.	
G-ASYL	Cessna 150E	British Skysports	
G-ASYP	Cessna 150E	T. S. Quirk	
G-ASYV	Cessna 310G	Chromecourt Ltd.	
G-ASYW	Bell 47G-2	Bristow Helicopters Ltd.	
G-ASYZ	Victa Airtourer 100	J. B. Cave	
G-ASZB	Cessna 150E	H. J. Cox	
G-ASZE	A.61 Terrier 2	P. J. Moore	
G-ASZF	Boeing 707-336C	British Airways	
G-ASZG	Boeing 707-336C	British Airways	
G-ASZJ	S.C.7 Skyvan 3A-100	Short Bros Ltd.	
G-ASZR	Fairtravel Linnet	H. C. D. & F. J. Garner	
G-ASZS	GY.80 Horizon 160	I. B. Willis	
G-ASZU	Cessna 150E	C. C. Dobson	
G-ASZV	T.66 Nipper 2	G. Thompson	
G-ASZW	Cessna F.172F	Glos-Air (Sales) Ltd.	
G-ASZX	A.61 Terrier	W. D. Hill	
G-ASZY	FRED Srs. 2	E. Clutton & A. J. F. Tabenor	
G-ASZZ	Cessna 310J	European Steel Sheets Ltd.	
G-ATAA	PA-28 Cherokee 180	Brendair	
G-ATAD	Mooney M.20C	H. W. Walker	
G-ATAF	Cessna F.172F	G. Bush	
G-ATAG	Jodel DR. 1050	J. L. Banks & ptnrs.	
G-ATAH	Cessna 336 Skymaster	Alderney Air Charter	
G-ATAI	D.H.104 Dove 8	Centrax Ltd. & E.C.C. Quarries Ltd.	
G-ATAO	PA-24 Comanche 260	C. V. Jones	
G-ATAS	PA-28 Cherokee 180	D. R. Wood	
G-ATAT	Cessna 150E	Ron Webster (Midlands) Ltd.	
G-ATAU	D.62B Condor	The Tiger Club	
G-ATAV	D.62C Condor	Lasham Gliding Soc. Ltd.	
G-ATAW	A.109 Airedale	Jean Dalton	
G-ATBF	*F-86E Sabre 4 (XB733)	Historic Aircraft Preservation Soc.	
G-ATBG	Nord 1002	L. M. Walton	
G-ATBH	Aero 145	Colpak Aviation Ltd.	
G-ATBI	Beech A.23 Musketeer	R. F. G. Dent	
G-ATBJ	Sikorsky S-61N	British Airways Helicopters Ltd.	
G-ATBK	Cessna F.172F	F. W. Sherrell	
G-ATBL	D.H.60G Moth	A. Haigh-Thomas	
G-ATBN	PA-28 Cherokee 140	Stapleford Flying Club Ltd.	
G-ATBP	Fournier RF-3	C. Jacques & ptnrs.	
G-ATBS	D.31 Turbulent	J. A. Thomas	
G-ATBU	A.61 Terrier 2	P. R. Anderson	
G-ATBV	PA-23 Aztec 250	Cabair Ltd.	

Notes	Reg.	Type	Owner or Operator
	G–ATBW	T.66 Nipper 2	D. Marshall
	G–ATBX	PA-20 Pacer 135	R. I. Souch
	G–ATBZ	W.S-58 Wessex 60	Bristow Helicopters Ltd.
	G–ATCC	A.109 Airedale	A. Robinson
	G–ATCD	D.5/180 Husky	Oxford Flying & Gliding Group
	G–ATCE	Cessna U.206	J. J. Aviation Ltd.
	G–ATCI	Victa Airtourer 100	Sunderland Flying Club Ltd.
	G–ATCJ	Luton L.A.4A Minor	R. M. Sharphouse
	G–ATCL	Victa Airtourer 100	Sunderland Flying Club Ltd.
	G–ATCN	Luton L.A.4A Minor	J. C. Gates & C. Neilson
	G–ATCR	Cessna 310	Staverton Apache Ltd.
	G–ATCU	Cessna 337	University of Cambridge
	G–ATCX	Cessna 182H Skylane	I. H. Sugden
	G–ATCY	PA-23 Aztec 250	Eastern-Air Executive Ltd.
	G–ATDA	PA-28 Cherokee 160	D. E. Siviter (Motors) Ltd.
	G–ATDB	Nord 1101 Noralpha	J. B. Jackson
	G–ATDC	PA-23 Aztec 250	Edinburgh Flying Services Ltd.
	G–ATDN	A.61 Terrier 2	J. F. Moore
	G–ATDO	Bo 208C Junior	H. Swift
	G–ATDS	HPR.7 Herald 209	Express Air Services Ltd.
	G–ATDV	PA-24 Comanche 400	Ugrain Aviation (Jersey) Ltd.
	G–ATDZ	Z-326 Trener Master	M. D. Popoff
	G–ATED	Hiller UH-12E	North Scottish Helicopters Ltd.
	G–ATEF	Cessna 150E	M. R. Nichols & ptnrs.
	G–ATEG	Cessna 150E	150 Group
	G–ATEM	PA-28 Cherokee 180	G. Wyles & W. Adams
	G–ATEP	EAA Biplane	E. L. Martin
	G–ATES	PA-32 Cherokee Six 260	Aeroscot Ltd.
	G–ATET	PA-30 Twin Comanche 160	Brenda Bayliss
	G–ATEV	Jodel DR. 1050	B. A. Mills & G. W. Payne
	G–ATEW	PA-30 Twin Comanche 160	Air Northumbria Group
	G–ATEX	Victa Airtourer 100	Landenair (GB) Ltd.
	G–ATEZ	PA-28 Cherokee 140	J. A. Burton
	G–ATFA	Bensen B-8	J. Butler
	G–ATFD	Jodel DR.1050	H. Fawcett & ptnrs.
	G–ATFF	PA-23 Aztec 250	Topflight Aviation Co. Ltd.
	G–ATFG	Brantly B.2B	R. J. Chapman Ltd.
	G–ATFK	PA-30 Twin Comanche 160	L. J. Martin
	G–ATFL	Cessna F.172F	R. L. Beverley
	G–ATFM	Sikorsky S-61N	British Airways Helicopters Ltd.
	G–ATFR	PA-25 Pawnee 150	J. F. Pelham-Born & R. V. Miller
	G–ATFU	D.H.85 Leopard Moth	A. H. Carrington & C. D. Duthy-James
	G–ATFV	Agusta-Bell 47J-2A	GSM Helicopters Ltd.
	G–ATFW	Luton L.A.4A Minor	A. J. Hammond
	G–ATFX	Cessna F.172G	Bristol & Wessex Aeroplane Club Ltd.
	G–ATFY	Cessna F.172G	W. Linskill
	G–ATGC	Victa Airtourer 100	Barnside Flying Group
	G–ATGE	Jodel DR.1050	J. R. Roberts
	G–ATGF	M.S.892A Rallye Commodore 150	E. G. Bostock & R. A. Punter
	G–ATGG	M.S.885 Super Rallye	N. Turner & ptnrs.
	G–ATGH	Brantly B.2B	R. Crook
	G–ATGO	Cessna F.172G	Bristol & Wessex Aeroplane Club Ltd.
	G–ATGP	Jodel DR.1050	W. M. Haley
	G–ATGY	GY.80 Horizon	P. W. Gibberson
	G–ATGZ	GH-4 Gyroplane	G. Griffiths
	G–ATHA	PA-23 Apache 235	A. Tucker
	G–ATHD	D.H.C.I Chipmunk 22	Spartan Flying Group Ltd.
	G–ATHF	Cessna 150F	Cambridge Aero Club Ltd.
	G–ATHG	Cessna 150F	G. T. Williams
	G–ATHH	Nord 1101 Noralpha	J. G. Kay
	G–ATHJ	PA-23 Aztec 250	H. I. Williams & ptnrs.
	G–ATHK	Aeronca 7AC Champion	A. Corran
	G–ATHL	Wallis WA-116/F	G. V. Wallis
	G–ATHM	Wallis WA-116 Srs. 1	Wallis Autogyros Ltd.
	G–ATHN	Nord 1101 Noralpha	E. L. Martin
	G–ATHR	PA-28 Cherokee 180	Britannia Airways Ltd.
	G–ATHT	Victa Airtourer 115	Alouette Leasing
	G–ATHU	A.61 Terrier 1	Marquess of Salisbury & Hon. C. E. V. Cecil
	G–ATHV	Cessna 150F	A. W. Pyle

Reg.	Type	Owner or Operator	Notes
G-ATHW	Mooney Mk. 20E	F. J. L. Aran	
G-ATHX	Jodel DR. 100A	J. W. West	
G-ATHZ	Cessna 150F	Solitair Flight Management Ltd.	
G-ATIA	PA-24 Comanche 260	A. P. Holmes & ptnrs.	
G-ATIB	Bensen B.8M	K. J. Atkins	
G-ATIC	Jodel DR.1050	M. A. George & F. G. Thorogood	
G-ATID	Cessna 337	C. S. Bishop Ltd.	
G-ATIE	Cessna 150F	Staverton Flying School Ltd.	
G-ATIG	HPR-7 Herald 214	Brymon Aviation Ltd.	
G-ATII	Beech S.35 Bonanza	Yankee Rose Air Charters Ltd.	
G-ATIN	Jodel D.117	Mrs. M. J. Underhill	
G-ATIR	Stampe SV.4C	Mitchell Aviation	
G-ATIS	PA-28 Cherokee 160	Leabond Ltd.	
G-ATIX	Nord 1101 Noralpha	B. J. Cope & others	
G-ATIZ	Jodel D.117	N. Chandler	
G-ATJA	Jodel DR.1050	Lesley Dee Fashions (Groby) Ltd.	
G-ATJC	Victa Airtourer 100	G. F. Richardson & G. J. Cresswell	
G-ATJD	PA-28 Cherokee 140	Novair Ltd.	
G-ATJF	PA-28 Cherokee 140	Alpha Flying Club	
G-ATJG	PA-28 Cherokee 140	A. Nayyar & C. Lindesay	
G-ATJL	PA-24 Comanche 260	M. W. Webb	
G-ATJN	Jodel D.119	A. L. Wickens	
G-ATJP	PA-23 Apache 160	DMR Computer Ltd.	
G-ATJR	PA-E23 Aztec 250	Bonaire	
G-ATJT	GY.80 Horizon 160	M. Chamberlain	
G-ATJU	Cessna 150F	Sunderland Flying Club Ltd.	
G-ATJW	Nord 1101 Noralpha	H. W. Elkin	
G-ATJX	Bucker Bu 131 Jungmann	J. E. Fricker & G. H. A. Bird	
G-ATJZ	PA-E23 Aztec 250	Automatic Finance Ltd.	
G-ATKC	Stampe S.V.4B	The Tiger Club	
G-ATKD	Cessna 150F	Cambridge Aero Club Ltd.	
G-ATKE	Cessna 150F	Skegness Air Taxi Services Ltd.	
G-ATKF	Cessna 150F	Cambridge Aero Club Ltd.	
G-ATKG	Hiller UH-12B	Bristow Helicopters Ltd.	
G-ATKH	Luton L.A.4A Minor	L. Hepper	
G-ATKI	Piper J-3C-65 Cub	A. C. Netting	
G-ATKS	Cessna F.172G	Derek Crouch (Contractors) Ltd.	
G-ATKT	Cessna F.172G	N. Y. Souster	
G-ATKU	Cessna F.172G	S. E. Ward & Sons (Engineers) Ltd.	
G-ATKW	PA-23 Aztec 250	Larchspeed Ltd.	
G-ATKX	Jodel D.140C	The Tiger Club	
G-ATKY	Cessna 150F	R. L. Beverley	
G-ATKZ	T.66-2 Nipper	M. W. Knights	
G-ATLA	Cessna 182J Skylane	Shefford Transport Engineers Ltd.	
G-ATLB	Jodel DR.1050-M1	The Tiger Club	
G-ATLC	PA-23 Aztec 250	Alderney Air Charter Ltd.	
G-ATLD	Cessna E-310K	Centreline Air Services Ltd.	
G-ATLG	Hiller UH-12B	Bristow Helicopters Ltd.	
G-ATLH	Fewsdale Gyro-Glider	F. Fewsdale	
G-ATLM	Cessna F.172G	Yorkshire Flying Services Ltd.	
G-ATLN	Cessna F.172G	Ceejay Aviation	
G-ATLO	Brantly 305	Freemans of Bewdley (Aviation) Ltd.	
G-ATLP	Bensen B.8M	C. D. Julian	
G-ATLR	Cessna F.172G	A. Wood & R. F. Patmore	
G-ATLT	Cessna U-206A	North Denes Aerodrome Ltd.	
G-ATLV	Jodel D.120	G. Douglas	
G-ATLW	PA-28 Cherokee 180	Links Systems Ltd.	
G-ATMB	Cessna F.150F	J. Rutherford	
G-ATMC	Cessna F.150F	Cambridge Aero Club Ltd.	
G-ATMG	M.S.893 Rallye Commodore 180	F. W. Fay & ptnrs.	
G-ATMH	Beagle D.5/180 Husky	Devon & Somerset Gliding Club Ltd.	
G-ATMI	H.S.748 Srs. 2A	Dan-Air Services Ltd.	
G-ATMJ	H.S.748 Srs. 2A	Dan-Air Services Ltd.	
G-ATMM	Cessna F.150F	Cambridge Technical Developments (Leasing) Ltd.	
G-ATMN	Cessna F.150F	J. H. Pickering & J. F. Thurlow	
G-ATMP	Cessna 210F	J. L. Way	
G-ATMT	PA-30 Twin Comanche 160	D. H. T. Bain	
G-ATMU	PA-23 Apache 160	Randy International Flying Associates Ltd.	
G-ATMW	PA-28 Cherokee 140	Bencray Ltd.	
G-ATMX	Cessna F.150F	Shropshire Aero Club Ltd.	

Notes	Reg.	Type	Owner or Operator
	G–ATMY	Cessna F.150F	S. E. Fellows
	G–ATNB	PA-28 Cherokee 180	Chaplin Auto Preparation Ltd.
	G–ATNE	Cessna F.150F	Bristol & Wessex Aeroplane Club Ltd.
	G–ATNI	Cessna F.150F	Air Service Training Ltd.
	G–ATNK	Cessna F.150F	Air Service Training Ltd.
	G–ATNL	Cessna F.150F	R. & Mrs. P. R. Budd
	G–ATNU	Cessna 182A	A. Bennett & W. W. Willis
	G–ATNV	PA-24 Comanche 260	Self-Fly Europe
	G–ATNX	Cessna F.150F	Mooney Aviation Ltd.
	G–ATNY	Cessna 337A	Ron Webster (Midlands) Ltd.
	G–ATOA	PA-23 Apache 160	Segadaloes Ltd.
	G–ATOD	Cessna F.150F	W. T. Tummon
	G–ATOE	Cessna F.150F	Argo Air Services Ltd.
	G–ATOG	Cessna F.150F	Manpower Grampian
	G–ATOH	D.62B Condor	G. H. Daniels
	G–ATOI	PA-28 Cherokee 140	O. & E. Flying Ltd.
	G–ATOJ	PA-28 Cherokee 140	O. T. Kernahan
	G–ATOK	PA-28 Cherokee 140	Link-Hampson Ltd.
	G–ATOL	PA-28 Cherokee 140	Tamar Flying Group
	G–ATOM	PA-28 Cherokee 140	T. Darlingwater & ptnrs.
	G–ATON	PA-28 Cherokee 140	Newcastle-upon-Tyne Aero Club Ltd.
	G–ATOO	PA-28 Cherokee 140	W. Davies
	G–ATOP	PA-28 Cherokee 140	T. R. & Mrs. E. A. Wiltshire
	G–ATOR	PA-28 Cherokee 140	J. P. Taylor
	G–ATOS	PA-28 Cherokee 140	AFT Craft Ltd.
	G–ATOT	PA-28 Cherokee 180	Roy Peplow & Co.
	G–ATOU	Mooney M.20E Super 21	B. Houghton
	G–ATOY	*PA-24 Comanche 260	Museum of Flight, E. Fortune
	G–ATPD	H.S. 125 Srs 1B	Juliett Aviation Ltd.
	G–ATPE	H.S. 125-1B	Moseley Group (PSV) Ltd.
	G–ATPJ	BAC One-Eleven 301	Dan-Air Services Ltd.
	G–ATPK	BAC One-Eleven 301	Laker Airways Ltd.
	G–ATPL	BAC One-Eleven 301	Dan-Air Services Ltd.
	G–ATPM	Cessna F.150F	Dan-Air Flying Club
	G–ATPN	PA-28 Cherokee 140	D. A. Thompson & L. Martin
	G–ATPR	PA-E23 Aztec 250	Flight Refuelling Ltd.
	G–ATPT	Cessna 182J Skylane	M. J. Baird-Smith
	G–ATPV	JB.01 Minicab	S. Russell
	G–ATRC	Beech B.95A Travel Air	A. G. Christmas Ltd.
	G–ATRG	PA-18 Super Cub 150	Lasham Gliding Soc. Ltd.
	G–ATRI	Bo 208C Junior	W. H. Jones
	G–ATRK	Cessna F.150F	Ascombe Distributors
	G–ATRL	Cessna F.150F	R. J. Jackson
	G–ATRM	Cessna F.150F	W. G. Fisher
	G–ATRN	Cessna F.150F	J. G. Coop & ptnrs.
	G–ATRO	PA-28 Cherokee 140	390th Flying Group
	G–ATRP	PA-28 Cherokee 140	T. R. E. Cook
	G–ATRR	PA-28 Cherokee 140	Woodgate Aviation
	G–ATRU	PA-28 Cherokee 180	Britannia Airways Ltd.
	G–ATRW	PA-32 Cherokee Six 260	Kent Messenger Ltd.
	G–ATRX	PA-32 Cherokee Six 260	R. F. Gibbs
	G–ATRY	Alon A-2 Aircoupe	B. W. George
	G–ATSB	PA-E23 Aztec 250	A. M. Stol Air Ltd. (G–ATZJ)
	G–ATSI	Bo 208C Junior	T. M. H. Paterson
	G–ATSJ	Brantly 305	R. H. Ryan
	G–ATSL	Cessna F.172G	H. G. Le Cheminant
	G–ATSM	Cessna 337A	Tremletts (Skycraft) Ltd.
	G–ATSR	Beech M.35 Bonanza	Alstan Aviation Ltd.
	G–ATST	M.S.893A Rallye Commodore	G. B. Instrument Panel Co. Ltd.
	G–ATSU	Jodel D.140B	J. S. Burnett Ltd.
	G–ATSX	Bo 208C Junior	Cotswold Aero Club Ltd.
	G–ATSY	Wassmer WA41 Super Baladou IV	Gilbey Warren Co. Ltd.
	G–ATSZ	PA-30 Twin Comanche 160	Air Kilroe Ltd.
	G–ATTB	Wallis WA.116-1	D. A. Wallis
	G–ATTD	Cessna 182J Skylane	Hanro Aviation Ltd.
	G–ATTF	PA-28 Cherokee 140	Mooney Aviation Ltd.
	G–ATTG	PA-28 Cherokee 140	J. F. Thurlow & J. H. Pickering
	G–ATTI	PA-28 Cherokee 140	A. S. Dredge & A. Gayden
	G–ATTK	PA-28 Cherokee 140	Andrewsfield Flying Club
	G–ATTM	Jodel DR. 250-160	D. H. Smith
	G–ATTP	BAC One-Eleven 207	Dan-Air Services Ltd.

Reg.	Type	Owner or Operator	Notes
G–ATTR	Bo 208C Junior 3	S. Luck	
G–ATTU	PA-28 Cherokee 140	Leith Air Ltd.	
G–ATTV	PA-28 Cherokee 140	Air South	
G–ATTX	PA-28 Cherokee 180	Violet M. Lambeth	
G–ATTY	PA-32 Cherokee Six 260	L. A. Dingemans & D. J. Everett	
G–ATUA	PA-25 Pawnee 235	A.D.S. (Aerial) Ltd.	
G–ATUB	PA-28 Cherokee 140	British Airways	
G–ATUC	PA-28 Cherokee 140	Airways Aero Associations Ltd.	
G–ATUD	PA-28 Cherokee 140	British Airways	
G–ATUF	Cessna F.150F	Devair Aviation Services Ltd.	
G–ATUG	D.62B Condor	The Tiger Club	
G–ATUH	T.66 Nipper	V. H. Hallam	
G–ATUI	Bo 208C Junior	Cotswold Aero Club	
G–ATUL	PA-28 Cherokee 180	R. J. Roberts & Sons	
G–ATVF	D.H.C.1 Chipmunk 22	RAFGSA	
G–ATVG	Hiller UH-12E	Management Aviation Ltd.	
G–ATVH	BAC One-Eleven 207	Dan-Air Services Ltd. *City of Newcastle-upon-Tyne*	
G–ATVI	SIPA 903	J. Martin	
G–ATVK	PA-28 Cherokee 140	E. A. Clack	
G–ATVL	PA-28 Cherokee 140	West London Aero Services	
G–ATVO	PA-28 Cherokee 140	Herts & Essex Aero Club	
G–ATVP	*F.B.5 Gunbus (2345)	RAF Museum	
G–ATVS	PA-28 Cherokee 180	Marshalls Woodflakes Ltd.	
G–ATVW	D.62B Condor	J. R. Stainer & D. W. Evernden	
G–ATVX	Bo 208C Junior	G. W. Stanmore	
G–ATWA	Jodel DR.1050	Llamedoes Flying Group	
G–ATWE	M.S.892A Rallye Commodore	D. I. Murray	
G–ATWG	PA-30 Twin Comanche 160	L. R. Davies	
G–ATWJ	Cessna F.172F	C. J. & J. Freeman	
G–ATWL	Jodel D.120	T. A. S. Rosie	
G–ATWO	PA-28 Cherokee 180	Vectaphone Manufacturing Ltd.	
G–ATWP	Alon A-2 Aircoupe	F. Bolton	
G–ATWR	PA-30 Twin Comanche 160	Lubenham Fidelities & Investments Co. Ltd.	
G–ATWS	Luton L.A.4A Minor	K. J. Hazelwood	
G–ATWV	Boeing 707-336C	British Airways	
G–ATWZ	M.S.892 Rallye Commodore	R. G. Allen & K. F. Olivier	
G–ATXA	PA-22 Tri-Pacer 150	J. W. Holmes	
G–ATXD	PA-30 Twin Comanche 160	Southwark Estates Ltd.	
G–ATXF	GY-80 Horizon 150	G. A. Hinchcliffe	
G–ATXM	PA-28 Cherokee 180	J. Khan	
G–ATXN	Mitchell-Proctor Kittiwake	D. W. Kent	
G–ATXO	SIPA 903	M. Hillam	
G–ATXR	AFB 1 gas balloon	Mrs. C. M. Bulmer *Omega One*	
G–ATXW	WHE Airbuggy	W. H. Ekin	
G–ATXY	WHE Airbuggy	W. H. Ekin (Engineering) Ltd.	
G–ATXZ	Bo 208C Junior	J. A. Taylor	
G–ATYA	PA-25 Pawnee 235	Skegness Air Taxi Services Ltd.	
G–ATYM	Cessna F.150G	J. T. Tyer	
G–ATYN	Cessna F.150G	Skegness Air Taxi Services Ltd.	
G–ATYS	PA-28 Cherokee 180	E. K. Chalke	
G–ATYV	Bell 47G	Heliwork Ltd.	
G–ATYW	Beagle B.206 Srs. 1	W. Lancs Aero Club Ltd.	
G–ATYZ	M.S.880B Rallye Club	Nylo Flying Group	
G–ATZA	Bo 208C Junior	W. C. Roberts	
G–ATZB	Hiller UH-12B	Bristow Helicopters Ltd.	
G–ATZC	Boeing 707-365C	British Caledonian Airways *Loch Katrine*	
G–ATZD	Boeing 707-365C	British Airways	
G–ATZG	AFB2 gas balloon	F./Lt. S. Cameron *Aeolis*	
G–ATZK	PA-28 Cherokee 180	Aerosport Ltd.	
G–ATZL	Jodel DR.250	C. G. Gray	
G–ATZM	Piper J-3C-65 Cub	B. Brooks	
G–ATZO	Beagle B.206 Srs. 1	Arnos Exporting Co. Ltd.	
G–ATZS	Wassmer WA41 Super IV Baladou	J. R. MacAlpine-Dounie & P. A. May	
G–ATZU	PA-30 Twin Comanche 160	Rotair Flying Services Ltd.	
G–ATZV	PA-30 Twin Comanche 160	Via Nova International Ltd.	
G–ATZY	Cessna F.150G	Air Service Training Ltd.	
G–ATZZ	Cessna F.150G	Argo Air Services Ltd.	
G–AUTO	Cessna 441 Conquest	Automobile Association Ltd.	

Notes	Reg.	Type	Owner or Operator
	G–AVAA	Cessna F.150G	Argo Air Services Ltd.
	G–AVAI	H.S.125 Srs. 3B	Brown and Root (U.K.) Ltd.
	G–AVAJ	Hiller UH-12B	Bristow Helicopters Ltd.
	G–AVAO	PA-30 Twin Comanche 160	Heath Street Car Hirings Ltd.
	G–AVAP	Cessna F.150G	Seawing Flying Club Ltd.
	G–AVAR	Cessna F.150G	West Wales Flying Club
	G–AVAS	Cessna F.172H	Birmingham Aviation Ltd.
	G–AVAU	PA-30 Twin Comanche 160	L. Batin
	G–AVAW	D.62B Condor	Avato Flying Group
	G–AVAX	PA-28 Cherokee 180	College of Air Training
	G–AVAY	PA-28 Cherokee 180	College of Air Training
	G–AVAZ	PA-28 Cherokee 180	College of Air Training
	G–AVBA	PA-28 Cherokee 180	College of Air Training
	G–AVBB	PA-28 Cherokee 180	College of Air Training
	G–AVBC	PA-28 Cherokee 180	College of Air Training
	G–AVBE	PA-28 Cherokee 180	College of Air Training
	G–AVBG	PA-28 Cherokee 180	College of Air Training
	G–AVBH	PA-28 Cherokee 180	College of Air Training
	G–AVBJ	PA-28 Cherokee 180	College of Air Training
	G–AVBL	PA-30 Twin Comanche 160	Fleet Cameras Ltd.
	G–AVBP	PA-28 Cherokee 140	Bristol & Wessex Aeroplane Club Ltd.
	G–AVBS	PA-28 Cherokee 180	I. J. James
	G–AVBT	PA-28 Cherokee 180	Cardinal Engineering Ltd.
	G–AVBU	PA-32 Cherokee Six 260	Tempus Aviation (Holdings) Ltd.
	G–AVBW	BAC One-Eleven 320	Laker Airways
	G–AVBX	BAC One-Eleven 320	Laker Airways
	G–AVBY	BAC One-Eleven 320	Laker Airways
	G–AVBZ	Cessna F.172H	J. Seville
	G–AVCA	Brantly B.2B	M. J. & Mrs. G. M. Page
	G–AVCC	Cessna F.172H	S. B. Price
	G–AVCD	Cessna F.172H	Toon Ghose Aviation Ltd.
	G–AVCE	Cessna F.172H	Cleco Electrical Industries Ltd.
	G–AVCM	PA-24 Comanche 260	F. Smith & Sons Ltd.
	G–AVCP	PA-30 Twin Comanche 160	Chartair Ltd.
	G–AVCS	A.61 Terrier I	L. M. Farrell & A. R. C. Hunter
	G–AVCT	Cessna F.150G	Sierra Aviation Services
	G–AVCU	Cessna F.150G	P. R. Moss
	G–AVCV	Cessna 182J Skylane	J. A. Moores
	G–AVCW	PA-30 Twin Comanche 160	Manro Transport Ltd.
	G–AVCX	PA-30 Twin Comanche 160	F. J. Stevens
	G–AVCY	PA-30 Twin Comanche 160	T. S. Grimshaw Ltd.
	G–AVDA	Cessna 182K Skylane	J. W. Grant
	G–AVDE	Turner Gyroglider Mk. I	J. S. Smith
	G–AVDF	*Beagle Pup 100	Brighton Transport Museum
	G–AVDG	Wallis WA-116 Srs. I	K. H. Wallis
	G–AVDR	Beech B80 Queen Air	Automatic Finance Corporation Ltd.
	G–AVDS	Beech B80 Queen Air	Automatic Finance Corporation Ltd.
	G–AVDT	Aeronca 7AC Champion	W. R. Prescott
	G–AVDW	D.62B Condor	N. H. Jones
	G–AVDX	H.S.125 Srs. 3B	Civil Aviation Authority
	G–AVDY	Luton L.A.4A Minor	D. E. Evans & ptnrs.
	G–AVDZ	PA-25 Pawnee 235	Skegness Air Taxi Services Ltd.
	G–AVEB	Morane MS 230 Et 2	Hon. P. Lindsay
	G–AVEC	Cessna F.172H	W. H. Ekin (Engineering) Co. Ltd.
	G–AVEF	Jodel D.150	The Tiger Club
	G–AVEG	SIAI-Marchetti S.205	Hanway Car Sales Ltd.
	G–AVEH	SIAI-Marchetti S.205	South Lancs. Flyers Ltd.
	G–AVEM	Cessna F.150G	Airwork Services Ltd.
	G–AVEN	Cessna F.150G	Airwork Services Ltd.
	G–AVEO	Cessna F.150G	Airwork Services Ltd.
	G–AVER	Cessna F.150G	Reedy Supplies Ltd.
	G–AVET	Beech C55 Baron	Spline Gauges Ltd.
	G–AVEU	Wassmer WA.41 Baladou	Baladou Flying Group
	G–AVEX	D.62B Condor	Cotswold Roller Hire Ltd.
	G–AVEY	Currie Super Wot	A. Eastelow
	G–AVEZ	HPR-7 Herald 210	Air UK
	G–AVFA	H.S.121 Trident 2E	British Airways
	G–AVFB	H.S.121 Trident 2E	British Airways
	G–AVFC	H.S.121 Trident 2E	British Airways
	G–AVFD	H.S.121 Trident 2E	British Airways
	G–AVFE	H.S.121 Trident 2E	British Airways
	G–AVFF	H.S.121 Trident 2E	British Airways

Reg.	Type	Owner or Operator	Notes
G–AVFG	H.S.121 Trident 2E	British Airways	
G–AVFH	H.S.121 Trident 2E	British Airways	
G–AVFI	H.S.121 Trident 2E	British Airways	
G–AVFJ	H.S.121 Trident 2E	British Airways	
G–AVFK	H.S.121 Trident 2E	British Airways	
G–AVFL	H.S.121 Trident 2E	British Airways	
G–AVFM	H.S.121 Trident 2E	British Airways	
G–AVFN	H.S.121 Trident 2E	British Airways	
G–AVFO	H.S.121 Trident 2E	British Airways	
G–AVFP	PA-28 Cherokee 140	H. D. Vince Ltd.	
G–AVFR	PA-28 Cherokee 140	B. A. Tuson & G. N. Morrow	
G–AVFS	PA-32 Cherokee Six 300	Headcorn Parachute Club Ltd.	
G–AVFU	PA-32 Cherokee Six 300	S. L. H. Construction Ltd.	
G–AVFW	PA-30 Twin Comanche 160	Woodlands Investments Ltd.	
G–AVFX	PA-28 Cherokee 140	J. E. Lawson	
G–AVFY	PA-28 Cherokee 140	F. Spencer-Jones	
G–AVFZ	PA-28 Cherokee 140	Keenair Services Ltd.	
G–AVGA	PA-24 Comanche 260	G. G. Asmore & F. Mumford	
G–AVGB	PA-28 Cherokee 140	G. Abbot	
G–AVGC	PA-28 Cherokee 140	B. A. Bennett	
G–AVGD	PA-28 Cherokee 140	A. D. Wren	
G–AVGE	PA-28 Cherokee 140	A. G. Branch Contractors Ltd.	
G–AVGH	PA-28 Cherokee 140	Mooney Aviation Ltd.	
G–AVGI	PA-28 Cherokee 140	Bencray Ltd.	
G–AVGJ	Jodel DR.1050	H. A. N. Crocker	
G–AVGK	PA-28 Cherokee 180	Liverpool Aero Club Ltd.	
G–AVGL	Cessna F.150G	F .A. E. Pyle	
G–AVGN	PA-24 Comanche 260	Viscount Chelsea	
G–AVGP	BAC One-Eleven 408	British Airways	
G–AVGU	Cessna F.150G	J. A. & Mrs. J. M. C. Pothecary	
G–AVGV	Cessna F.150G	British Skysports	
G–AVGX	Bo 208C Junior	R. J. Pascoe	
G–AVGY	Cessna 182K Skylane	H. E. Peacock	
G–AVGZ	Jodel DR.1050	W. H. Gilchrist & A. T. Howarth	
G–AVHC	Brooklands Mosquito II	R. H. Ryan	
G–AVHF	Beech A.23 Musketeer	J. G. Stewart & I. M. S. Ferriman	
G–AVHH	Cessna F.172H	R. T. Pritchard	
G–AVHJ	Wassmer WA.41 Baladou	D. G. Pickering & ptnrs.	
G–AVHL	Jodel DR.105A	Girls Venture Corps	
G–AVHM	Cessna F.150G	Municipal Product and Services Ltd.	
G–AVHN	Cessna F.150G	Bristol and Wessex Aero Club Ltd.	
G–AVHT	Auster AOP.9 (WZ711)	M. Somerton-Rayner	
G–AVHY	Fournier RF.4D	R. Swinn & J. Conolly	
G–AVHZ	PA-30 Twin Comanche 160	P. S. Bubbear & J. M. Glanville	
G–AVIA	Cessna F.150G	Laarbruch Flying Club	
G–AVIB	Cessna F.150G	Warwickshire Aero Club	
G–AVIC	Cessna F.172H	W. Wales Flying Club Ltd.	
G–AVID	Cessna 182J	Inch Farming Co. Ltd.	
G–AVIE	Cessna F.172H	N. Denes Aerodrome Ltd.	
G–AVIG	A-B 206B JetRanger	Bristow Helicopters Ltd.	
G–AVII	A-B 206A JetRanger	Bristow Helicopters Ltd.	
G–AVIL	Alon A.2 Aircoupe	D. W. Vernon	
G–AVIN	M.S.880B Rallye Club	A. P. Locke & D. R. F. Sapte	
G–AVIO	M.S.880B Rallye Club	Ms. J. M. Elliott	
G–AVIP	Brantly B.2B	Cosworth Engineering Ltd.	
G–AVIR	Cessna F.172H	W. Lancashire Aero Club Ltd.	
G–AVIS	Cessna F.172H	Jon Paul Photography Ltd.	
G–AVIT	Cessna F.150G	Shropshire Aero Club Ltd.	
G–AVIU	Jodel DR.1050	P. R. Cremer	
G–AVIZ	Scheibe SF.25A Motorfalke	D. C. Pattison & D. A. Wilson	
G–AVJB	V.815 Viscount	Jersey European Airways	
G–AVJE	Cessna F.150G	J. H. Pickering	
G–AVJF	Cessna F.172H	W. R. & Mrs. B. M. Young	
G–AVJG	Cessna 337B	P. R. Moss	
G–AVJH	D.62 Condor	O. G. Baum	
G–AVJI	Cessna F.172H	Royal Artillery Aero Club Ltd.	
G–AVJJ	PA-30 Twin Comanche 160	Peregrine Air Services Ltd.	
G–AVJK	Jodel DR.1050 M.1	J. H. B. Urmston	
G–AVJN	Brantly B.2B	John Berry Ltd.	
G–AVJO	Fokker E.111 Replica (422)	Personal Plane Services Ltd.	
G–AVJU	PA-24 Comanche 260	Anchor Finance Ltd.	
G–AVJV	Wallis WA-117 Srs. 1	K. H. Wallis (G–ATCV)	
G–AVJW	Wallis WA-118 Srs. 2	K. H. Wallis (G–ATPW)	

Notes	Reg.	Type	Owner or Operator
	G–AVKB	MB.50 Pipistrelle	R. K. Haldenby & T. S. Warren
	G–AVKD	Fournier RF.4D	Lasham RF4 Group
	G–AVKE	Gadfly HDW.1	British Rotorcraft Museum
	G–AVKG	Cessna F. 172H	W. Lancs Aero Club Ltd.
	G–AVKI	Nipper T.66 Srs. 3	R. J. Corbett & ptnrs.
	G–AVKJ	Nipper T.66 Srs. 3	P. W. Hunter
	G–AVKM	D.62B Condor	N. H. Jones
	G–AVKN	Cessna 401	Lewcol Enterprises
	G–AVKP	A.109 Airedale	St. Pirans Flying Group
	G–AVKR	Bo 208C Junior	D. F. Barley & D. A. Bishop
	G–AVKS	Bell 47G-2	Bristow Helicopters Ltd.
	G–AVKX	Hiller UH-12E	Management Aviation Ltd.
	G–AVKY	Hiller UH-12E	Agricopters Ltd.
	G–AVKZ	PA-23 Aztec 250	Hunnable Holdings Ltd.
	G–AVLA	PA-28 Cherokee 140	Mrs. J. I. & P. Sprent
	G–AVLB	PA-28 Cherokee 140	D. C. Tyne & Co.
	G–AVLC	PA-28 Cherokee 140	Swansea & District Flying School & Club Ltd.
	G–AVLD	PA-28 Cherokee 140	P. V. Evans
	G–AVLE	PA-28 Cherokee 140	P. A. Johnstone & E. J. Morgan
	G–AVLF	PA-28 Cherokee 140	W. London Aero Services Ltd.
	G–AVLG	PA-28 Cherokee 140	D. Golding & P. J. Pearce
	G–AVLH	PA-28 Cherokee 140	T. L. Wilkinson
	G–AVLI	PA-28 Cherokee 140	R. H. Neale
	G–AVLJ	PA-28 Cherokee 140	Barbers Animal Products Ltd. & E. D. Hughes
	G–AVLN	B.121 Pup 2	C. B. G. Masefield
	G–AVLO	Bo 208C Junior	P. J. Thompson & R. G. Williams
	G–AVLP	PA-23 Aztec 250	Survey Flights Ltd.
	G–AVLR	PA-28 Cherokee 140	G. H. & Mrs. K. M. Wylde
	G–AVLS	PA-28 Cherokee 140	K. Dando
	G–AVLT	PA-28 Cherokee 140	E. A. Clack & M. T. Pritchard
	G–AVLU	PA-28 Cherokee 140	London Transport (Central Road Services) Sports Association Flying Club
	G–AVLV	PA-23 Aztec 250	Cabair Ltd.
	G–AVLW	Fournier RF 4D	T. M. W. Webster
	G–AVLY	Jodel D.120A	R. Arbon & ptnrs.
	G–AVMA	GY.80 Horizon 180	B. R. & S. Hildick
	G–AVMB	D.62B Condor	N. H. Jones
	G–AVMD	Cessna 150G	IIP Associates
	G–AVMF	Cessna F. 150G	J. D. Palfreman
	G–AVMH	BAC One-Eleven 510	British Airways
	G–AVMI	BAC One-Eleven 510	British Airways
	G–AVMJ	BAC One-Eleven 510	British Airways
	G–AVMK	BAC One-Eleven 510	British Airways
	G–AVML	BAC One-Eleven 510	British Airways
	G–AVMM	BAC One-Eleven 510	British Airways
	G–AVMN	BAC One-Eleven 510	British Airways
	G–AVMO	BAC One-Eleven 510	British Airways
	G–AVMP	BAC One-Eleven 510	British Airways
	G–AVMR	BAC One-Eleven 510	British Airways
	G–AVMS	BAC One-Eleven 510	British Airways
	G–AVMT	BAC One-Eleven 510	British Airways
	G–AVMU	BAC One-Eleven 510	British Airways
	G–AVMV	BAC One-Eleven 510	British Airways
	G–AVMW	BAC One-Eleven 510	British Airways
	G–AVMX	BAC One-Eleven 510	British Airways
	G–AVMY	BAC One-Eleven 510	British Airways
	G–AVMZ	BAC One-Eleven 510	British Airways
	G–AVNB	Cessna F.150G	G. A. J. Bowles
	G–AVNC	Cessna F.150G	Northampton Aviation Services Ltd.
	G–AVNG	Beech A80 Queen Air	Vernair Transport Services
	G–AVNI	PA-30 Twin Comanche	D.P. Aviation
	G–AVNL	PA-23 Aztec 250	Cabair Ltd.
	G–AVNM	PA-28 Cherokee 180	College of Air Training
	G–AVNN	PA-28 Cherokee 180	College of Air Training
	G–AVNO	PA-28 Cherokee 180	College of Air Training
	G–AVNP	PA-28 Cherokee 180	College of Air Training
	G–AVNR	PA-28 Cherokee 180	College of Air Training
	G–AVNS	PA-28 Cherokee 180	College of Air Training
	G–AVNT	PA-28 Cherokee 180	College of Air Training
	G–AVNU	PA-28 Cherokee 180	College of Air Training

Reg.	Type	Owner or Operator	Notes
G–AVNV	PA-28 Cherokee 180	College of Air Training	
G–AVNW	PA-28 Cherokee 180	College of Air Training	
G–AVNX	Fournier RF-4D	O. C. Harris & C. G. Masterman	
G–AVNY	Fournier RF-4D	A. N. Mavrogordato	
G–AVNZ	Fournier RF-4D	Cobra Group	
G–AVOA	Jodel DR.1050	I. MacPherson	
G–AVOD	Beagle D5/180 Husky	J. Hockton	
G–AVOH	D.62B Condor	N. H. Jones	
G–AVOI	H.S.125 Srs. 3B	Markheath Securities Ltd.	
G–AVOM	Jodel DR.221	M. A. Mountford	
G–AVON	Luton LA.5A Major	G. R. Mee	
G–AVOO	PA-18-150 Super Cub	T. A. McMullin	
G–AVOR	Lockspeiser Land Development Aircraft	D. Lockspeiser	
G–AVOZ	PA-28 Cherokee 180	Downley Garages Ltd.	
G–AVPA	Sopwith Pup Replica	C. J. Warrilow	
G–AVPB	Boeing 707-336C	British Airways	
G–AVPC	D.31 Turbulent	J. Sharp	
G–AVPD	D.92 Bebe	S. W. McKay	
G–AVPE	H.S.125 Srs. 3B	British Aerospace	
G–AVPH	Cessna F.150G	W. Lancashire Aero Club	
G–AVPI	Cessna F.172H	R. Jones & J. Chancellor	
G–AVPJ	D.H.82A Tiger Moth	The Barnstormers Ltd.	
G–AVPK	M.S.892A Rallye Commodore	D. A. Gray	
G–AVPL	M.S.892 Rallye Commodore	J. H. Atkinson	
G–AVPM	Jodel D.117	J. Houghton	
G–AVPN	HPR.7 Herald 213	Air UK	
G–AVPR	PA-30 Twin Comanche 160	Cold Storage (Jersey) Ltd.	
G–AVPS	PA-30 Twin Comanche 160	Pegasus Aviation Ltd.	
G–AVPT	PA-18 Super Cub 150	N. H. Jones	
G–AVPU	PA-18 Super Cub 150	Scottish Gliding Union Ltd.	
G–AVPV	PA-28 Cherokee 180	E. A. Clack	
G–AVPX	Taylor JT.1 Monoplane	S. M. Smith	
G–AVPY	PA-25 Pawnee 235	Farm Aviation Services Ltd.	
G–AVRF	H.S.125 Srs. 3B	BAC (Holdings) Ltd.	
G–AVRG	H.S.125 Srs. 3A	Shell Aircraft Ltd.	
G–AVRK	PA-28 Cherokee 180	Dollar Air Services Ltd.	
G–AVRL	Boeing 737-204	Britannia Airways Ltd. *Sir Ernest Shackleton*	
G–AVRM	Boeing 737-204	Britannia Airways Ltd. *James Watt*	
G–AVRN	Boeing 737-204	Britannia Airways Ltd. *Capt. James Cook*	
G–AVRO	Boeing 737-204	Britannia Airways Ltd. *Sir Francis Drake*	
G–AVRP	PA-28 Cherokee 140	M. L. Rhodes	
G–AVRS	GY.80 Horizon 180	Rogers Autos Ltd.	
G–AVRT	PA-28 Cherokee 140	F. Clarke	
G–AVRU	PA-28 Cherokee 180	Fenland Tractors Ltd.	
G–AVRW	GY-20 Minicab	R. B. Pybus	
G–AVRX	PA-23 Aztec 250	The Cronite Group Ltd.	
G–AVRY	PA-28 Cherokee 180	Roses Flying Group	
G–AVRZ	PA-28 Cherokee 180	Briglea Engineering Ltd.	
G–AVSA	PA-28 Cherokee 180	Alliance Aviation Ltd.	
G–AVSB	PA-28 Cherokee 180	White House Garage Ltd.	
G–AVSC	PA-28 Cherokee 180	W. London Aero Services Ltd.	
G–AVSD	PA-28 Cherokee 180	Tracair Ltd.	
G–AVSE	PA-28 Cherokee 180	S E Aviation Ltd.	
G–AVSF	PA-28 Cherokee 180	Goodwood Terrena Ltd.	
G–AVSH	PA-28 Cherokee 180	Elken Ltd.	
G–AVSI	PA-28 Cherokee 180	E. P. van Mechelen	
G–AVSK	Bell 47G-4A	Autair Helicopter Services Ltd.	
G–AVSP	PA-28 Cherokee 180	Trig Engineering Ltd.	
G–AVSR	Beagle D 5/180 Husky	A. L. Young	
G–AVTB	Nipper T.66 Srs. 3	S. Stride & J. Hobday	
G–AVTE	Bell 206A JetRanger	W. R. Finance Ltd.	
G–AVTH	Jodel DR.1051	T. Knowles	
G–AVTJ	PA-32 Cherokee Six 260	G. Jacobs	
G–AVTK	PA-32 Cherokee Six 260	Rodney Saunders Associates Ltd.	
G–AVTM	Cessna F.150H	Falcon Flying Services	
G–AVTO	Cessna F.150H	Swanton Morley Flying Group	
G–AVTP	Cessna F.172H	Western Air Training Ltd.	
G–AVTS	PA-23 Aztec 250	Tyler Goodchild Aviation Ltd.	
G–AVTT	Ercoupe 415D	A. H. Cullimore	

Notes	Reg.	Type	Owner or Operator
	G-AVTV	M.S.893 Rallye Commodore	A. Lister
	G-AVTX	Taylor JT.1 Monoplane	P. Lockwood
	G-AVUA	Cessna F.172H	Recreational Flying Centre (Popham) Ltd.
	G-AVUD	PA-30 Twin Comanche 160B	F. M. Aviation
	G-AVUG	Cessna F.150H	Dukeries Aviation Ltd.
	G-AVUH	Cessna F.150H	Fly-Hire Ltd.
	G-AVUI	Cessna F.150H	Dukeries Aviation Ltd.
	G-AVUJ	F.8L Falco 4	M. Shield
	G-AVUK	Enstrom F-28A	Spooner Aviation Ltd.
	G-AVUL	Cessna F.172H	D. H. Stephens & J. A. Blythe
	G-AVUM	Hughes 269B	P. Berriman
	G-AVUN	PA-30 Twin Comanche 160B	Overseas Aero Leasing
	G-AVUS	PA-28 Cherokee 140	R. Matthews
	G-AVUT	PA-28 Cherokee 140	W. B. Bateson
	G-AVUU	PA-28 Cherokee 140	Goodwood Estate Co. Ltd.
	G-AVUV	Cessna 310N	Airwork Services Ltd.
	G-AVUZ	PA-32 Cherokee Six 300	P. & Mrs. M. E. Biggins
	G-AVVB	H.S.125 Srs. 3B	Brown & Root (UK) Ltd.
	G-AVVC	Cessna F.172H	Kestrel Air Ltd.
	G-AVVE	Cessna F.150H	Practavia Ltd.
	G-AVVF	D.H.104 Dove 8	Martin Baker (Engineering) Ltd.
	G-AVVG	PA-28 Cherokee 180	604 Squadron Flying Club
	G-AVVI	PA-30 Twin Comanche 160B	Steepletone Products Ltd.
	G-AVVJ	M.S.893 Rallye Commodore	F. A. O. Gaze
	G-AVVL	Cessna F.150H	R. J. & Mrs. B. Ford-Sagers
	G-AVVM	Jodel D.117	R. R. Corker
	G-AVVN	D.62C Condor	Avato Flying Group
	G-AVVO	*Avro 652A Anson 19 (VL348)	Newark Air Museum
	G-AVVS	Hughes 269B	W. Holmes
	G-AVVT	PA-23 Aztec 250	Kayglynn Investment Holdings Ltd.
	G-AVVU	Beech A.23A Musketeer	A. Rennie
	G-AVVV	PA-28 Cherokee 180	Alba Motels
	G-AVVW	Cessna F.150H	Bristol & Wessex Aeroplane Club Ltd.
	G-AVVX	Cessna F.150H	J. C. & Mrs. M. Abberley
	G-AVVY	Cessna F.150H	W. S. Davies
	G-AVWA	PA-28 Cherokee 140	T. W. Electric (Gretton) Ltd.
	G-AVWB	PA-28 Cherokee 140	A. E. James & P. Houghton
	G-AVWD	PA-28 Cherokee 140	Apache Aircraft Services Ltd.
	G-AVWE	PA-28 Cherokee 140	W. C. C. Meyer
	G-AVWF	PA-28 Cherokee 140	Liverpool Aero Club Ltd.
	G-AVWG	PA-28 Cherokee 140	Bencray Ltd.
	G-AVWH	PA-28 Cherokee 140	B. P. W. Faithfull
	G-AVWI	PA-28 Cherokee 140	J. C. Smith & Miss L. M. Veitch
	G-AVWJ	PA-28 Cherokee 140	E.F.G. Flying Services Ltd.
	G-AVWL	PA-28 Cherokee 140	Achilles School of Flying Ltd.
	G-AVWM	PA-28 Cherokee 140	The Misses A. J. & C. B. Speight
	G-AVWN	PA-28R Cherokee Arrow 180	Seven City Overseas Ltd.
	G-AVWO	PA-28R Cherokee Arrow 180	C & S Controls Ltd.
	G-AVWR	PA-28R Cherokee Arrow 180	Mono-Construction Ltd.
	G-AVWT	PA-28R Cherokee Arrow 180	G. W. Barker & ptnrs.
	G-AVWU	PA-28R Cherokee Arrow 180	J. Neilson
	G-AVWV	PA-28R Cherokee Arrow 180	Mapair Ltd.
	G-AVWW	Mooney M.20F	A. & J. & G. Cullen
	G-AVWY	Fournier RF-4D	T. G. Hoult
	G-AVWZ	Fournier RF-4D	Mrs. L. P. Holmes & D. Bartlett
	G-AVXA	PA-25 Pawnee 235	Howard Avis (Aviation) Ltd.
	G-AVXB	PL-1 Gyroplane	C. Mowat
	G-AVXC	Nipper T.66 Srs. 3	W. G. Wells & ptnrs.
	G-AVXD	Nipper T.66 Srs. 3	J. M. Murrie
	G-AVXF	PA-28R Cherokee Arrow 180	Allzones Travel Ltd.
	G-AVXI	H.S.748 Srs. 2	Civil Aviation Authority
	G-AVXJ	H.S.748 Srs. 2	Civil Aviation Authority
	G-AVXK	H.S.125 Srs. 3B-RA	Shell Aircraft Ltd.
	G-AVXV	Bleriot XI (BAPC 104)	L. D. Goldsmith
	G-AVXW	D.62B Condor	Medway Flying Group Ltd.
	G-AVXX	Cessna FR.172E	Hadrian Flying Group
	G-AVXY	Auster AOP.9 (XK417)	R. Windley
	G-AVYF	Beech A.23-24 Musketeer	Wearside Aviation Group
	G-AVYK	A.61 Terrier 3	Airways Aero Associations Ltd.
	G-AVYL	PA-28 Cherokee 180	Miller Aerial Spraying Ltd.
	G-AVYM	PA-28 Cherokee 180	Border Piper Aviation Ltd.

Reg.	Type	Owner or Operator	Notes
G–AVYO	PA-28 Cherokee 140	Goodwood Estate Co. Ltd.	
G–AVYP	PA-28 Cherokee 140	T. D. Reid (Braids) Ltd.	
G–AVYR	PA-28 Cherokee 140	E. Drew (Construction & Plant Hire) Ltd.	
G–AVYS	PA-28R Cherokee Arrow 180	Leisure Sport Ltd.	
G–AVYT	PA-28R Cherokee Arrow 180	H. Stephenson	
G–AVYV	Jodel D.120	J. B. J. Berrow	
G–AVYX	AB-206 JetRanger	S.W. Electricity Board	
G–AVYZ	BAC One-Eleven 320L	Laker Airways	
G–AVZA	IMCO Callair A-9	Arable & Bulb Chemicals Ltd.	
G–AVZB	Aero Z-37 Cmelak	ADS (Aerial) Ltd.	
G–AVZC	Hughes 269B	Ocean Rangers Charters Ltd.	
G–AVZE	D.62B Condor	A. J. M. Trowbridge & J. Harris	
G–AVZI	Bo 208C Junior	C. F. Rogers	
G–AVZM	Beagle B.121 Pup I	ARAZ Group	
G–AVZN	Beagle B.121 Pup I	W. E. Cro & Sons Ltd.	
G–AVZO	Beagle B.121 Pup I	Dungenair Ltd.	
G–AVZP	Beagle B.121 Pup I	Northampton School of Flying Co. Ltd.	
G–AVZR	PA-28 Cherokee 180	G. Knowles & ptnrs.	
G–AVZS	Cessna 310B	J & B's Aviation Ltd.	
G–AVZT	PA-31 Navajo	Cabair Ltd.	
G–AVZU	Cessna F.150H	Miss I. Sammons	
G–AVZV	Cessna F.172H	Hull Aero Club	
G–AVZW	EAA Model P Biplane	R. G. Maidment & G. R. Edmundson	
G–AVZX	M.S.880B Rallye Club	L. Clutton & ptnrs.	
G–AVZY	M.S.880B Rallye Club	R. McLindsay	
G–AWAA	M.S.880B Rallye Club	P. A. Cairns	
G–AWAC	GY-80 Horizon 180	A. E. Chapman	
G–AWAD	Beech D 55 Baron	College of Air Training	
G–AWAE	Beech D 55 Baron	College of Air Training	
G–AWAF	Beech D 55 Baron	College of Air Training	
G–AWAG	Beech D 55 Baron	College of Air Training	
G–AWAH	Beech D 55 Baron	College of Air Training	
G–AWAI	Beech D 55 Baron	College of Air Training	
G–AWAJ	Beech D 55 Baron	College of Air Training	
G–AWAK	Beech D 55 Baron	College of Air Training	
G–AWAL	Beech D 55 Baron	College of Air Training	
G–AWAM	Beech D 55 Baron	College of Air Training	
G–AWAN	Beech D 55 Baron	College of Air Training	
G–AWAO	Beech D 55 Baron	College of Air Training	
G–AWAP	SA 318B Alouette Astazou	Helicopter Hire Ltd.	
G–AWAT	D.62B Condor	N. H. Jones	
G–AWAU	*Vickers F.B.27A Vimy (replica) (F8614)	RAF Museum	
G–AWAV	Cessna F.150F	J. F. Thurlow & J. H. Pickering	
G–AWAW	Cessna F.150F	S.B. International Aviation Ltd.	
G–AWAX	Cessna 150D	Cambridge Technical Developments (Leasing) Ltd.	
G–AWAZ	PA-28R Cherokee Arrow 180	Wells Timbercraft Ltd.	
G–AWBA	PA-28R Cherokee Arrow 180	W. London Aero Services Ltd.	
G–AWBB	PA-28R Cherokee Arrow 180	M. H. Cundey	
G–AWBC	PA-28R Cherokee Arrow 180	Astrapoint Ltd.	
G–AWBE	PA-28 Cherokee 140	Achilles School of Flying Ltd.	
G–AWBG	PA-28 Cherokee 140	EFG Flying Services Ltd.	
G–AWBH	PA-28 Cherokee 140	R. C. A. Mackworth	
G–AWBJ	Fournier RF.4D	The BJ Group	
G–AWBL	BAC One-Eleven 416	British Airways	
G–AWBM	D.31 Turbulent	W. B. Gunn & W. B. Paige	
G–AWBN	PA-30 Twin Comanche 160	Stourfield Investments Ltd.	
G–AWBP	Cessna 182L Skylane	A. H. & Mrs. P. M. Butcher	
G–AWBS	PA-28 Cherokee 140	W. London Aero Services Ltd.	
G–AWBT	PA-30 Twin Comanche 160	Express Aviation Services Ltd.	
G–AWBU	Morane-Saulnier N (replica)	Personal Plane Services Ltd.	
G–AWBV	Cessna 182L Skylane	Airviews (Manchester) Ltd.	
G–AWBW	*Cessna F.172H	Brunel Technical College, Lulsgate	
G–AWBX	Cessna F.150H	Western Air Training Ltd.	
G–AWCD	CEA DR.253	M. L. Desoutter & Scoba Ltd.	
G–AWCH	Cessna F.150H	M. Bua	
G–AWCJ	Cessna F.150H	Transknight Ltd.	
G–AWCK	Cessna F.150H	Wycombe Air Centre Ltd.	
G–AWCL	Cessna F.150H	Civil Service Flying Club	

Notes	Reg.	Type	Owner or Operator
	G-AWCM	Cessna F.150H	Peterborough Aero Club Ltd.
	G-AWCN	Cessna FR.172E	LEC Refrigeration Ltd.
	G-AWCO	Cessna F.150H	K. A. Learmonth
	G-AWCP	Cessna F.150H	Herefordshire Aero Club Ltd.
	G-AWCR	Piccard Ax6 balloon	The London Balloon Club *London Pride I*
	G-AWCW	Beech E.95 Travel Air	H. W. Astor
	G-AWCX	PA-32 Cherokee Six 260	Kansey Ltd.
	G-AWCY	PA-32 Cherokee Six 260	Robinson & Carr Ltd.
	G-AWDA	Nipper T.66 Srs. 3	C. T. Storey & ptnrs.
	G-AWDD	Nipper T.66 Srs. 3	T. D. G. Roberts
	G-AWDH	D.31 Turbulent	J. H. Tetley
	G-AWDI	PA-23 Aztec 250	Air Foyle Ltd.
	G-AWDJ	Cessna 411	Choucell Ltd.
	G-AWDL	PA-25 Pawnee 235	Peter Charles (Airfarmers) Ltd.
	G-AWDO	D.31 Turbulent	R. Watling-Greenwood
	G-AWDP	PA-28 Cherokee 180	Brian Ilston Ltd.
	G-AWDR	Cessna FR.172E	Largoair Ltd.
	G-AWDU	Brantly B.2B	Coleraine Landscape Services
	G-AWDW	Bensen B.8	M. R. Langton
	G-AWDX	Beagle B.121 Pup 1	J. Pearse & O. D. West
	G-AWDZ	Beagle B.121 Pup 1	Smithwell Ltd.
	G-AWED	PA-31-310 Navajo	Cabair Ltd.
	G-AWEF	Stampe SV.4B	N. H. Jones
	G-AWEG	Cessna 172G	H. Lawson
	G-AWEI	D.62B Condor	N. H. Jones
	G-AWEL	Fournier RF.4D	A. B. Clymo
	G-AWEM	Fournier RF.4D	B. J. Griffin
	G-AWEN	Jodel DR.1050	L. G. Earnshaw & ptnrs.
	G-AWEO	Cessna F.150H	Banbury Plant Hire Ltd.
	G-AWEP	JB-01 Minicab	F. S. Jackson
	G-AWER	PA-23 Aztec 250	Woodgate Aviation
	G-AWES	Cessna 150H	Ralair Ltd.
	G-AWET	PA-28 Cherokee 180	Broadland Flying Group Ltd.
	G-AWEU	PA-28 Cherokee 140	Liverpool Aero Club Ltd.
	G-AWEV	PA-28 Cherokee 140	J. T. Garrod
	G-AWEX	PA-28 Cherokee 140	Northampton Heating & Ventilating Ltd.
	G-AWEY	PA-28R Cherokee Arrow 180	Peak Trailers Ltd.
	G-AWEZ	PA-28R Cherokee Arrow 180	D. O. Hooper & K. Erbsmail
	G-AWFB	PA-28R Cherokee Arrow 180	Luke Aviation Ltd.
	G-AWFC	PA-28R Cherokee Arrow 180	J. A. Clarke
	G-AWFD	PA-28R Cherokee Arrow 180	D. E. Roberts & J. G. Fisher
	G-AWFE	Jodel D.140E	Airscooters Ltd.
	G-AWFF	Cessna F.150H	Herefordshire Aero Club Ltd.
	G-AWFH	Cessna F.150H	Norfolk & Norwich Aero Club Ltd.
	G-AWFJ	PA-28R Cherokee Arrow 180	R. Watt
	G-AWFK	PA-28R Cherokee Arrow 180	J. A. Rundle (Holdings) Ltd.
	G-AWFN	D.62B Condor	J. Guy
	G-AWFO	D.62B Condor	Cornwall Flying Group
	G-AWFP	D.62B Condor	Wingsouth Ltd.
	G-AWFR	D.31 Turbulent	S. W. Usherwood
	G-AWFT	Jodel D.9 Bebe	W. H. Cole
	G-AWFW	Jodel D.117	F. H. Greenwell
	G-AWFX	Sikorsky S-61N	British Airways Helicopters Ltd.
	G-AWFY	SA.318C Alouette Astazou	Dollar Air Services Ltd.
	G-AWFZ	Beech A23 Musketeer	R. Sweet & B. D. Corbett
	G-AWGA	A.109 Airedale	A. C. W. Day
	G-AWGC	Cessna F.172H	Surrey & Kent Flying Club Ltd.
	G-AWGD	Cessna F.172H	A. A. Mattacks
	G-AWGE	Cessna F.172H	R. A. Gray
	G-AWGJ	Cessna F.172H	J. & C. J. Freeman
	G-AWGK	Cessna F.150H	G. R. Brown
	G-AWGM	Mitchell-Procter Kittiwake 2	Arkle Research & Development
	G-AWGN	Fournier RF.4D	P. J. Foreman
	G-AWGP	Cessna T210H	Gledhill Water Storage Ltd.
	G-AWGR	Cessna F.172H	P. Bushell
	G-AWGU	AB-206B JetRanger II	British Airways Helicopters Ltd.
	G-AWGX	Cessna F.172H	Aberdeen Aero Club
	G-AWGY	Cessna F.150H	Exeter Flying Club Ltd.
	G-AWGZ	Taylor JT.1 Monoplane	A. Hill
	G-AWHB	*CASA 2.111 (6J+PR)	Historic Aircraft Museum
	G-AWHU	Boeing 707-379C	British Airways

Reg.	Type	Owner or Operator	Notes
G–AWHV	Rollason Beta B.2A	Rollason Aircraft & Engines Ltd.	
G–AWHW	Rollason Beta B.2A	C. E. Bellhouse	
G–AWHX	Rollason Beta B.2	J. A. H. Chadwick	
G–AWIF	Brookland Mosquito	L. Chiappi	
G–AWIG	Jodel D.112	K. R. Nunn	
G–AWII	V.S.349 Spitfire VC (AR501)	Shuttleworth Trust	
G–AWIJ	V.S.329 Spitfire 11A (P7350)	RAF Battle of Britain Historic Aircraft Flight	
G–AWIK	Beech 23 Musketeer	Resource Investors Management Ltd.	
G–AWIN	Campbell-Bensen B.8MC	M. J. Cuttel & J. Deane	
G–AWIO	Brantly B.2B	G. J. Ward & E. J. Roche	
G–AWIP	Luton L.A.4A Minor	J. Houghton	
G–AWIR	Midget Mustang	K. E. Sword	
G–AWIT	PA-28 Cherokee 180	C. D. Linton & ptnrs.	
G–AWIV	Storey TSR.3	C. J. Lesson	
G–AWIW	Stampe SV.4B	Rothmans Aerobatic Team	
G–AWIY	PA-23 Aztec 250	Queen's University of Belfast	
G–AWJA	Cessna 182L Skylane	German Tourist Facilities Ltd.	
G–AWJC	Brighton gas balloon	P. D. Furlong *Slippery William*	
G–AWJE	Nipper T.66 Srs. 3	Jubilee Group	
G–AWJF	Nipper T.66 Srs. 3	R. Wilcock	
G–AWJI	M.S.880B Rallye Club	Thames Estaury Flying Services	
G–AWJN	Tigercraft Tiger Mk. II	Frederick Fewsdale	
G–AWJO	Tigercraft Tiger Mk. II	Frederick Fewsdale	
G–AWJP	Tigercraft Tiger Mk. III	Frederick Fewsdale	
G–AWJR	Tigercraft Tiger Mk. I	Frederick Fewsdale	
G–AWJS	Tigercraft Mosquito Mk. I	Frederick Fewsdale	
G–AWJT	Tigercraft Tiger Mk. I	J. H. Turner	
G–AWJV	*D.H.98 Mosquito TT Mk. 35 (TA634)	Mosquito Aircraft Museum	
G–AWJW	AB-206B JetRanger II	British Caledonian Helicopters Ltd.	
G–AWJX	Z.526 Akrobat	Aerobatics International Ltd.	
G–AWJY	Z.526 Akrobat	Ello Manufacturing Co.	
G–AWJZ	Cessna F.150H	Westshells Ltd.	
G–AWKB	M.J.5 Sirocco F.2 39	G. D. Claxton	
G–AWKD	PA-17 Vagabond	A. T. & Mrs. M. R. Dowie	
G–AWKM	B.121 Pup I	G. J. Morgan & P. P. Kneath	
G–AWKO	B.121 Pup I	D. A. Sadler	
G–AWKP	Jodel DR.253	R. C. Chandless	
G–AWKS	M.S.880B Rallye Club	P. R. Newell	
G–AWKT	M.S.880B Rallye Club	D. C. Strain	
G–AWKW	PA-24 Comanche 180	F. J. Bellamy	
G–AWKX	Beech A65 Queen Air	Ralli Bros & Comex Ltd.	
G–AWKZ	PA-23 Apache 160	E. A. Clack & T. Pritchard	
G–AWLA	Cessna F.150H	F. B. Reynolds	
G–AWLB	D.31 Turbulent	A. E. Shouler	
G–AWLE	Cessna F.172H	Sunderland Flying Club Ltd.	
G–AWLF	Cessna F.172H	Clement Spring Co. Ltd.	
G–AWLG	SIPA 903	T. G. Turner	
G–AWLI	PA-22 Tri-Pacer 150	D. T. Cheetham	
G–AWLJ	Cessna F.150H	D. S. Watts	
G–AWLL	AB-206B JetRanger 2	Gleneagles Helicopters Ltd.	
G–AWLM	Bensen B.8MS	C. J. E. Ashby	
G–AWLP	Mooney M.20F	Siminco Ltd.	
G–AWLR	Nipper T.66 Srs. 3	J. D. Lowther	
G–AWLS	Nipper T.66 Srs. 3	S. W. Brown	
G–AWLW	Hawker Hurricane 11B (P3308)	Strathallan Aircraft Collection	
G–AWLY	Cessna F.150H	Birmingham Aviation Ltd.	
G–AWLZ	Fournier RF.4D	E. V. Goodwin & C. R. Williamson	
G–AWMC	Campbell-Bensen B.8MS	M. E. Sykes-Hankinson	
G–AWMD	Jodel D.11	D. A. Lord	
G–AWMF	PA-18-150 Super Cub	Airways Aero Associations Ltd.	
G–AWMI	Glos-Airtourer 115	V. C. Birch	
G–AWMK	AB-206A JetRanger	Bristow Helicopters Ltd.	
G–AWMM	M.S.893A Rallye Commodore 180	Callow Aviation	
G–AWMN	Luton L.A.4A Minor	R. E. R. Wilks	
G–AWMP	Cessna F.172H	W. Rennie-Roberts	
G–AWMR	D.31 Turbulent	P. R. M. Bowlan	
G–AWMS	H.S.125 Srs. 3B	Rio Tinto Zinc Services Ltd.	
G–AWMT	Cessna F.150H	J. A. Wright	

Notes	Reg.	Type	Owner or Operator
	G–AWMU	Cessna F.172H	Tee-side Transport Commercial Services Ltd.
	G–AWNA	Boeing 747-136	British Airways *Sir Richard Grenville*
	G–AWNB	Boeing 747-136	British Airways
	G–AWNC	Boeing 747-136	British Airways
	G–AWND	Boeing 747-136	British Airways *Christopher Marlowe*
	G–AWNE	Boeing 747-136	British Airways *Sir Francis Drake*
	G–AWNF	Boeing 747-136	British Airways
	G–AWNG	Boeing 747-136	British Airways
	G–AWNH	Boeing 747-136	British Airways *Sir Walter Raleigh*
	G–AWNI	Boeing 747-136	British Airways
	G–AWNJ	Boeing 747-136	British Airways *John Donne*
	G–AWNK	Boeing 747-136	British Airways *William Shakespeare*
	G–AWNL	Boeing 747-136	British Airways
	G–AWNM	Boeing 747-136	British Airways
	G–AWNN	Boeing 747-136	British Airways *Sebastian Cabot*
	G–AWNO	Boeing 747-136	British Airways *Sir Francis Bacon*
	G–AWNP	Boeing 747-136	British Airways *Sir John Hawkins*
	G–AWNT	BN-2A Islander	Survey Flights Ltd.
	G–AWOA	M.S.880B Rallye Club	Oakley Motor Units Ltd.
	G–AWOE	Aero Commander 680E	J. M. Houlder
	G–AWOF	PA-15 Vagabond	D. S. Morgan
	G–AWOH	PA-17 Vagabond	K. M. Bowen
	G–AWOJ	Cessna F.172H	Aberdeen Aero Club Ltd.
	G–AWOL	Bell 206B JetRanger 2	Gleneagles Helicopters Ltd.
	G–AWOT	Cessna F.150H	Coventry Air Training School Ltd.
	G–AWOU	Cessna 170B	North Denes Aerodrome Ltd.
	G–AWPA	D.31A Turbulent	J. T. Heaton
	G–AWPG	Z.526 Akrobat	R. E. Legg
	G–AWPH	P.56 Provost T Mk. I	J. A. D. Bradshaw
	G–AWPJ	Cessna F.150H	M. L. Sarjeant
	G–AWPK	PA-23 Aztec 250	Air Atlantique Ltd.
	G–AWPL	Bensen B-8	N. F. Higgins
	G–AWPN	Shield Xyla	T. Worrall
	G–AWPP	Cessna F.150H	D. Williams
	G–AWPS	PA-28 Cherokee 140	J. D. Widdicombe & C. Young
	G–AWPU	Cessna F.150J	Light Planes (Lancashire) Ltd.
	G–AWPW	PA-12 Super Cruiser	P. Ligertwood
	G–AWPX	Cessna 150E	W. R. Emberton
	G–AWPY	Bensen B.8M	J. M. Deane
	G–AWPZ	Andreasson BA-4B	S. E. & P. J. C. Phillips
	G–AWRB	B.121 Pup I	P. O. P Pulvermacher
	G–AWRK	Cessna F.150J	T. D. Close
	G–AWRL	Cessna F.172H	K. W. Harris & ptnrs.
	G–AWRP	Grasshopper CR.LTH-I	Cierva Rotorcraft Ltd.
	G–AWRS	*Avro 19 Srs. 2	Strathallan Aircraft Collection
	G–AWRT	Glos-Airtourer 115	Hollybush Investments Ltd.
	G–AWRW	B.121 Pup Srs. 2	Ambrian Aviation Ltd.
	G–AWRZ	Bell 47G-5	Heliwork Ltd.
	G–AWSA	*Avro 652A Anson 19 (N5054)	Norfolk & Suffolk Aviation Museum
	G–AWSD	Cessna F.150J	B. D. & W. Phillips
	G–AWSF	Cessna 401	Con-Mech Engineers Ltd.
	G–AWSH	Z.526 Akrobat	Aerobatics International Ltd.
	G–AWSK	Agusta-Bell 47G-2	Bristow Helicopters Ltd.
	G–AWSL	PA-28 Cherokee 180D	Fascia Ltd.
	G–AWSM	PA-28 Cherokee 235	C.S.E. Aviation Ltd.
	G–AWSN	D.62B Condor	J. Leader
	G–AWSO	D.62B Condor	N. H. Jones
	G–AWSP	D.62B Condor	R. Q. & A. S. Bond
	G–AWSR	D.62B Condor	N. H. Jones
	G–AWSS	D.62C Condor	N. H. Jones
	G–AWST	D.62B Condor	Humberside Aviation
	G–AWSU	F.8L Falco Srs. 4	M. Slazenger
	G–AWSV	Skeeter 12 (XM553)	Maj. M. Somerton-Rayner
	G–AWSY	Boeing 737-204	Britannia Airways Ltd. *General James Wolfe*
	G–AWSZ	M.S.894A Rallye Minerva 220	D. Quinn & J. McCloskey
	G–AWTA	Cessna 310N	A. H. Wiltshire
	G–AWTJ	Cessna F.150J	IIP Associates
	G–AWTL	PA-28 Cherokee 180D	Leston Aviation
	G–AWTM	PA-28 Cherokee 140	Keenair Services Ltd.
	G–AWTR	Beech A.23 Musketeer	R. B. Aviation Ltd.
	G–AWTU	Beech A.23 Musketeer	Air Haven

Reg.	Type	Owner or Operator	Notes
G–AWTV	Beech A.23 Musketeer	R. Hammond	
G–AWTW	Beech B.55 Baron	Worldwide Wheels Ltd.	
G–AWTX	Cessna F.150J	Tango X-Ray Flying Group	
G–AWUA	Cessna P.206D	G. B. Grant & Sons (Farmers) Ltd.	
G–AWUB	GY.201-Minicab	H. P. Burrill	
G–AWUE	Jodel DR.1050	S. Bichan	
G–AWUF	H.S.125 Srs. 1B	Guinness Peat Group Services Ltd.	
G–AWUG	Cessna F.150H	Norfolk & Norwich Aero Club Ltd.	
G–AWUH	Cessna F.150H	K. A. Learmonth	
G–AWUJ	Cessna F.150H	R. J. Jones & J. M. Allen	
G–AWUL	Cessna F.150H	M. Cairns & ptnrs.	
G–AWUM	Cessna F.150H	Western Air Training Ltd.	
G–AWUN	Cessna F.150H	Northamptonshire School of Flying Ltd.	
G–AWUO	Cessna F.150H	W. Todd	
G–AWUP	Cessna F.150H	R. H. Timmis	
G–AWUR	Cessna F.150J	K. A. Learmonth	
G–AWUS	Cessna F.150J	Birmingham Aviation Ltd.	
G–AWUT	Cessna F.150J	J. M. Kirk	
G–AWUU	Cessna F.150J	Enniskillen Flying Club	
G–AWUW	Cessna F.172H	E. Shipley & H. Wilson	
G–AWUX	Cessna F.172H	J. D. A. Shields & ptnrs.	
G–AWUY	Cessna F.172H	J. & B. Powell (Printers) Ltd.	
G–AWUZ	Cessna F. 172H	K. Wickenden	
G–AWVA	Cessna F. 172H	W. Lipka	
G–AWVB	Jodel D.117	C. M. & T. R. C. Griffin	
G–AWVC	B.121 Pup 1	Middleton Music Stores Ltd.	
G–AWVE	Jodel DR.1050M.1	D. J. Leo & ptnrs.	
G–AWVF	P.56 Provost T.1 (XF877)	M. W. Stow	
G–AWVH	Glos-Airtourer 115	Mardenair Ltd.	
G–AWVK	H.P.137 Jetstream	Decca Navigator Co. Ltd.	
G–AWVN	Aeronca 7AC Champion	Bowker Air Services Ltd.	
G–AWVP	Brookland Hornet	Brookland Rotorcraft Ltd.	
G–AWVR	Skyship non-rigid airship	Skyships Ltd.	
G–AWVS	Cessna 337D	W. H. Crispe & Sons Ltd.	
G–AWVW	PA-23 Aztec 250D	Heath Street Car Hirings Ltd.	
G–AWVY	BN-2A Islander	Euroair Transport Ltd.	
G–AWVZ	Jodel D.112	D. C. Stokes	
G–AWWE	B.121 Pup 2	G. J. Bunting	
G–AWWF	B.121 Pup 1	Three Counties Aero Club Ltd.	
G–AWWI	Jodel D.117	J. H. Kirkham	
G–AWWL	H.S.125 Srs. 3B-RA	McAlpine Aviation Ltd.	
G–AWWM	GY-201 Minicab	J. S. Brayshaw	
G–AWWN	Jodel DR.1051	Jodel Flying Group	
G–AWWO	Jodel DR.1050	D. R. Gray & ptnrs.	
G–AWWP	Woody Pusher III	M. S. Bird & Mrs. R. D. Bird	
G–AWWS	SC.7 Skyvan Srs. 3	Vernair Transport Services	
G–AWWT	D.31 Turbulent	J. G. Alderton & J. Martendale	
G–AWWU	Cessna FR.172F	Dowdeswell Engineering Co. Ltd.	
G–AWWV	Cessna FR.172F	I. R. Hamilton & J. P. M. Stewart	
G–AWWW	Cessna 401	Westair Flying Services Ltd.	
G–AWWX	BAC One-Eleven 509	Dan-Air Services Ltd.	
G–AWWZ	BAC One-Eleven 509	Monarch Airlines Ltd.	
G–AWXA	Cessna 182M Skylane	R. E. & Mrs. U. C. Markelow	
G–AWXH	Cessna F.150H	Bristol & Wessex Aeroplane Club Ltd.	
G–AWXO	H.S.125 Srs. 400B	McAlpine Aviation Ltd.	
G–AWXR	PA-28 Cherokee 180D	J. D. Williams	
G–AWXS	PA-28 Cherokee 180D	D. F. Smyth & ptnrs.	
G–AWXU	Cessna F.150J	Mona Aviation Ltd.	
G–AWXV	Cessna F.172H	D. N. Forrest	
G–AWXW	PA-23 Aztec 250D	Thurston Aviation Ltd.	
G–AWXY	M.S.885 Super Rallye	J. & B. Fowler	
G–AWXZ	SNCAN SV-4C	Personal Plane Services Ltd.	
G–AWYB	Cessna FR.172F	C. W. Larkin	
G–AWYE	H.S.125 Srs. 1B	Rolls-Royce (1971) Ltd.	
G–AWYF	G.159 Gulfstream 1	Ford Motor Co. Ltd.	
G–AWYH	Aero Commander 200D	W. J. D. Roberts	
G–AWYJ	B.121 Pup 2	B. Richardson & M. J. Holland	
G–AWYL	Jodel DR.253B	Clarville Ltd.	
G–AWYO	B.121 Pup 1	B. R. C. Wild	
G–AWYR	BAC One-Eleven 501	British Caledonian Airways *Isle of Tiree*	

Notes	Reg.	Type	Owner or Operator
	G–AWYS	BAC One-Eleven 501	British Caledonian Airways *Isle of Bute*
	G–AWYT	BAC One-Eleven 501	British Caledonian Airways *Isle of Barra*
	G–AWYU	BAC One-Eleven 501	British Caledonian Airways *Isle of Colonsay*
	G–AWYV	BAC One-Eleven 501	British Caledonian Airways *Isle of Harris*
	G–AWYX	M.S.880B Rallye Club	Joy M. L. Edwards
	G–AWYY	T.57 Camel replica (C1701)	Leisure Sport Ltd.
	G–AWYZ	H.S.121 Trident 3B	British Airways
	G–AWZA	H.S.121 Trident 3B	British Airways
	G–AWZB	H.S.121 Trident 3B	British Airways
	G–AWZC	H.S.121 Trident 3B	British Airways
	G–AWZD	H.S.121 Trident 3B	British Airways
	G–AWZE	H.S.121 Trident 3B	British Airways
	G–AWZF	H.S.121 Trident 3B	British Airways
	G–AWZG	H.S.121 Trident 3B	British Airways
	G–AWZH	H.S.121 Trident 3B	British Airways
	G–AWZI	H.S.121 Trident 3B	British Airways
	G–AWZJ	H.S.121 Trident 3B	British Airways
	G–AWZK	H.S.121 Trident 3B	British Airways
	G–AWZL	H.S.121 Trident 3B	British Airways
	G–AWZM	H.S.121 Trident 3B	British Airways
	G–AWZN	H.S.121 Trident 3B	British Airways
	G–AWZO	H.S.121 Trident 3B	British Airways
	G–AWZP	H.S.121 Trident 3B	British Airways
	G–AWZR	H.S.121 Trident 3B	British Airways
	G–AWZS	H.S.121 Trident 3B	British Airways
	G–AWZU	H.S.121 Trident 3B	British Airways
	G–AWZV	H.S.121 Trident 3B	British Airways
	G–AWZW	H.S.121 Trident 3B	British Airways
	G–AWZX	H.S.121 Trident 3B	British Airways
	G–AWZZ	H.S.121 Trident 3B	British Airways
	G–AXAA	Canadair CL-44D-4	
	G–AXAB	PA-28 Cherokee 140	Bencray Ltd.
	G–AXAJ	Glos-Airtourer 150	E. N. Struys
	G–AXAK	M.S.880B Rallye Club	R. L. & Mrs. C. Stewart
	G–AXAN	D.H.82A Tiger Moth	A. J. Cheshire
	G–AXAO	Omega 56 balloon	P. D. Furlong *Renatus Cartesius*
	G–AXAS	Wallis WA-116T	K. H. Wallis (G–AVDH)
	G–AXAT	Jodel D.117A	M. I. Goss
	G–AXAU	PA-30 Twin Comanche 160C	RHM Investments Ltd.
	G–AXAV	PA-30 Twin Comanche 160C	Nottingham Aviation Ltd.
	G–AXAW	Cessna 421A	Bembridge Air Hire Ltd.
	G–AXAX	PA-23 Aztec 250D	Euroair Transport Ltd.
	G–AXAZ	PA-31 Navajo	Meridian Airmaps Ltd.
	G–AXBB	BAC One-Eleven 409	Air UK *Island Envoy*
	G–AXBD	PA-25 Pawnee 235C	Farm Aviation Services Ltd.
	G–AXBG	Bensen B.8M	R. Curtis
	G–AXBH	Cessna F.172H	Farrowcrest Ltd.
	G–AXBJ	Cessna F.172H	Sussex Marine Aviation Charters Ltd.
	G–AXBU	Cessna FR.172F	Sir P. Grant-Suttie
	G–AXBW	D.H.82A Tiger Moth (T5879)	R. Venning
	G–AXBY	Cessna 401A	Hawtal Whiting (Design & Engineering) Co. Ltd.
	G–AXBZ	D.H.82A Tiger Moth	D. H. McWhir
	G–AXCA	PA-28R Cherokee Arrow 200	J. A. Tooth
	G–AXCC	Bell 47G-2	Bristow Helicopters Ltd.
	G–AXCD	Agusta-Bell 47G-2	Bristow Helicopters Ltd.
	G–AXCF	Agusta-Bell 47G-2	Bristow Helicopters Ltd.
	G–AXCG	Jodel D.117	J. W. Hollingsworth & M. J. Doherty
	G–AXCK	BAC One-Eleven 401	Dan-Air Services Ltd.
	G–AXCL	M.S.880B Rallye Club	A. Cope & F. Ratcliffe
	G–AXCM	M.S.880B Rallye Club	P. A. & W. A. Benjamin
	G–AXCN	M.S.880B Rallye Club	Severn Valley Aero Club
	G–AXCP	BAC One-Eleven 401	Dan Air Services Ltd.
	G–AXCX	B.121 Pup 2	Deltair Ltd.
	G–AXCY	Jodel D.117	R. M. Rennoldson
	G–AXCZ	Stampe SV-4C	K. L. Adam
	G–AXDB	Piper L-4 Cub	N. D. Norman
	G–AXDC	PA-23 Aztec 250D	Trago Mills (South Devon) Ltd.

Reg.	*Type*	*Owner or Operator*	*Notes*
G–AXDD	PA-31-300 Navajo	H. A. Stradling & Sons Ltd.	
G–AXDE	Bensen B.8	T. J. Hartwell	
G–AXDH	BN-2A Islander	Parachute Regiment Freefall Club	
G–AXDI	Cessna F.172H	Jim Russell International Racing Drivers Ltd.	
G–AXDK	Jodel DR.315	W. B. Wright & Sons Ltd.	
G–AXDL	PA-30 Twin Comanche 160C	Northern Executive Aviation Ltd.	
G–AXDM	H.S.125 Srs. 400B	Ferranti Ltd.	
G–AXDN	*BAC-Sud Concorde 01	Duxford Aviation Soc.	
G–AXDU	B.121 Pup 2	Deltair Ltd.	
G–AXDV	B.121 Pup 1	R. A. Chappell	
G–AXDW	B.121 Pup 1	Cranfield Institute of Technology	
G–AXDX	Wassmer Jodel D.120	G. D. Pybus	
G–AXDY	Falconar F-11	G. K. Ellis	
G–AXDZ	Cassutt Racer Srs. 111M	A. Chadwick	
G–AXEB	Cassutt Racer Srs. 111M	G. E. Horder	
G–AXEC	Cessna 182M	R. F. Fox	
G–AXED	PA-25 Pawnee 235	Sprayfields (Scothern) Ltd.	
G–AXEI	*Ward Gnome	Lincolnshire Aviation Museum	
G–AXEJ	Hughes 369HS	Ryburn Air Ltd.	
G–AXEO	Scheibe SF.25B Falke	D. Collinson	
G–AXER	PA-30 Twin Comanche 160C	Astrojet Ltd. & Danro Dental Products Ltd.	
G–AXET	B.121 Pup 2	J. C. Furneaux	
G–AXEU	B.121 Pup 2	Wings Flying Group	
G–AXEV	B.121 Pup 2	B. Richardson	
G–AXEW	B.121 Pup 1	C. J. Spicer & A. A. Gray	
G–AXEX	B.121 Pup 1	Lubenham Fidelities & Investment Co. Ltd.	
G–AXFA	PA-23 Aztec 250D	Herts & Essex Aero Club	
G–AXFB	BN-3 Nymph	BN (Bembridge) Ltd.	
G–AXFD	PA-25 Pawnee 235	J.E.F. Aviation Ltd.	
G–AXFE	Beech B.90 King Air	GKN Group Services Ltd.	
G–AXFF	Cessna A.188 Agwagon	Ag-Air	
G–AXFG	Cessna 337D	Alfred Smith & Son (Penzance) Ltd.	
G–AXFM	Servotec Grasshopper 3	Cierva Rotorcraft Ltd.	
G–AXFN	Jodel D.119	R. O. Harper	
G–AXGA	PA-19 Super Cub 95	Felthorpe Flying Group Ltd.	
G–AXGC	M.S.880B Rallye Club	K. M. & H. Bowen	
G–AXGD	M.S.880B Rallye Club	J. G. R. Read & ptnrs.	
G–AXGE	M.S.880B Rallye Club	A. R. T. Banks	
G–AXGG	Cessna F.150J	G. W. G. C. Sudlow	
G–AXGP	Piper L-4B Cub	W. K. Butler	
G–AXGR	Luton L.A.4A Minor	T. M. W. Webster	
G–AXGS	D.62B Condor	N. H. Jones	
G–AXGT	D.62B Condor	P. Simpson & ptnrs.	
G–AXGU	D.62B Condor	N. H. Jones	
G–AXGV	D.62B Condor	C. N. Bonneywell	
G–AXGW	Boeing 707-336C	British Airways	
G–AXGX	Boeing 707-336C	British Airways	
G–AXGZ	D.62B Condor	Lines Condor Group	
G–AXHA	Cessna 337A	E. A. Pitcher	
G–AXHC	Stampe SV-4C	K. L. Hawse & ptnrs.	
G–AXHE	BN-2A Islander	Peterborough Parachute Centre Ltd.	
G–AXHG	M.S.880B Rallye Club	D. M. Leonard	
G–AXHI	M.S.880B Rallye Club	P. Newsham	
G–AXHO	B.121 Pup 2	Harvair Ltd.	
G–AXHP	Piper L-4H Cub	W. F. Barnes & R. W. Griffin	
G–AXHR	Piper L-4H Cub (329601)	D. E. Elphick	
G–AXHS	M.S.880B Rallye Club	A. L. Langley	
G–AXHT	M.S.880B Rallye Club	V. C. A. Fisher	
G–AXHV	Jodel D.117A	D. M. Cashmore & K. R. Payne	
G–AXHX	M.S.892 Rallye Commodore	Lord Davies	
G–AXIA	B.121 Pup 1	Cranfield Institute of Technology	
G–AXIE	B.121 Pup 2	Ellan Vannin Flying Group	
G–AXIF	B.121 Pup 2	Stadrest Ltd.	
G–AXIG	B.125 Bulldog 104	George House (Holdings) Ltd.	
G–AXIH	Bu 133 Jungmeister	R. E. Legg	
G–AXIO	PA-28 Cherokee 140B	W. London Aero Services Ltd.	
G–AXIP	PA-28 Cherokee 140B	R. J. Stevens	
G–AXIR	PA-28 Cherokee 140B	C. T. Brinson	
G–AXIT	M.S.893A Rallye Commodore 180	South Wales Gliding Club Ltd.	

Notes	Reg.	Type	Owner or Operator
	G–AXIV	PA-23 Aztec 250D	Houston Management Services Ltd.
	G–AXIW	Scheibe SF.25B Falke	Herefordshire Gliding Club Ltd.
	G–AXIX	Glos-Airtourer 150	R. A. Sareen
	G–AXIY	Bird Gyrocopter	Gerald Bird
	G–AXJB	Omega 84 balloon	Hot-air Group *Jester*
	G–AXJH	B.121 Pup 2	J. Pearce
	G–AXJI	B.121 Pup 2	E. A. Clack
	G–AXJJ	B.121 Pup 2	P. Hirst & ptnrs.
	G–AXJK	BAC One-Eleven 501	British Caledonian Airways *Isle of Staffa*
	G–AXJM	BAC One-Eleven 501	British Caledonian Airways *Isle of Islay*
	G–AXJN	B.121 Pup 2	Toon Ghose Aviation Ltd.
	G–AXJO	B.121 Pup 2	Lowe (Paddock Wood) Ltd.
	G–AXJR	Scheibe SF.25B Falke	M. J. Munday & N. E. M. Coombe
	G–AXJV	PA-28 Cherokee 140B	Mona Aviation Ltd.
	G–AXJW	PA-28 Cherokee 140B	I. Goodchild
	G–AXJX	PA-28 Cherokee 140B	R. Gilkes
	G–AXJY	Cessna U-206D Super Skywagon	A. I. Walgate & Son
	G–AXKD	PA-23 Aztec 250D	Jones & Bailey Contractors Ltd.
	G–AXKH	Luton L.A.4A Minor	M. E. Vaisey
	G–AXKI	Jodel D.9 Bebe	T. A. Hodges
	G–AXKJ	Jodel D.9 Bebe	C. H. Morris
	G–AXKK	Westland Bell 47G-4A	Bristow Helicopters Ltd.
	G–AXKL	Westland Bell 47G-4A	Bristow Helicopters Ltd.
	G–AXKM	Westland Bell 47G-4A	Bristow Helicopters Ltd.
	G–AXKN	Westland Bell 47G-4A	Bristow Helicopters Ltd.
	G–AXKO	Westland Bell 47G-4A	Bristow Helicopters Ltd.
	G–AXKR	Westland Bell 47G-4A	Bristow Helicopters Ltd.
	G–AXKS	Westland Bell 47G-4A	Bristow Helicopters Ltd.
	G–AXKT	Westland Bell 47G-4A	Bristow Helicopters Ltd.
	G–AXKU	Westland Bell 47G-4A	Bristow Helicopters Ltd.
	G–AXKV	Westland Bell 47G-4A	Bristow Helicopters Ltd.
	G–AXKW	Westland Bell 47G-4A	Bristow Helicopters Ltd.
	G–AXKX	Westland Bell 47G-4A	Bristow Helicopters Ltd.
	G–AXKY	Westland Bell 47G-4A	Bristow Helicopters Ltd.
	G–AXKZ	Westland Bell 47G-4A	Bristow Helicopters Ltd.
	G–AXLA	Westland Bell 47G-4A	Bristow Helicopters Ltd.
	G–AXLG	Cessna 310K	Smiths (Outdrives) Ltd.
	G–AXLI	Nipper T.66 Srs. 3	P. W. Thomas & K. Richter
	G–AXLK	SE.3160 Alouette III	United Helicopters Ltd.
	G–AXLS	Jodel DR.105A	T. L. Giles
	G–AXLZ	PA-19 Super Cub 95	J. C. Quantrell
	G–AXMA	PA-24 Comanche 180	Tegrel Products Ltd.
	G–AXMB	Slingsby Motor Cadet	I. G. Smith
	G–AXMD	Omega 20 balloon	Nimble Bread Ltd. *Nimble*
	G–AXMG	BAC One-Eleven 518	Monarch Airlines Ltd.
	G–AXMM	Bell 206A JetRanger	R. G. Woodward
	G–AXMN	J/5B Autocar	I. R. F. Hammond
	G–AXMP	PA-28 Cherokee 180	Concorde Garage (Elmsawell) Ltd.
	G–AXMR	PA-31-300 Navajo	Ron Webster (Midlands) Ltd.
	G–AXMS	PA-30 Twin Comanche 160C	Ernest Green International Ltd.
	G–AXMT	Bu 133 Jungmeister	S. R. Flack
	G–AXMU	BAC One-Eleven 432	Air UK *Island Spirit*
	G–AXMW	B.121 Pup 1	DJP Engineering (Knebworth) Ltd.
	G–AXMX	B.121 Pup 2	Susan A. Jones
	G–AXMY	PA-30 Twin Comanche 160C	Air & General Services Ltd.
	G–AXNA	Boeing 737-204C	Britannia Airways Ltd. *Robert Clive of India*
	G–AXNB	Boeing 737-204C	Britannia Airways Ltd. *Charles Darwin*
	G–AXNC	Boeing 737-204	Britannia Airways Ltd. *Sir Frederick Handley Page*
	G–AXNJ	Wassmer Jodel D.120	Clive Flying Group
	G–AXNK	Cessna F.150J	Mona Aviation Ltd.
	G–AXNL	B.121 Pup 1	Mike Bennett Ltd.
	G–AXNM	B.121 Pup 1	Majabay Ltd.
	G–AXNN	B.121 Pup 2	Knights Aviation Ltd.
	G–AXNP	B.121 Pup 2	Deltair Ltd.
	G–AXNR	B.121 Pup 2	Specialised Mouldings Ltd. & G. Broadley
	G–AXNS	B.121 Pup 2	R. Light & ptnrs.

Reg.	*Type*	*Owner or Operator*	*Notes*
G-AXNW	SNCAN SV-4C	E. N. Grace	
G-AXNX	Cessna 182M	Air Tows Ltd.	
G-AXNY	Fixter Pixie	J. van Geest	
G-AXNZ	Pitts S.1C Special	D. W. Leech & D. Parfrey	
G-AXOG	PA-23 Aztec 250D	R. W. Diggens	
G-AXOH	M.S.894 Rallye Minerva	Bristol Cars Ltd.	
G-AXOI	Jodel D.9 Bebe	P. W. Thomas	
G-AXOJ	B.121 Pup 2	Gordon King (Aviation) Ltd.	
G-AXOL	Currie Wot	D. R. Campbell & A. Kay	
G-AXOR	PA-28 Cherokee 180D	J. A. Butterfield & ptnrs.	
G-AXOS	M.S.894A Rallye Minerva	Seven Flying Group	
G-AXOT	M.S.893 Rallye Commodore 180	Col. J. F. Williams-Wynne	
G-AXOV	Beech B55A Baron	S. Brod	
G-AXOX	BAC One-Eleven 432	Air UK *Island Endeavour*	
G-AXOZ	B.121 Pup 1	Tees-side Aviation Ltd.	
G-AXPB	B.121 Pup 1	D. Smith	
G-AXPD	B.121 Pup 1	Surrey & Kent Flying Club Ltd.	
G-AXPF	Cessna F.150K	Westair Flying Services Ltd.	
G-AXPG	Mignet HM-293	W. H. Cole (Historic Aircraft Museum)	
G-AXPM	B.121 Pup 1	D. Taylor	
G-AXPN	B.121 Pup 2	Starline Elms Coaches	
G-AXRA	Campbell Cricket	L. E. Schnurr	
G-AXRB	Campbell Cricket	J. C. P. Thomas	
G-AXRC	Campbell Cricket	K. W. Hayr	
G-AXRD	Campbell Cricket	Glyndwr Rees	
G-AXRK	Practavia Sprite 115	E. G. Thale	
G-AXRL	PA 28 Cherokee 160	T. W. Clark	
G-AXRO	PA-30 Twin Comanche 160C	Havelet Aviation	
G-AXRP	SNCAN SV-4C	M. D. Tweedie & ptnrs.	
G-AXRR	Auster AOP.9 (XR241)	Shuttleworth Trust	
G-AXRS	Boeing 707-355C	British Caledonian Airways *Loch Lomond*	
G-AXRT	Cessna FA.150K	Hornet Aviation Ltd.	
G-AXRU	Cessna FA.150K	Civil Service Flying Club	
G-AXSC	B.121 Pup 1	J. Hawkins	
G-AXSD	B.121 Pup 1	Surrey & Kent Flying Club Ltd.	
G-AXSF	Procter Petrel	Procter Aircraft Associates	
G-AXSG	PA-28 Cherokee 180	Shropshire Aero Club Ltd.	
G-AXSH	PA-28 Cherokee 140B	Brailey & Co. (Aviation) Ltd.	
G-AXSI	Cessna F.172H	B. M. Godward	
G-AXSJ	Cessna FA.150K	Staverton Flying School Ltd.	
G-AXSL	Cessna 310P	Air Service Training Ltd.	
G-AXSM	Jodel DR.1051	C. Cousten	
G-AXSP	PA-30 Twin Comanche 160C	Henfrey Engineering & Developments Ltd.	
G-AXSV	Jodel DR.340	Beamville Ltd.	
G-AXSW	Cessna FA.150K	Furness Aviation Ltd.	
G-AXSX	Beech C.23 Sundowner	D. M. Balfour	
G-AXSZ	PA-28 Cherokee 140B	A. R. Powley	
G-AXTA	PA-28 Cherokee 140B	Carlisle Aviation Co. Ltd.	
G-AXTB	PA-28 Cherokee 140B	Stapleford Flying Club	
G-AXTC	PA-28 Cherokee 140B	Airways Aero Associations Ltd.	
G-AXTD	PA-28 Cherokee 140B	Vincent-Walker Engineering Ltd.	
G-AXTE	PA-28 Cherokee 140B	N. Clayton	
G-AXTF	PA-28 Cherokee 140B	Webb & Co.	
G-AXTG	PA-28 Cherokee 140B	M. J. Cowham	
G-AXTH	PA-28 Cherokee 140B	W. London Aero Services Ltd.	
G-AXTI	PA-28 Cherokee 140B	LT (Central Road Services) Sports Association	
G-AXTJ	PA-28 Cherokee 140B	TK Aero Enterprises Ltd.	
G-AXTK	PA-28 Cherokee 140B	Andrewsfield Flying Club Ltd.	
G-AXTL	PA-28 Cherokee 140B	L. Williams	
G-AXTM	PA-28 Cherokee 140B	Cormack (Aircraft Services) Ltd.	
G-AXTO	PA-24 Comanche 260	Wolverhouse Ltd.	
G-AXTP	PA-28 Cherokee 180	D. I. L. Butler	
G-AXTX	Jodel D.112	J. J. Penney	
G-AXUA	B.121 Pup 1	F. R. Blennerhassett & C. Wedlake	
G-AXUB	BN-2A Islander	Bristow Helicopters Ltd.	
G-AXUC	PA-12 Super Cruiser	V. N. Mukaloff	
G-AXUE	Jodel DR.105A	Carlton Flying Group	
G-AXUF	Cessna FA.150K	Airwork Services Ltd.	

Notes	Reg.	Type	Owner or Operator
	G–AXUI	H.P.137 Jetstream 1	Cranfield Institute of Technology
	G–AXUJ	J/1 Autocrat	R. G. Earp & J. W. H. Lee
	G–AXUK	Jodel DR.1050	R. Pidcock & ptnrs.
	G–AXUL	Canadair CL-44D-4	British Cargo Airlines Ltd.
	G–AXUM	H.P.137 Jetstream 1	Cranfield Institute of Technology
	G–AXUV	Cessna F.172H	F. A. & Mrs. E. M. Smith
	G–AXUW	Cessna FA.150K	Coventry Air Training School
	G–AXUX	Beech B95 Travel Air	J. H. Southern
	G–AXUZ	Practavia Sprite 125	C. B. Healey
	G–AXVB	Cessna F.172H	Staverton Flying School Ltd.
	G–AXVC	Cessna FA.150K	M. J. Godwin & Sons Ltd.
	G–AXVG	H.S.748 Srs. 2	Dan-Air Services Ltd.
	G–AXVK	Campbell Cricket	Campbell Gyroplanes Ltd.
	G–AXVM	Campbell Cricket	C. H. T. Cliff
	G–AXVN	McCandless M.4	R. McCandless
	G–AXVS	Jodel DR.1050	F. W. Tilley
	G–AXVU	Omega 84 balloon	Brede Balloons Ltd. *Henry VII*
	G–AXVV	Piper L-4H Cub	J. MacCartny
	G–AXVW	Cessna F.150K	Alan Lord Photography
	G–AXVX	Cessna F.172H	D. A. Doust
	G–AXWA	Auster AOP 9	T. Platt
	G–AXWB	Omega 65 balloon	A. Robinson & M. J. Moore *Ezekiel*
	G–AXWD	Jurca MJ 10	F.P.A. Group
	G–AXWE	Cessna F.150K	Light Planes (Lancashire) Ltd.
	G–AXWF	Cessna F.172H	Redfir Aviation Ltd.
	G–AXWG	BN-2A Islander	Bristow Helicopters Ltd.
	G–AXWP	BN-2A Islander	Aurigny Air Services
	G–AXWR	BN-2A Islander	Aurigny Air Services
	G–AXWT	Jodel D.11	R. Owen
	G–AXWV	Jodel DR.253	Murray Motors
	G–AXWY	Taylor JT.2 Titch	S. E. O. Tomlinson
	G–AXWZ	PA-28R Cherokee Arrow 200	Melbourns Brewery Ltd.
	G–AXXA	PA-28 Cherokee 180E	Sicon Hydraulics Ltd.
	G–AXXC	CP.301B Emeraude	J. R. R. Gale & J. Tetley
	G–AXXD	Hughes H.269B	Twyford Moors (Helicopters) Ltd.
	G–AXXG	BN-2A Islander	G.K.N. Group Services Ltd.
	G–AXXH	BN-2A Islander	M. P. Data Prep Ltd.
	G–AXXJ	BN-2A Islander	Haywards Aviation Ltd.
	G–AXXM	CP.301A Emeraude	Mrs. P. H. Wren
	G–AXXN	WHE Airbuggy	W. H. Ekin
	G–AXXR	Beech 95-B55A Baron	Firth Plant Ltd.
	G–AXXV	D.H.82A Tiger Moth (DE992)	L. B. Jefferies
	G–AXXW	Jodel D.117	J. P. Bell
	G–AXXY	Boeing 707-336B	British Airways
	G–AXXZ	Boeing 707-336B	British Airways
	G–AXYA	PA-31-300 Navajo	C.S.E. (Aircraft Services) Ltd.
	G–AXYD	BAC One-Eleven 509	Dan-Air Services Ltd.
	G–AXYK	Taylor JT.1 Monoplane	C. Oakins
	G–AXYO	PA-25 Pawnee 235	Westwick Distributors Ltd.
	G–AXYU	Jodel D.9 Bebe	D. P. Jones & S. R. Sissons
	G–AXYV	Luton Beta Srs. 2	D. G. Wiggins
	G–AXYX	WHE Airbuggy	R. T. Ginn
	G–AXYY	WHE Airbuggy	M. P. Chetwynd-Talbot
	G–AXYZ	WHE Airbuggy	W. H. Ekin
	G–AXZA	WHE Airbuggy	J. C. Kitchin
	G–AXZB	WHE Airbuggy	W. H. Ekin
	G–AXZC	PA-28 Cherokee 180E	College of Air Training
	G–AXZD	PA-28 Cherokee 180E	College of Air Training
	G–AXZE	PA-28 Cherokee 180E	College of Air Training
	G–AXZF	PA-28 Cherokee 180E	College of Air Training
	G–AXZJ	Cessna F.172H	Smiths' Aviation
	G–AXZM	Nipper Mk. III	S. J. Booth & D. A. Young
	G–AXZO	Cessna 180	R.S.A. Parachute Club Ltd.
	G–AXZP	PA-23 Aztec 250	White House Garage, Ashford Ltd.
	G–AXZR	Taylor JT.2 Titch	A. J. Fowler & D. E. Evans
	G–AXZT	Jodel D.117	H. W. Baines
	G–AXZU	Cessna 182N	R. Taylor & ptnrs.
	G–AYAA	PA-28 Cherokee 180E	College of Air Training
	G–AYAB	PA-28 Cherokee 180E	College of Air Training
	G–AYAC	PA-28R Cherokee Arrow 200	Steer Aviation Ltd.
	G–AYAD	PA-30 Twin Comanche 160C	Apache Aircraft Services Ltd.
	G–AYAE	Bell 47G-4A	Helicopter Hire Ltd.

Reg.	Type	Owner or Operator	Notes
G–AYAF	PA-30 Twin Comanche 160C	Arrow Air Services (Charter) Ltd.	
G–AYAI	Fournier RF-5	Exeter RF Group	
G–AYAJ	Cameron O-84 balloon	E. T. Hall *Flaming Pearl*	
G–AYAK	Yak-11	A. E. Hutton	
G–AYAL	Omega 56 balloon	Nimble Bread Ltd. *Nimble II*	
G–AYAN	Slingsby Motor Cadet Mk. III	I. Stevenson	
G–AYAO	Cessna F.172H	Transmatic Fyllan Ltd.	
G–AYAP	PA-28 Cherokee 180E	College of Air Training	
G–AYAR	PA-28 Cherokee 180E	College of Air Training	
G–AYAS	PA-28 Cherokee 180E	College of Air Training	
G–AYAT	PA-28 Cherokee 180E	College of Air Training	
G–AYAU	PA-28 Cherokee 180E	College of Air Training	
G–AYAV	PA-28 Cherokee 180E	College of Air Training	
G–AYAW	PA-28 Cherokee 180E	College of Air Training	
G–AYBD	Cessna F.150K	Trehaven Aviation Ltd.	
G–AYBG	Scheibe SF.25B Falke	Doncaster Sailplane Services	
G–AYBK	PA-28 Cherokee 180E	College of Air Training	
G–AYBO	PA-23 Aztec 250D	Twinguard Leasing Ltd.	
G–AYBP	Jodel D.112	A. P. Davies & ptnrs.	
G–AYBT	PA-28 Cherokee 180E	College of Air Training	
G–AYBU	Omega 84 balloon	D. R. Gibbons	
G–AYBV	Chasle Tourbillon	B. A. Mills	
G–AYCC	Campbell Cricket	K. W. E. Denson	
G–AYCE	CP.301C Emeraude	K. Webb	
G–AYCF	Cessna FA.150K	Leo Designs	
G–AYCG	SNCAN SV-4C	Colton Aviation Services Ltd.	
G–AYCJ	Cessna TP.206D	Brymon Aviation Ltd.	
G–AYCM	Bell 206A JetRanger	Express Helicopters Ltd.	
G–AYCN	Piper L-4H Cub	F. J. Cox	
G–AYCO	CEA DR.360	L. M. Gould	
G–AYCP	Jodel D.112	W. Hutchings	
G–AYCT	Cessna F.172H	Kontrox Ltd.	
G–AYDG	M.S.894A Rallye Minerva	R. Vaughan & F. T. Skipper (Electronics) Ltd.	
G–AYDI	D.H.82A Tiger Moth	R. B. Woods	
G–AYDJ	Campbell Cricket	A. van Preussen	
G–AYDR	SNCAN SV-4C	R. A. Phillips	
G–AYDU	AJEP Tailwind	AJEP Development Ltd.	
G–AYDV	Coates SA.11-1 Swalesong	J. R. Coates	
G–AYDW	A.61 Terrier 2	J. S. Harwood	
G–AYDX	A.61 Terrier 2	R. G. Ford	
G–AYDY	Luton L.A.4A Minor	L. J. E. Goldfinch	
G–AYDZ	Jodel DR.200	Don Martin (Car Sales) Ltd.	
G–AYEB	Jodel D.112	F. W. & G. F. T. Taylor	
G–AYEC	CP.301A Emeraude	A. P. Docherty & J. S. Barker	
G–AYED	PA-24 Comanche 260	Patgrove Ltd.	
G–AYEE	PA-28 Cherokee 180E	College of Air Training	
G–AYEF	PA-28 Cherokee 180E	College of Air Training	
G–AYEG	Falconer F-9	A. G. Thelwall	
G–AYEH	Jodel DR.1050	R. O. F. Harper & P. R. Skeels	
G–AYEI	PA-31-300 Navajo	Keller, Bryant & Co. Ltd.	
G–AYEJ	Jodel DR.1050	G. Weaver	
G–AYEK	Jodel DR.1050	I. Shaw & B. Hanson	
G–AYEL	Bell 47G-5	Dollar Air Services Ltd.	
G–AYEN	Piper L-4H Cub	P. Warde & C. F. Morris	
G–AYER	H.S.125 Srs. 400B	Lazer Investments Ltd.	
G–AYES	M.S.892A Rallye Commodore 150	Waveney Flying Group	
G–AYET	M.S.892A Rallye Commodore 150	Cleanacres Ltd.	
G–AYEU	Brookland Hornet	A. J. Philpotts	
G–AYEV	Jodel DR.1050	M. R. Ireland	
G–AYEW	Jodel DR.1051	D. G. Hammersley & R. E. Kendal	
G–AYEX	Boeing 707-355C	British Caledonian Airways *Loch Leven*	
G–AYEY	Cessna F.150K	A. G. C. Finance Ltd.	
G–AYFA	SA Twin Pioneer 3	Flight One Ltd.	
G–AYFC	D.62B Condor	J. B. Randle	
G–AYFD	D.62B Condor	N. H. Jones	
G–AYFE	D.62C Condor	R. G. Bearcroft	
G–AYFF	D.62B Condor	W. C. C. Meyer	
G–AYFG	D.62C Condor	Wolds Gliding Club	
G–AYFH	D.62B Condor	B. J. Collins	
G–AYFJ	M.S.880B Rallye Club	K. Hutson & ptnrs.	

Notes	Reg.	Type	Owner or Operator
	G–AYFM	H.S.125 Srs. 400B	Ford Motor Co.
	G–AYFP	Jodel D.140	S. K. Minocha
	G–AYFS	Brookland Hornet	Brookland Rotorcraft Ltd.
	G–AYFT	PA-39 Twin Comanche C/R	F. Kirby
	G–AYFV	Crosby BA-4B	W. B. Limb
	G–AYFX	AA-1 Yankee	D. W. F. Willard & B. Refson
	G–AYFY	EAA Biplane	H. Kuehling
	G–AYFZ	PA-31-300 Navajo	Anglo-Normandy Aviation Ltd.
	G–AYGA	Jodel D.117	E. J. Baxter & ptnrs.
	G–AYGB	Cessna 310Q	Airwork Services Ltd.
	G–AYGC	Cessna F.150K	D. W. Barron
	G–AYGD	Jodel DR.1051	L. H. Oakins
	G–AYGE	SNCAN SV-4C	Owledge Ltd.
	G–AYGG	Jodel D.120	J. E. Hobbs
	G–AYGM	Cessna T.210K	Rogers Aviation Sales Ltd.
	G–AYGN	Cessna 210K	Cheshire Scaffold Co. Ltd.
	G–AYGO	Cessna FR.172G	Airviews (Manchester) Ltd.
	G–AYGX	Cessna FR.172G	Ranelagh Garage Ltd.
	G–AYGZ	Beech 58 Baron	General Engineering Co. (Ilford) Ltd.
	G–AYHA	AA-1 Yankee	P. Chambers
	G–AYHH	Campbell Cricket	A. J. Philpotts
	G–AYHI	Campbell Cricket	W. H. Beevers
	G–AYHX	Jodel D.117A	A. J. S. Bullen
	G–AYHY	Fournier RF-4D	N. H. Jones
	G–AYIA	Hughes 369HS	G. D. E. Briton
	G–AYIB	Cessna 182N Skylane	Emair Bridlington Ltd.
	G–AYIF	PA-28 Cherokee 140C	C.S.E. (Aircraft Services) Ltd.
	G–AYIG	PA-28 Cherokee 140C	Harris Aviation Services Ltd.
	G–AYIH	PA-28 Cherokee 140C	Emilio Ferrari Ltd.
	G–AYII	PA-28R Cherokee Arrow 200	Devon Growers Ltd. & A. L. Bacon
	G–AYIJ	SNCAN SV-4B	R. J. Maxey & ptnrs.
	G–AYIL	Scheibe SF.25B Falke	S. Evans & ptnrs.
	G–AYIO	PA-28 Cherokee 140C	S. Grant & ptnrs.
	G–AYIP	PA-39 Twin Comanche C/R	P. D. Lees & G. Pinkus
	G–AYIT	D.H.82A Tiger Moth	R. L. H. Alexander & ptnrs.
	G–AYIU	Cessna 182N Skylane	Moatair Service Ltd.
	G–AYIZ	H.S.125 Srs. 400B	Peaktape Ltd.
	G–AYJA	Jodel DR. 1050	J. R. Surbey
	G–AYJB	SNCAN SV-4C	N. J. Robertson & ptnrs.
	G–AYJD	Alpavia-Fournier RF-3	C. Wren
	G–AYJP	PA-28 Cherokee 140C	RAF Brize Norton Flying Club Ltd.
	G–AYJR	PA-28 Cherokee 140C	RAF Brize Norton Flying Club Ltd.
	G–AYJS	PA-28 Cherokee 140C	Thames Estuary Flying Services Ltd.
	G–AYJT	PA-28 Cherokee 140C	Jennifer M. Whitaker
	G–AYJU	Cessna TP-206A	Berrard Ltd.
	G–AYJW	Cessna FR.172G	Alpine Press Ltd.
	G–AYJY	Isaacs Fury II	A. V. Francis
	G–AYKA	Beech 95-B55A Baron	R. M. & C. T. Harlock (Thorney) Ltd.
	G–AYKC	D.H.82A Tiger Moth	G. Freeman & ptnrs.
	G–AYKD	Jodel DR.1050	R. G. E. Armfield
	G–AYKF	M.S.880B Rallye Club	D. W. Harris & ptnrs.
	G–AYKJ	Jodel D.117A	G. D. M. Wynne
	G–AYKK	Jodel D.117	P. Cawkwell & ptnrs.
	G–AYKL	Cessna F.150L	Aero Group 78
	G–AYKS	Leopoldoff L-7	D. J. Elliott
	G–AYKT	Jodel D.117	D. R. C. Hunter
	G–AYKU	PA-E23 Aztec 250D	Simulated Flight Training Ltd.
	G–AYKV	PA-28 Cherokee 140C	J. M. Whitaker
	G–AYKW	PA-28 Cherokee 140C	R. E. & R. A. Yallop
	G–AYKX	PA-28 Cherokee 140C	J. A. Bowers & P. Jones
	G–AYKZ	SAI KZ-8	R. E. Mitchell
	G–AYLA	Glos-Airtourer 115	Vagabond Flying Group
	G–AYLB	PA-39 Twin Comanche C/R	E. A. Radnall & Co. Ltd.
	G–AYLC	Jodel DR.1051	E. W. B. Trollope
	G–AYLE	M.S.880B Rallye Club	J. E. Stephenson
	G–AYLF	Jodel DR.1051	C. A. Robbins & D. Saunders
	G–AYLG	H.S.125 Srs. 400B	British Steel Corporation
	G–AYLJ	PA-31 Navajo	Northern Executive Aviation Ltd.
	G–AYLK	Stampe SV-4C	R. W. & P. R. Budge
	G–AYLL	Jodel DR.1050	V. E. Hanning-Lee
	G–AYLM	AA-1 Yankee	G. W. Thompson & ptnrs.
	G–AYLN	AA-1 Yankee	Coleraine Landscape Services
	G–AYLO	AA-1 Yankee	M. Brown

G–ATFU (Top) D.H.85 Leopard Moth/*A. S. Wright*

G–AYMG (Centre) Handley Page HPR-7 Herald 213 of Air UK

G–BEKR (Bottom) Rand KR-2

Reg.	Type	Owner or Operator	Notes
G–AYLP	AA-1 Yankee	Genoa Precision Engineers Ltd.	
G–AYLT	Boeing 707-336C	British Airways	
G–AYLU	Pitts S-1D Special	I. M. G. Senior & J. G. Harper	
G–AYLV	Jodel D.120	R. E. Wray	
G–AYLX	Hughes 269C	Transtar Haulage Ltd.	
G–AYLY	PA-23 Aztec 250	Air UK	
G–AYLZ	Super Aero 45 Srs. 2	J. R. B. Aviation Ltd.	
G–AYMA	Stolp Starduster Too	P. J. Leggo	
G–AYME	Fournier RF.5	R. D. Goodger	
G–AYMG	HPR-7 Herald 213	Air UK	
G–AYMK	PA-28 Cherokee 140C	The Piper Flying Group	
G–AYML	PA-28 Cherokee 140C	J. M. Bendle	
G–AYMN	PA-28 Cherokee 140C	K. Fletcher	
G–AYMO	PA-23 Aztec 250	Goldstar Publications Ltd.	
G–AYMP	Currie Wot Special	H. F. Moffatt	
G–AYMR	Lederlin 380L Ladybug	J. S. Brayshaw	
G–AYMT	Jodel DR.1050	Merlin Flying Club Ltd.	
G–AYMU	Jodel D.112	R. M. White	
G–AYMV	Western 20 balloon	G. F. Turnbull & ptnrs. *Tinkerbelle*	
G–AYMX	Bell 206A JetRanger	W. Holmes	
G–AYMY	Bell 47G-5	B.E.A.S. Ltd.	
G–AYMZ	PA-28 Cherokee 140C	Carlisle Aviation Co. Ltd.	
G–AYNA	Currie Wot	R. W. Hart	
G–AYNB	PA-31 300 Navajo	B. Mendes	
G–AYND	Cessna 310Q	Morgan Bentley & ptnrs.	
G–AYNF	PA-28 Cherokee 140C	C.S.E. (Aircraft Services) Ltd	
G–AYNG	PA-28 Cherokee 140C	C.S.E. (Aircraft Services) Ltd	
G–AYNJ	PA-28 Cherokee 140C	J. O. Carlisle	
G–AYNN	Cessna 185B Skywagon	Bencray Ltd.	
G–AYNP	Westland S.55 Srs. 3	Bristow Helicopters Ltd.	
G–AYNR	H.S.125 Srs. 400B	McAlpine Aviation Ltd.	
G–AYNS	Airmaster H2-B1	D. J. Fry	
G–AYOD	Cessna 172	J. Vicary	
G–AYOJ	H.S.125 Srs. 400B	Bakerloo Investments Co. Ltd.	
G–AYOL	GY-80 Horizon 180	J.B.D.R. Flying Group Ltd.	
G–AYOM	Sikorsky S-61N Mk. 2	British Airways Helicopters Ltd.	
G–AYOP	BAC One-Eleven 530	British Caledonian Airways *Isle of Hoy*	
G–AYOU	Cessna 401B	Zaar International Cinema & TV Programmes Ltd.	
G–AYOW	Cessna 182N Skylane	D. P. H. Lennox	
G–AYOX	V.814 Viscount	British Midland Airways Ltd.	
G–AYOY	Sikorsky S-61N Mk. 2	British Airways Helicopters Ltd.	
G–AYOZ	Cessna FA.150L	Exeter Flying Club Ltd.	
G–AYPA	Beech A-24R Sierra	D. R. Brown & A. E. Cohen	
G–AYPB	Beech C-23 Sundowner	B. F. Bloomfield	
G–AYPC	Beech 70 Queen Air	Vernair Transport Services	
G–AYPD	Beech 95 B.55 Baron	Sir W. S. Dugdale	
G–AYPE	Bo 209 Monsun	R. E. W. Wheeler & A. C. Weedon	
G–AYPF	Cessna F.177RG	Paine Electrics Ltd.	
G–AYPG	Cessna F.177RG	Smith Aviation	
G–AYPH	Cessna F.177RG	G. A. Witherington	
G–AYPI	Cessna F.177RG	Cardinal Aviation Ltd.	
G–AYPJ	PA-28 Cherokee 180E	M. J. F. Aviation Ltd.	
G–AYPM	PA-19 Super Cub 95	C. H. A. Bott	
G–AYPO	PA-19 Super Cub 95	Mrs. J. E. Mavrogordato	
G–AYPP	PA-19 Super Cub 95	McAully Flying Group	
G–AYPR	PA-19 Super Cub 95	T. E. C. Cushing Ltd.	
G–AYPS	PA-19 Super Cub 95	K. F. J. Gardner	
G–AYPT	PA-19 Super Cub 95	Laarbruch Flying Club	
G–AYPU	PA-28R Cherokee Arrow 200	Alpine Ltd.	
G–AYPV	PA-28 Cherokee 140D	Meeting Point Ltd.	
G–AYPZ	Campbell Cricket	A. Melody	
G–AYRA	Campbell Cricket	R. C. Thomas	
G–AYRB	Campbell Cricket	G. L. Clarke	
G–AYRC	Campbell Cricket	G. A. Coventry	
G–AYRE	Campbell Cricket	Campbell Aircraft Ltd.	
G–AYRF	Cessna F.150L	Northern Auto Salvage	
G–AYRG	Cessna F.172K	D. J. C. Turnbull	
G–AYRH	M.S.892A Rallye Commodore 150	J. D. Watt	
G–AYRI	PA-28R Cherokee Arrow 200	Delta Motor Co. (Windsor) Ltd.	
G–AYRK	Cessna 150J	K. A. Learmonth	

Notes	Reg.	Type	Owner or Operator
	G–AYRL	Fournier SFS.31 Milan	W. A. L. Mitchell
	G–AYRM	PA-28 Cherokee 140D	E. S. Dignam
	G–AYRN	Schleicher ASK-14	V. J. F. Falconer
	G–AYRO	Cessna FA.150L Aerobat	Buddale Ltd.
	G–AYRP	Cessna FA.150L Aerobat	Pegasus School of Flying Ltd.
	G–AYRS	Jodel D.120	J. H. Tetley & G. C. Smith
	G–AYRT	Cessna F.172K	Fly-Gay Ltd.
	G–AYRU	BN-2A-6 Islander	Joint Services Parachute Centre
	G–AYRY	H.S.125 Srs. 1B	McAlpine Aviation Ltd.
	G–AYSA	PA-23 Aztec 250C	E. W. Noakes
	G–AYSB	PA-30 Twin Comanche 160C	Sandcliffe Aviation
	G–AYSG	Cessna F.172K	Coventry (Civil) Aviation Ltd.
	G–AYSH	Taylor JT.1 Monoplane	C. J. Lodge
	G–AYSK	Luton L.A.4A Minor	P. F. Bennison & ptnrs.
	G–AYSX	Cessna F.177RG	Nasaire Ltd.
	G–AYSY	Cessna F.177RG	Wells (Barrow) Ltd.
	G–AYSZ	Cessna FA.150L Aerobat	Taxi Truck Services
	G–AYTA	M.S.880B Rallye Club	Willoughby Farm Ltd.
	G–AYTB	M.S.880B Rallye Club	The Newcastle-upon-Tyne Aero Club Ltd.
	G–AYTC	PA-E23 Aztec 250C	Automatic Finance Ltd.
	G–AYTD	PA-23 Aztec 250C	Interland Air Services Ltd.
	G–AYTF	Bell 206B JetRanger 2	Group Lotus Car Co. Ltd.
	G–AYTH	Cessna FR.172H	Zonex Ltd.
	G–AYTJ	Cessna 207 Super Skywagon	Foxair
	G–AYTM	G.164A Ag-Cat	Miller Aerial Spraying Ltd.
	G–AYTN	Cameron O-65 balloon	P. H. Hall & R. F. Jessett *Prometheus*
	G–AYTP	PA-23E Aztec 250E	J. Traynor
	G–AYTR	CP.301A Emeraude	D. A. Cuttriss & A. H. Jefferson
	G–AYTT	Phoenix LA-4A Duet	Gp. Capt. A. S. Knowles
	G–AYTV	MJ.2A Tempete	P. Russell
	G–AYTY	Bensen Autogyro	J. H. Wood
	G–AYUB	CEA DR.253B	D. M. E. Rawling
	G–AYUC	Cessna F.150L	R. C. Ashford
	G–AYUD	PA-25 Pawnee 235	Farmair Ltd.
	G–AYUF	PA-31-300 Navajo	Cabair Ltd.
	G–AYUG	PA-28 Cherokee 140D	P. W. Ward
	G–AYUH	PA-28 Cherokee 180F	M. S. Bayliss
	G–AYUI	PA-28 Cherokee 180	Routair Aviation Services Ltd.
	G–AYUJ	Evans VP.1 Volksplane	J. H. Blake & S. G. D. Rithie
	G–AYUL	PA-23 Aztec 250E	Kattan (GB) Ltd.
	G–AYUM	Slingsby T.61 Falke	Doncaster & District Gliding Club
	G–AYUN	Slingsby T.61 Falke	C. W. Vigar & R. J. Watts
	G–AYUR	Slingsby T.61 Falke	W. A. Urwin
	G–AYUS	Taylor JT.1 Monoplane	D. G. J. Barker
	G–AYUT	Jodel DR.1050	R. Norris
	G–AYUV	Cessna F.172H	Arch Motors Manufacturing Ltd.
	G–AYUX	D.H.82A Tiger Moth (PG651)	P. R. Harris
	G–AYUY	Cessna FA.150L Aerobat	Forgehurst Ltd.
	G–AYVA	Cameron O-84 balloon	A. Kirk *April Fool*
	G–AYVB	Cessna F.172K	Arden Farms
	G–AYVC	PA-23 Aztec 250E	McAlpine Aviation Ltd.
	G–AYVF	H.S.121 Trident 3B	British Airways
	G–AYVG	Boeing 707-321	British Midland Airways Ltd.
	G–AYVI	Cessna T.210H	Trident Marine Ltd.
	G–AYVJ	PA-23 Aztec 250D	Kilby Bros. (Property) Ltd.
	G–AYVM	PA-31-300 Navajo	Casair Aviation Services Ltd.
	G–AYVN	Luton L.A.5A Major	T. M. W. Webster
	G–AYVO	Wallis WA120 Srs. 1	K. H. Wallis
	G–AYVP	Woody Pusher	J. R. Wraight
	G–AYVT	Brochet MB.84	Dunelm Flying Group
	G–AYVU	Cameron O-56 balloon	Shell-Mex & B.P. Ltd. *Hot Potato*
	G–AYVV	ST.10 Diplomate	DMJ Travel
	G–AYVX	M.S.893A Rallye Commodore	D. Morris
	G–AYVY	D.H.82A Tiger Moth (PG617)	G. Smith
	G–AYWA	*Avro 19 Srs. 2	Strathallan Aircraft Collection
	G–AYWD	Cessna 182N	Trans Para Aviation Ltd.
	G–AYWE	PA-28 Cherokee 140C	Lake Green Garage Ltd.
	G–AYWF	PA-23 Aztec 250C	Peregrine Air Services Ltd.
	G–AYWG	PA-E23 Aztec 250C	David Knott (Plant) Ltd.
	G–AYWH	Jodel D.117A	J. M. Knapp & ptnrs.
	G–AYWI	BN-2A Mk. III-1 Trislander	Aurigny Air Services
	G–AYWL	Taylor JT.1 Monoplane	D. G. Wiggins

Reg.	Type	Owner or Operator	Notes
G–AYWM	Glos-Airtourer Super 150	F. B. Miles	
G–AYWS	Beech C23 Sundowner	Waygrand Ltd.	
G–AYWT	Stampe SV-4C	B. K. Lecomber	
G–AYWU	Cessna 150G	C. L. Duke	
G–AYWV	PA-39 Twin Comanche 160 C/R	C.S.E. (Aircraft Services) Ltd.	
G–AYWW	PA-28R Cherokee Arrow 200D	B. Walker & Co. (Dursley) Ltd.	
G–AYWZ	PA-39 Twin Comanche 160 C/R	C.S.E. (Aircraft Services) Ltd.	
G–AYXA	PA-39 Twin Comanche 160 C/R	C.S.E. (Aircraft Services) Ltd.	
G–AYXO	Luton L.A.5 Major	A. C. T. Broomcroft	
G–AYXP	Jodel D.117A	G. N. Davies	
G–AYXS	SIAI-Marchetti S205-18R	W. F. South	
G–AYXT	Westland S.55 Srs. 2	Autair International Ltd.	
G–AYXU	Champion 7KCAB Citabria	H. Fould & ptnrs.	
G–AYXV	Cessna FA.150L	Leo Designs	
G–AYXW	Evans VP.1 Volksplane	J. S. Penny	
G–AYXX	Cessna F.172H	Robinaire Ltd.	
G–AYXY	PA-39 Twin Comanche 160 C/R	C.S.E. (Aircraft Services) Ltd.	
G–AYXZ	PA-39 Twin Comanche 160 C/R	C.S.E. (Aircraft Services) Ltd.	
G–AYYC	Taylor JT.1 Monoplane	F. J. Hoysted	
G–AYYD	M.S.894A Rallye Minerva	J. E. Dyson & T. E. O'Connors	
G–AYYF	Cessna F.150L	Staverton Flying School	
G–AYYG	H.S.748 Srs. 2A	Dan-Air Services Ltd.	
G–AYYK	Slingsby T.61A Falke	Polish Flying Club Ltd.	
G–AYYL	Slingsby T.61A Falke	Airways Aero Associations Ltd.	
G–AYYN	PA-28R Cherokee Arrow 200B	International Ski-Sales Ltd. & W. & R. Leggott Ltd.	
G–AYYO	Jodel DR.1050/M1	Bustard Flying Club	
G–AYYT	Jodel DR.1050/M1	T. S. Warren & ptnrs.	
G–AYYU	Beech C23 Sundowner	A. F. Clements	
G–AYYX	M.S.880B Rallye Club	79 Flying Group	
G–AYYY	M.S.880B Rallye Club	P. J. Cottle	
G–AYYZ	M.S.880B Rallye Club	J. V. White & R. D. Steele	
G–AYZC	PA-E23 Aztec 250D	Casair Aviation Services Ltd.	
G–AYZE	PA-39 Twin Comanche 160 C/R	J. R. Fuller	
G–AYZH	Taylor JT.2 Titch	K. J. Munro	
G–AYZI	Stampe SV-4C	F. M. Barrett	
G–AYZJ	Westland Sikorsky S-55	Autair International Ltd.	
G–AYZK	Jodel DR.1050/M1	G. S. Claybourn	
G–AYZN	PA-E23 Aztec 250	Central Air Services (Air Envoy) Ltd.	
G–AYZO	PA-23 Aztec 250	Aviation Enterprises Ltd.	
G–AYZS	D.62B Condor	R. K. G. Hannington	
G–AYZT	D.62B Condor	R. G. Bearcroft	
G–AYZU	Slingsby T.61A Falke	The Falcon Gliding Group	
G–AYZW	Slingsby T.61A Falke	Portmoak Falke Syndicate	
G–AYZX	Fournier RF-5	R. D. Goodger	
G–AYZY	PA-39 Twin Comanche 160C/R	Dormit Timber Industry Ltd.	
G–AZAB	PA-30 Twin Comanche 160	T. W. P. Sheffield	
G–AZAD	Jodel DR.1051	I. C. Young & J. S. Paget	
G–AZAG	AB 206A JetRanger	Donald MacQueen Ltd.	
G–AZAJ	PA-28R Cherokee Arrow 200B	McKenzie & Tapp Ltd.	
G–AZAU	Grasshopper Type 02	Cierva Rotorcraft Ltd.	
G–AZAV	Cessna 337F	W. T. Johnson & Sons (Huddersfield) Ltd.	
G–AZAW	GY-80 Horizon 160	Nuberg Ltd.	
G–AZAZ	Bensen B-8M	FAA Museum	
G–AZBA	T.66 Nipper 3	E. N. Simmons	
G–AZBB	MBB Bo 209 Monsun	Cheyne Motors Ltd.	
G–AZBC	PA-39 Twin Comanche 160 C/R	C.S.E. (Aircraft Services) Ltd.	
G–AZBD	PA-39 Twin Comanche 160 C/R	C.S.E. (Aircraft Services) Ltd.	
G–AZBE	Glos-Airtourer Super 150	T. C. Edwards	
G–AZBF	PA-39 Twin Comanche 160 C/R	C.S.E. (Aircraft Services) Ltd.	
G–AZBH	Cameron O-84 balloon	Serendipity Balloon Group *Serendipity*	

Notes	Reg.	Type	Owner or Operator
	G-AZBI	Jodel D.150	T. A. Rawson & K. H. Siorpaes
	G-AZBK	PA-E23 Aztec 250E	Qualitair Engineering Ltd.
	G-AZBL	Jodel D.9 Bebe	West Midlands Flying Group
	G-AZBN	AT-16 Harvard 2B (FT391)	Strathallan Aircraft Collection
	G-AZBT	Western O-65 balloon	D. J. Harris *Hermes*
	G-AZBW	PA-39 Twin Comanche 160 C/R	Rijory Ltd.
	G-AZBX	Western O-65 balloon	Jasper Balloon Group *Thursday's Child*
	G-AZBY	W.S-58 Wessex 60 Srs. I	Bristow Helicopters Ltd.
	G-AZCB	Stampe SV-4C	J. Tilzey & ptnrs.
	G-AZCE	Pitts S-I Special	R. J. Oulton
	G-AZCF	Sikorsky S-6IN	British Airways Helicopters Ltd.
	G-AZCH	H.S.125 Srs. 3B/RA	Shell Aircraft Ltd.
	G-AZCI	Cessna 320A Skyknight	Landsurcon (Air Survey) Ltd.
	G-AZCK	B.121 Pup 2	Cosalt Ltd.
	G-AZCL	B.121 Pup 2	Cameron Rainwear Ltd.
	G-AZCP	B.121 Pup I	Three Counties Aero Club Ltd.
	G-AZCT	B.121 Pup I	D. R. Rolfe
	G-AZCU	B.121 Pup I	Surrey & Kent Flying Club
	G-AZCZ	B.121 Pup 2	Forest Aviation Ltd.
	G-AZDA	B.121 Pup I	G. H. G. Bishop & K. E. Fehrenbach
	G-AZDC	Sikorsky S-61N	Bristow Helicopters Ltd. *Dunnotar*
	G-AZDD	MBB Bo 209 Monsun	E. M. Emerson & N. Hughes-Narborough
	G-AZDE	PA-28R Cherokee Arrow 200B	Electro-Motion UK (Export) Ltd.
	G-AZDF	Cameron O-84 balloon	M. C. Abram *Hannibal*
	G-AZDH	PA-31-300 Navajo	Casair Aviation Services Ltd.
	G-AZDK	Beech B55 Baron	Burton Metal Fabrications Ltd.
	G-AZDW	PA-28 Cherokee 180F	J. Todd
	G-AZDX	PA-28 Cherokee 180F	Anglo-Dansk Marine Engineering Co. Ltd.
	G-AZDY	D.H.82A Tiger Moth	B. A. Mills
	G-AZDZ	Cessna 172K	Stansted Fluid Power Ltd.
	G-AZEA	Cessna 182N	Forth Flying Group Ltd.
	G-AZED	BAC One-Eleven 414	Dan-Air Services Ltd.
	G-AZEE	M.S.880B Rallye Club	P. L. Clements
	G-AZEF	Jodel D.120	P. Cawkwell & G. Firth
	G-AZEG	PA-28 Cherokee 140D	M. I. Cranston & P. A. Lewis
	G-AZER	Cameron O-42 balloon	M. P. Dokk-Olsen & P. L. Jaye *Shy Tot*
	G-AZEU	B.121 Pup 2	J. N. Russell
	G-AZEV	B.121 Pup 2	G. P. Martin
	G-AZEW	B.121 Pup 2	Deltair Ltd.
	G-AZFA	B.121 Pup 2	K. F. Plummer
	G-AZFB	Boeing 720-051B	Monarch Airlines Ltd.
	G-AZFC	PA-28 Cherokee 140D	CFS Aircraft Ltd.
	G-AZFE	PA-23 Aztec 250D	Air Charter (Scotland)
	G-AZFF	Jodel D.112	M. K. Field
	G-AZFI	PA-28R Cherokee Arrow 200B	Hawksworth Garage Ltd.
	G-AZFL	Cessna 310P	Salan Productions Ltd.
	G-AZFM	PA-28R Cherokee Arrow 200B	Lincs Poultry Machinery Ltd.
	G-AZFO	PA-39 Twin Comanche 160 C/R	G. Firbank
	G-AZFP	Cessna F.177RG	Selflock Ltd.
	G-AZFR	Cessna 401B	Johnson Group Management Services Ltd.
	G-AZFS	Beech B80 Queen Air	Globetrotter Survey Co. Ltd.
	G-AZFZ	Cessna 414	C. H. Taylor & Co. Ltd.
	G-AZGA	Jodel D.120	A. J. Smith & G. E. Williamson
	G-AZGB	PA-E23 Aztec 250D	Qualitair Engineering Ltd.
	G-AZGC	Stampe SV-4C	The Hon. Patrick Lindsay
	G-AZGD	Stampe SV-4C	F. H. Bateman
	G-AZGE	Stampe SV-4A	M. R. L. Astor
	G-AZGF	B.121 Pup 2	J. A. Macreadie
	G-AZGG	Beech C90 King Air	The Plessey Co. Ltd.
	G-AZGH	M.S.880B Rallye Club	R. G. Moore
	G-AZGI	M.S.880B Rallye Club	G. E. M. Hallett & ptnrs.
	G-AZGJ	M.S.880B Rallye Club	S. C. Howes & T. E. French
	G-AZGL	M.S.894A Rallye Minerva	The Cambridge Aero Club Ltd.
	G-AZGY	CP.301B Emeraude	Rodingair Flying Group
	G-AZGZ	D.H.82A Tiger Moth (NM181)	F. R. Manning
	G-AZHA	PA-E23 Aztec 250E	Air Charter Scotland Ltd.
	G-AZHB	Robin HR 100-200	W. H. Everett & Son Ltd.

Reg.	Type	Owner or Operator	Notes
G–AZHC	Jodel D.112	Sywell Skyraiders Flying Group	
G–AZHD	Slingsby T.61A Falke	West Wales Gliding Co. Ltd.	
G–AZHF	Cessna 150L	Coventry Air Training School Ltd.	
G–AZHH	SA 102.5 Cavalier	D. W. Buckle *Time*	
G–AZHI	Glos-Airtourer Super 150	A. W. Jenner & ptnrs	
G–AZHJ	S.A. Twin Pioneer Srs. 3	Flight One Ltd.	
G–AZHK	Robin HR 100-200	E. A. & W. M. C. Payton	
G–AZHL	PA-31-300 Navajo	BAC Windows Ltd.	
G–AZHM	Cassutt Racer	M. S. Crossley *Razor Blade*	
G–AZHO	Jodel DR. 1050	S. Alexander	
G–AZHR	Piccard Ax6 balloon	J. W. Moss *Happiness*	
G–AZHT	Glos-Airtourer T.3	M. J. Cuttell	
G–AZHU	Luton L.A.4A Minor	A. E. Morris	
G–AZIA	PA-39 Twin Comanche 160 C/R	Toton Plant Hire Ltd.	
G–AZIB	ST-10 Diplomate	Wilmslow Audio Ltd.	
G–AZID	Cessna FA.150L	Oldment Ltd.	
G–AZIE	PA-25 Pawnee 235	Aerocare Agricultural Services	
G–AZIG	Fournier RF-4D	A. H. R. Stansfield	
G–AZIH	J/IN Alpha	I. R. F. Hammod & L. A. Groves	
G–AZII	Jodel D.117A	J. S. Brayshaw	
G–AZIJ	Jodel DR.360	K. H. Tostevin	
G–AZIK	PA-34-200 Seneca	C.S.E. (Aircraft Services) Ltd.	
G–AZIL	Slingsby T.61B Falke	I. Jamieson	
G–AZIM	PA-31 Navajo	IDS Aircraft Ltd.	
G–AZIN	Canadair CL-44D-4	British Cargo Airlines Ltd.	
G–AZIO	SNCAN SV-4C	Rollason Aircraft & Engines Ltd.	
G–AZIP	Cameron O-65 balloon	Dante Balloon Group *Dante*	
G–AZIR	Stampe SV-4C	Rollason Aircraft & Engines Ltd.	
G–AZJA	BN-2A-1 Mk. III Trislander	Aurigny Air Services	
G–AZJB	PA-34-200 Seneca	W. S. Churchill	
G–AZJC	Fournier RF-5	D. W. Sutherland	
G–AZJD	AT-6D Harvard III	Gladaircraft Ltd.	
G–AZJE	JB-01 Minicab	J. B. Evans	
G–AZJI	Western O-65 balloon	W. Davison *Peek-a-Boo*	
G–AZJK	Agusta-Bell 47G-2	Twyford Moors (Helicopters) Ltd.	
G–AZJL	Agusta-Bell 47G-2	Twyford Moors (Helicopters) Ltd.	
G–AZJM	Boeing 707-324C	British Midland Airways Ltd.	
G–AZJN	Robin DR 300/140	Wright Farm Eggs Ltd.	
G–AZJV	Cessna F.172L	The JV Group	
G–AZJW	Cessna F.150L	Seair	
G–AZJX	Cessna F.150L	Gordon King (Aviation) Ltd.	
G–AZJY	Cessna FRA.150L	Shropshire Aero Club Ltd.	
G–AZJZ	PA-23 Aztec 250E	Air Commuter Ltd.	
G–AZKA	M.S.880B Rallye Club	J. M. F. Groat & Sons Ltd.	
G–AZKC	M.S.880B Rallye Club	L. J. Martin	
G–AZKD	M.S.880B Rallye Club	C. B. Dew	
G–AZKE	M.S.880B Rallye Club	M. J. Powell	
G–AZKG	Cessna F.172L	Hartmann Ltd.	
G–AZKH	Cessna F.177RG	F. B. Spriggs	
G–AZKI	Noorduyn Harvard 2B (FT229)	A. E. Hutton	
G–AZKK	Cameron O-56 balloon	Gemini Balloon Group *Gemini*	
G–AZKM	Boeing 720-051B	Monarch Airlines Ltd.	
G–AZKN	Robin HR.100/200	P. T. Bolton	
G–AZKO	Cessna F.337F	Maidstone Investments Co. Ltd.	
G–AZKP	Jodel D.117	C. M. Fitton	
G–AZKR	PA-24 Comanche 180	B. J. Boughton	
G–AZKS	AA-1A Trainer	G. A. P. N. Barlow	
G–AZKT	Cessna F.177RG	Rogers Aviation Sales Ltd.	
G–AZKV	Cessna FRA.150L	Professional Programmers & Analysts (Yorks) Ltd.	
G–AZKW	Cessna F.172L	Banbury Plant Hire Ltd.	
G–AZKZ	Cessna F.172L	Rogers Aviation Sales Ltd.	
G–AZLA	Taylor JT.2 Titch	Jeffrey Chappell	
G–AZLE	Boeing N2S–5 Kaydet	A. E. Poulson	
G–AZLF	Jodel D.120	J. Brooks	
G–AZLH	Cessna F.150L	E. Midlands School of Flying Ltd.	
G–AZLJ	BN-2A-1 Mk. III Trislander	Aurigny Air Services	
G–AZLK	Cessna F.150L	Mercury Flying Club Ltd.	
G–AZLL	Cessna FRA.150L	Cleveland Flying School	
G–AZLM	Cessna F.172L	J. F. Davis & Q. L. Rigby	
G–AZLN	PA-28 Cherokee 180F	D. H. L. Wigan	
G–AZLO	Cessna F337F	Helicrops Ltd.	

Notes	Reg.	Type	Owner or Operator
	G–AZLP	V.813 Viscount	British Midland Airways Ltd.
	G–AZLR	V.813 Viscount	British Midland Airways Ltd.
	G–AZLS	V.813 Viscount	British Midland Airways Ltd.
	G–AZLT	V.813 Viscount	British Midland Airways Ltd.
	G–AZLV	Cessna 172K	J. Braithwaite (Aerial Photography) Ltd.
	G–AZLY	Cessna F.150L	Cleveland Flying School
	G–AZLZ	Cessna F.150L	J. Fricker
	G–AZMA	Jodel D.140B	W. A. Braim Ltd.
	G–AZMB	Bell 47G–3B	Helicopter Farming Ltd.
	G–AZMC	Slingsby T.61A Falke	Essex Gliding Club Ltd.
	G–AZMD	Slingsby T.61C Falke	J. Worth & ptnrs.
	G–AZME	PA-31-300 Navajo	All Seasons Aviation Ltd.
	G–AZMF	BAC One-Eleven 530	British Caledonian Airways
	G–AZMG	PA-23 Aztec 250	Commair Aviation Ltd.
	G–AZMH	Morane-Saulnier M.S.500	Hon. P. Lindsay
	G–AZMJ	AA-5 Traveler	A. A. Cansick
	G–AZMK	PA-23 Aztec 250	Executive Wings Ltd.
	G–AZMN	Glos-Airtourer T.5	R. G. Lowerson & T. Ellefson
	G–AZMO	PA-32 Cherokee Six 260	Dateline International Dating Systems Ltd. & Grangewood Press Ltd.
	G–AZMV	D.62C Condor	Ouse Gliding Club Ltd.
	G–AZMX	PA-28 Cherokee 140	Mooney Aviation Ltd.
	G–AZMY	SIAI-Marchetti SF-260	Miss W. M. Miller
	G–AZMZ	M.S.893A Rallye Commodore 150	G. Daryn
	G–AZNA	V.813 Viscount	British Midland Airways Ltd.
	G–AZNB	V.813 Viscount	British Midland Airways Ltd.
	G–AZNC	V.813 Viscount	British Midland Airways Ltd.
	G–AZNF	Stampe SV-4C	H. J. Smith
	G–AZNI	S.A.315B Lama	Dollar Air Services Ltd (G–AWLC)
	G–AZNJ	M.S.880B Rallye Club	Miss J. G. White
	G–AZNK	Stampe SV-4A	S. Donghi
	G–AZNL	PA-28R Cherokee Arrow 200D	C. R. Balls
	G–AZNO	Cessna 182P	Strathmarine Flying Group
	G–AZNT	Cameron O-84 balloon	Cameron Balloons Ltd. *Oberon*
	G–AZNX	Boeing 720-051B	Monarch Airlines Ltd.
	G–AZNY	PA-E23 Aztec 250E	P & B Investment Holdings Ltd.
	G–AZNZ	Boeing 737-222	Britannia Airways Ltd. *Henry Hudson*
	G–AZOA	MBB Bo 209 Monsun	Dr. G. R. Outwin
	G–AZOB	MBB Bo 209 Monsun	G. N. Richardson
	G–AZOD	PA-23 Aztec 250D	Peregrine Air Services Ltd.
	G–AZOE	Glos-Airtourer 115	P. C. Logsdon
	G–AZOF	Glos-Airtourer Super 150	V. H. Bellamy
	G–AZOG	PA-28R Cherokee Arrow 200D	Winchfield Enterprises Ltd.
	G–AZOH	Beech 65 B.80 Queen Air	Fairey Surveys Ltd.
	G–AZOL	PA-34-200 Seneca	Granpack Ltd.
	G–AZOM	MBB Bo 105D	B.E.A.S. Ltd.
	G–AZON	PA-34-200-2 Seneca	Alma Investments Ltd.
	G–AZOO	Western O-65 balloon	Southern Balloon Group *Carousel*
	G–AZOR	MBB Bo 105D	Management Aviation Ltd.
	G–AZOS	Jurca Sirocco	R. Wells
	G–AZOT	PA-34-200 Seneca	L. G. Payne
	G–AZOU	Jodel DR.1051	T. W. Jones & ptnrs.
	G–AZOZ	Cessna FRA.150L	Airwork Services Ltd.
	G–AZPA	PA-25 Pawnee 235	Farm Aviation Services Ltd.
	G–AZPC	Slingsby T.61C Falke	B. C. Dixon
	G–AZPF	Fournier RF-5	R. Pye
	G–AZPH	Craft-Pitts S-1S Special	Aerobatics International Ltd.
	G–AZPV	Luton L.A.4A Minor	J. Scott
	G–AZPX	Western O-31 balloon	E. R. McCosh *Nessie*
	G–AZRA	MBB Bo 209 Monsun	The BBC Club
	G–AZRB	Cessna 340	Otter Controls Ltd.
	G–AZRC	Cessna 340	Margate Motors (Plant & Aircraft) Ltd.
	G–AZRD	Cessna 401B	John Finlan Ltd.
	G–AZRF	Sikorsky S-61N	Bristow Helicopters *Pitcaple*
	G–AZRG	PA-23 Aztec 250D	Woodgate Aviation Ltd.
	G–AZRH	PA-28 Cherokee 140D	CFS Aircraft Ltd.
	G–AZRI	Payne balloon	G. F. Payne *Shoestring*

Reg.	Type	Owner or Operator	Notes
G–AZRK	Fournier RF-5	Strathtay Flying Group	
G–AZRL	PA-19 Super Cub 95	B. A. Dunlop	
G–AZRM	Fournier RF-5	Miss R. S. A. Lloyd-Bostock	
G–AZRN	Cameron O-84 balloon	M. Yarrow *Gravida II*	
G–AZRP	Glos-Airtourer 115	Torfaen Self Drive Hire Ltd.	
G–AZRR	Cessna 310Q	Ames Company (Transport) Ltd.	
G–AZRS	PA-22 Tri-Pacer 150	E. A. Harrhy	
G–AZRU	AB-206B JetRanger	Heliwork Ltd.	
G–AZRV	PA-28R Cherokee Arrow 200B	S. G. Daniel	
G–AZRW	Cessna T.337C	A.D.S. (Aerial) Ltd.	
G–AZRX	GY-80 Horizon 160	Fine Stitchers Ltd.	
G–AZRZ	Cessna U-206F	Army Parachute Association	
G–AZSA	Stampe SV-4B	Pinnacle Market Promotions	
G–AZSC	AT-16 Harvard IIB (FT323)	M. W. Stow	
G–AZSD	Slingsby T.29B Motor Tutor	R. G. Boynton	
G–AZSE	PA-28R Cherokee Arrow 200D	Amstrad Consumer Electronics Ltd.	
G–AZSF	PA-28R Cherokee Arrow 200D	P. Blamire	
G–AZSG	PA-28 Cherokee 180E	Scotia Safari	
G–AZSH	PA-28R Cherokee Arrow 180	S. H. Hayward	
G–AZSK	Taylor JT.1 Monoplane	R. R. Lockwood	
G–AZSL	M.S.890B Rallye Commodore	F. J. Shevill	
G–AZSM	PA-28R Cherokee Arrow 180	Drumgate Ltd.	
G–AZSN	PA-28R Cherokee Arrow 200	Burch (Engineering) Ltd.	
G–AZSP	Cameron O-84 balloon	Esso Petroleum Ltd. *Esso*	
G–AZSS	Jodel D.9 Bebe	M. W. Rice	
G–AZSU	H.S.748 Srs. 2A	Dan-Air Services Ltd.	
G–AZSV	Hiller UH-12E	Shawline Helicopters Ltd.	
G–AZSW	Beagle 121 Pup 1	P. Evans & P. J. C. Graves	
G–AZSX	Beagle 121 Pup 1	P. W. Hunter	
G–AZSY	PA-24 Comanche 260	Christopher Foyle Aviation Leasing Co.	
G–AZSZ	PA-23 Aztec 250	Air Kilroe	
G–AZTA	MBB Bo 209 Monsun	R. S. Perks	
G–AZTD	PA-32 Cherokee Six 300	Presshouse Publications Ltd.	
G–AZTF	Cessna F.177RG	Carentals Ltd.	
G–AZTH	Bensen Autogyro	E. Henshaw	
G–AZTI	Bolkow Bo 105C	North Scottish Helicopters Ltd.	
G–AZTK	Cessna F.172F	C. C. Donald	
G–AZTM	Glos-Airtourer 115	I. J. Smith	
G–AZTN	Glos-Airtourer 115	Bernell Aviation Ltd.	
G–AZTO	PA-34-200 Seneca	Holding & Barnes Ltd.	
G–AZTR	SNCAN SV.4C	D. J. Shires	
G–AZTS	Cessna F.172L	Transgap Ltd.	
G–AZTT	PA-28R Cherokee Arrow 200	Rivermill Pyrford Ltd.	
G–AZTV	Stolp A 500 Starlet	P. Russell	
G–AZTW	Cessna F.177RG	T. R. Waling	
G–AZUG	AA-5 Traveler	Air Service Training Ltd.	
G–AZUH	PA-31-300 Navajo	Ocaso (Reinsurance Services) Ltd.	
G–AZUL	Stampe SV-4B	R. A. Seeley	
G–AZUM	Cessna F.172L	R. B. Lewis	
G–AZUN	Cessna F.172L	G. Fraser	
G–AZUO	Cessna F.177RG	Newbury Sand and Gravel Co. Ltd.	
G–AZUP	Cameron O-65 balloon	C. M. G. Ellis & ptnrs.	
G–AZUT	M.S.893A Rallye Commodore 180	Rallye Flying Group	
G–AZUU	Fournier RF-4D	R. A. Hayne & A. J. Hopson	
G–AZUV	Cameron O-65 balloon	D. S. Bush *Icarus*	
G–AZUX	Western O-65 balloon	H. C. J. & Mrs. S. L. G. Williams *Slow Djinn*	
G–AZUY	Cessna E.310L	Euro Advertising Ltd.	
G–AZUZ	Cessna FRA.150L	D. J. Parker	
G–AZVA	MBB Bo 209 Monsun	K. H. Wallis	
G–AZVB	MBB Bo 209 Monsun	P. C. Logsdon	
G–AZVC	MBB Bo 209 Monsun	G. E. Horder & D. Cockcroft Ltd.	
G–AZVE	AA-5 Traveler	Mindon Engineering (Nottingham) Ltd.	
G–AZVF	M.S.894A Rallye Minerva	J. McCleary & T. S. Brown	
G–AZVG	AA-5 Traveler	R. B. Sandell & Co. Ltd.	
G–AZVH	M.S.894A Rallye Minerva	C. H. T. Trace	
G–AZVI	M.S.892A Rallye Commodore	A. A. Broughton & Son Ltd.	

Notes	Reg.	Type	Owner or Operator
	G–AZVJ	PA-34-200 Seneca	Business Air Travel Ltd.
	G–AZVL	Jodel D.119	A. B. Fisher
	G–AZVM	Hughes 369HS	Diagnostic Reagents Ltd.
	G–AZVP	Cessna F.177RG	R. W. Martin & R. G. Saunders
	G–AZVR	Cessna F.150L	Ann Pascoe
	G–AZVS	H.S.125 Srs. 3B	Eastern Airways
	G–AZVT	Cameron O-84 balloon	Sky Soarer Ltd. *Jules Verne*
	G–AZVV	PA-28 Cherokee 180G	M. R. Woodgate
	G–AZVW	Bell 47G-5A	Helicopter Hire Ltd.
	G–AZVX	Bell 47G-5A	Helicopter Hire Ltd.
	G–AZVY	Cessna 310Q	Centreline Air Services Ltd.
	G–AZVZ	PA-28 Cherokee 140	Gordon King (Aviation) Ltd.
	G–AZWB	PA-28 Cherokee 140	Apache Aircraft Services Ltd.
	G–AZWD	PA-28 Cherokee 140	Airways Aero Associations Ltd.
	G–AZWE	PA-28 Cherokee 140	Airways Aero Associations Ltd.
	G–AZWF	SAN Jodel DR.1050	Miss M. M. Truchet
	G–AZWS	PA-28R Cherokee Arrow 200D	A. W. Gibbs (Holdings) Ltd.
	G–AZWT	Westland Lysander III (V9441)	Strathallan Aircraft Collection
	G–AZWU	Cessna F.150L	D. W. Walton
	G–AZWW	PA-23 Aztec 250E	G. M. W. Starkey
	G–AZWY	PA-24 Comanche 260	Keymer Son & Co. Ltd.
	G–AZXA	Beechcraft 95-C55 Baron	J. D. Habin
	G–AZXB	Cameron O-65 balloon	London Balloon Club Ltd. *London Pride II*
	G–AZXC	Cessna F.150L	Brailsford Aviation Ltd.
	G–AZXD	Cessna F.172L	Dukdeed Ltd.
	G–AZXE	Jodel D.120A	Kestrel Flying Group
	G–AZXG	PA-23 Aztec 250	M. Priest & K. Lefevre
	G–AZXH	PA-34-200-2 Seneca	Tapehurst Ltd.
	G–AZXI	Hughes 269C	Rassler Aero Services
	G–AZXM	H.S.121 Trident 2E	British Airways
	G–AZYA	GY-80 Horizon 160	T. Poole & G. L. Newbrook
	G–AZYB	Bell 47H-1	G. Watt
	G–AZYC	Cessna A.188B Agwagon	Mindacre Ltd.
	G–AZYD	M.S.893A Rallye Commodore	Deeside Gliding Club
	G–AZYF	PA-28 Cherokee 180	J. C. Glynn
	G–AZYG	PA-E23 Aztec 250	F. M. Barrett
	G–AZYI	Cessna E-310Q	Aircraft Mart
	G–AZYJ	PZL-104 Srs. 6 Wilga	Worcestershire Gliding Ltd.
	G–AZYK	Cessna 310Q	F. Adam
	G–AZYL	Portslade School free balloon	R. M. Glover *Mercury*
	G–AZYM	Cessna E-310Q	Air Charter & Travel Ltd.
	G–AZYR	Cessna 340	Selflock Ltd.
	G–AZYS	CP.301C-1 Emeraude	K. S. V. Bass
	G–AZYU	PA-E23 Aztec 250	Service Station Services Ltd.
	G–AZYV	Burns O-77 balloon	B. F. G. Ribbans *Contrary Mary*
	G–AZYX	M.S.893A Rallye Commodore	Black Mountain Gliding Co. Ltd.
	G–AZYY	Slingsby T.61A Falke	J. A. Towers
	G–AZYZ	WA.51A Pacific	N. H. Jones
	G–AZZA	PA-E23 Aztec 250	Air Charter (Scotland)
	G–AZZB	AB-206B JetRanger 2	British Caledonian Helicopters Ltd.
	G–AZZC	Douglas DC-10-10	Laker Airways *Eastern Belle*
	G–AZZD	Douglas DC-10-10	Laker Airways *Western Belle*
	G–AZZE	Beech A.23-19 Musketeer	K. D. Price
	G–AZZF	M.S.880B Rallye Club	J. Meaden & ptnrs.
	G–AZZG	Cessna 188 Agwagon	W. P. Miller
	G–AZZH	Practavia Pilot Sprite	K. G. Stewart
	G–AZZK	Cessna 414	Unifix Air Ltd.
	G–AZZL	PA-E23 Aztec 250	Airclass Ltd.
	G–AZZO	PA-28 Cherokee 140	Stapleford Flying Club Ltd.
	G–AZZP	Cessna F.172H	K. A. Learmonth
	G–AZZR	Cessna F.150L	Herefordshire Aero Club
	G–AZZT	PA-28 Cherokee 180	Stapleford Flying Club Ltd.
	G–AZZV	Cessna F.172L	Rogers Aviation Sales Ltd.
	G–AZZW	Fournier RF-5	Gloster Aero Group
	G–AZZX	Cessna FRA.150L	J. E. Uprichard & ptnrs.
	G–AZZZ	D.H.82A Tiger Moth	S. W. McKay
	G–BAAD	Evans Super VP-1	R. W. Husband
	G–BAAF	Manning-Flanders MF 1 replica	D. E. Bianchi
	G–BAAG	Beechcraft B.55 Baron	Mannin Aviation Ltd.

Reg.	Type	Owner or Operator	Notes
G–BAAH	Coates SA111 Swalesong	J. R. Coates	
G–BAAI	M.S.893A Rallye Commodore	A. F. Butcher	
G–BAAK	Cessna 207	Sunderland Parachute Centre Ltd.	
G–BAAL	Cessna 172A	V. H. Bellamy	
G–BAAP	PA-28R Cherokee Arrow 200	Shirley A. Shelley	
G–BAAR	PA-28R Cherokee Arrow 200	R. W. Burchardt	
G–BAAT	Cessna 182P Skylane	S. J. Martin Ltd.	
G–BAAU	Enstrom F-28A	Group Lotus Car Companies	
G–BAAV	Cessna FRA.150L	Enniskillen Flying Club Ltd.	
G–BAAW	Jodel D.112	J. M. Alexander	
G–BAAX	Cameron O-84 balloon	The New Holker Estate Co. Ltd. *Holker Hall*	
G–BAAY	Valton Viima II	Shipping & Airlines Ltd.	
G–BAAZ	PA-28R Cherokee Arrow 200D	A. W. Rix	
G–BABA	D.H.82A Tiger Moth	S. W. McKay	
G–BABB	Cessna F.150L	George House Holdings Ltd.	
G–BABC	Cessna F.150L	E. P. Collier	
G–BABD	Cessna FRA.150L	Wycombe Air Centre Ltd.	
G–BABE	Taylor JT.2 Titch	J. Berry	
G–BABG	PA-28 Cherokee 180	Trehaven Aviation Ltd.	
G–BABH	Cessna F.150L	N. F. O'Neill & E. J. Leathem	
G–BABK	PA-34-200 Seneca	D. F. J. & N. R. Flashman	
G–BABW	Beech E90 King Air	The Rank Organisation	
G–BABY	Taylor JT.2 Titch	J. R. D. Bygraves *Barnstormer Two*	
G–BACA	BAC Petrel	BAC Military Aircraft Division, Apprentice Training Dept.	
G–BACB	PA-34-200 Seneca	IPC Business Press Ltd.	
G–BACC	Cessna FRA.150L	R. F. Development Co. Ltd.	
G–BACD	Cessna FRA.150L	Cheshire Air Training School Ltd.	
G–BACE	Fournier RF-5	R. W. K. Stead	
G–BACF	Cessna F.337F	Wilson Salt Co. Ltd.	
G–BACH	Enstrom F.28A	Dargen Ltd.	
G–BACI	H.S.125 Srs. 400B	Clartacrest Ltd.	
G–BACJ	Jodel D.120	Wearside Flying Association	
G–BACK	D.H.82A Tiger Moth (DF130)	G. R. French & ptnrs.	
G–BACL	Jodel D.150	G. R. French	
G–BACM	Cessna FRA.150L	Air Compton Ltd.	
G–BACN	Cessna FRA.150L	Regent Motors	
G–BACO	Cessna FRA.150L	Miss B. Kennett	
G–BACP	Cessna FRA.150L	Norfolk & Norwich Aero Club Ltd.	
G–BADC	Luton Beta	H. M. Mackenzie	
G–BADE	PA-23 Aztec 250	Thurston Aviation Ltd.	
G–BADF	PA-34-200-2 Seneca	Strata Surveys Ltd.	
G–BADH	Slingsby T.61A Falke	E. M. Andrew & ptnrs.	
G–BADI	PA-E23 Aztec 250	W. London Aero Services Ltd.	
G–BADJ	PA-E23 Aztec 250	J. G. Hogg	
G–BADK	BN-2A-8 Islander	Brymon Aviation Ltd.	
G–BADL	PA-34-200 Seneca	Apollo Aviation Ltd.	
G–BADM	D.62B Condor	Rollason Aircraft & Engines Ltd.	
G–BADO	PA-32 Cherokee Six 300	D. Russell	
G–BADP	Boeing 737-204	Britannia Airways Ltd. *Sir Arthur Whitten Brown*	
G–BADR	Boeing 737-204	Britannia Airways Ltd. *Capt. Robert Falconer Scott*	
G–BADT	Cessna 402B	British Aircraft Corp.	
G–BADU	Cameron O-56 balloon	J. Philp *Dream Machine*	
G–BADV	Brochet MB-50	P. A. Cairns	
G–BADW	Pitts S-2A Special	Rothmans International Ltd.	
G–BADY	Pitts S-2A Special	Rothmans International Ltd.	
G–BADZ	Pitts S-2A Special	Rothmans International Ltd.	
G–BAEB	Robin DR.400/160	Bracknell Refrigeration Services Ltd.	
G–BAED	PA-E23 Aztec 250	Andair International Ltd.	
G–BAEE	Jodel DR.1050/M1	Joan H. Martin	
G–BAEF	Boeing 727-46	Dan-Air Services Ltd.	
G–BAEG	PA-31-300 Navajo	Aviation Beauport (Finance) Ltd.	
G–BAEJ	AA-5 Traveler	A. Vikander	
G–BAEM	Robin DR.400/125	J. D. Rees	
G–BAEN	Robin DR.400/180	Trans Europe Air Charter Ltd.	
G–BAEP	Cessna FRA.150L	Airwork Services Ltd.	
G–BAER	Cosmic Wind	R. S. Voice	
G–BAES	Cessna 337A	Page & Moy Ltd. & High Voltage Applications Ltd.	

Notes	Reg.	Type	Owner or Operator
	G–BAET	Piper L-4H Cub	C. M. G. Ellis
	G–BAEU	Cessna F.150L	Skyviews & General Ltd.
	G–BAEV	Cessna FRA.150L	South Midland Communications Ltd.
	G–BAEW	Cessna F.172M	Rogers Aviation Sales Ltd.
	G–BAEX	Cessna F.172M	D. H. Stephenson & ptnrs.
	G–BAEY	Cessna F.172M	Cleansing Service (Southern Counties)
	G–BAEZ	Cessna FRA.150L	F. Butterfield
	G–BAFA	AA-5 Traveler	Lewis Flying Group Ltd.
	G–BAFD	MBB Bo 105D	British Caledonian Helicopters Ltd.
	G–BAFG	D.H.82A Tiger Moth	C. D. Cyster
	G–BAFH	Evans VP-1 Volksplane	R. H. W. Beath
	G–BAFI	Cessna F.177RG	WSM Aviation
	G–BAFL	Cessna 182P	Anglian Double Glazing Ltd.
	G–BAFM	AT-16 Harvard IIB	Hon. P. Lindsay
	G–BAFN	Bell 212	British Airways Helicopters Ltd.
	G–BAFP	Robin DR.400/160	Miss G. A. Habsey
	G–BAFS	PA-18 Super Cub 150	Doncaster & District Gliding Club
	G–BAFT	PA-18 Super Cub 150	Cambridge University Gliding Trust Ltd.
	G–BAFU	PA-28 Cherokee 140	R. S. Fenwick
	G–BAFV	PA-18 Super Cub 95	P. Elliott
	G–BAFW	PA-28 Cherokee 140	B. J. Poulten
	G–BAFX	Robin DR.400/140	Copthorne Precision Products Ltd.
	G–BAFZ	Boeing 727-46	Dan-Air Services Ltd.
	G–BAGA	Cessna 182A Skylane	N. W. Parachute Centre
	G–BAGB	SIAI-Marchetti SF.260	British Midland Airways Ltd.
	G–BAGC	Robin DR.400/140	Hempalm Ltd.
	G–BAGE	Cessna T.210L	Worksop Aircraft Sales Partnership Ltd.
	G–BAGF	Jodel D.92 Bebe	G. R. French & J. D. Watt
	G–BAGG	PA-32 Cherokee Six 300E	J. S. Horne
	G–BAGI	Cameron O-31 balloon	Cameron Balloons Ltd. *Vital Spark*
	G–BAGL	SA.341G Gazelle Srs. 1	Westland Helicopters Ltd.
	G–BAGM	Wassmer WA.41	Alderney Flying Services Ltd.
	G–BAGN	Cessna F.177RG	Dudley Industrial Services
	G–BAGO	Cessna 421B	Lewcol Enterprises
	G–BAGR	Robin DR.400/125	J. W. T. Burfitt
	G–BAGS	Robin DR.400 2+2	Headcorn Flying School Ltd.
	G–BAGT	Helio H.295 Courier	B. J. C. Woodhall Ltd.
	G–BAGU	Luton L.A.5A Major	J. Gawley
	G–BAGV	Cessna U.206F	Corbett Farms Ltd.
	G–BAGW	Cessna F.150J	Sherburn Aero Club Ltd.
	G–BAGX	PA-28 Cherokee 140	Bolton Air Training School Ltd.
	G–BAGY	Cameron O-84 balloon	P. G. Dunnington *Beatrice*
	G–BAGZ	PA-34-200 Seneca	T. Bittan
	G–BAHC	PA-23 Aztec 250	Berkeley Hotel Ltd.
	G–BAHD	Cessna 182P Skylane	S. Brunt (Silverdale Staffs) Ltd.
	G–BAHE	PA-28 Cherokee 140	Mooney Aviation Ltd.
	G–BAHF	PA-28 Cherokee 140	Mooney Aviation Ltd.
	G–BAHG	PA-24 Comanche 260	Friendly Aviation (Jersey) Ltd.
	G–BAHH	Wallis WA.121	K. H. Wallis
	G–BAHI	Cessna F.150H	Coventry Air Training School Ltd.
	G–BAHJ	PA-24 Comanche 250	Videovision
	G–BAHL	Robin DR.400/160	Norvett Electronics Ltd.
	G–BAHN	Beech 58 Baron	Research Consultants Ltd.
	G–BAHO	Beech C.23 Sundowner	Air & General Finance Ltd.
	G–BAHP	Volmer VJ.22 Sportsman	J. H. H. Turner
	G–BAHR	PA-28 Cherokee 140	N. R. Goodwin & ptnrs.
	G–BAHS	PA-28R Cherokee Arrow 200-II	A. A. Wild & ptnrs.
	G–BAHT	Cessna F.172F	C. B. Healey
	G–BAHW	Cessna 310Q	Air Charter & Travel Ltd.
	G–BAHX	Cessna 182P	Clifford Leasing Co.
	G–BAHZ	PA-28R Cherokee Arrow 200-II	J. Burgess
	G–BAIA	PA-32 Cherokee Six 300E	Langham International (Aircraft) Ltd.
	G–BAIB	Enstrom F-28A	J. Lloyd
	G–BAIC	Cessna FRA.150L	Wycombe Air Centre Ltd.
	G–BAIF	Western O-65 balloon	B. M. Smith *Captain Starlight*
	G–BAIG	PA-34-200 Seneca	Elverston Electronics Ltd.
	G–BAIH	PA-28R Cherokee Arrow 200-II	J. Pemberton
	G–BAII	Cessna FRA.150L	D. Nixon & E. C. Heathcote
	G–BAIK	Cessna F.150L	Wickenby Aviation Ltd.
	G–BAIL	Cessna FR.172J	E. A. Black

Reg.	Type	Owner or Operator	Notes
G–BAIM	Cessna 310Q	Airwork Services Ltd.	
G–BAIN	Cessna FRA.150L	Airwork Services Ltd.	
G–BAIO	Cessna F.150L	Gordon King Aviation Ltd.	
G–BAIP	Cessna F.150L	J. S. Jones	
G–BAIR	Thunder Ax7-77 balloon	P. A. & Mrs. M. Hutchins	
G–BAIS	Cessna F.177RG	Loughton Aviation Ltd.	
G–BAIU	Hiller UH-12E	Heliwork Ltd.	
G–BAIW	Cessna F.172M	R. D. Green & M. J. Dawkins	
G–BAIX	Cessna F.172M	Tiltbrook Ltd.	
G–BAIY	Cameron O-65 balloon	Budget Rent A Car (UK) Ltd. *Lady Budget*	
G–BAIZ	Slingsby T.61A Falke	Lasham Gliding Soc. Ltd.	
G–BAJA	Cessna F.177RG	Don Ward Productions Ltd.	
G–BAJB	Cessna F.177RG	Bank Farm Ltd.	
G–BAJC	Evans VP-1	J. R. Clements	
G–BAJE	Cessna 177 Cardinal	Triavia Ltd.	
G–BAJN	AA-5 Traveler	Janacrew Ltd.	
G–BAJO	AA-5 Traveler	J. B. Cuddy	
G–BAJR	PA-28 Cherokee 180	K. F. Davison	
G–BAJT	PA-28R Cherokee Arrow 200-II	DMR Computor Ltd.	
G–BAJU	PA-23 Aztec 250	Mediterranean Caravan Sales Ltd.	
G–BAJV	SA.102.5 Cavalier	A. J. Starkey	
G–BAJW	Boeing 727-46	Dan-Air Services Ltd.	
G–BAJX	PA-E23 Aztec 250	A. J. Walgate & Son Ltd.	
G–BAJY	Robin DR.400/180	J. F. Ingledew	
G–BAJZ	Robin DR.400/125	J. F. Durcan	
G–BAKA	Sikorsky S-61N	Bristow Helicopters Ltd. *West Sole*	
G–BAKB	Sikorsky S-61N	Bristow Helicopters Ltd. *Montrose*	
G–BAKC	Sikorsky S-61N	Bristow Helicopters Ltd. *Forties*	
G–BAKD	PA-34-200-2 Seneca	NIC Instruments Ltd.	
G–BAKF	Bell 206B JetRanger 2	M. J. K. Belmont	
G–BAKG	Hughes 269C	W. R. Finance Ltd.	
G–BAKH	PA-28 Cherokee 140	Woodgate Aviation Ltd.	
G–BAKJ	PA-30 Twin Comanche 160	Overdraft Aviation	
G–BAKK	Cessna F.172H	P. J. Brown	
G–BAKL	F.27 Friendship 200	Air UK	
G–BAKM	Robin DR.400/140	P. E. Scott	
G–BAKN	SNCAN SV-4C	M. Holloway	
G–BAKO	Cameron O-84 balloon	D. C. Dokk-Olsen *Pied Piper*	
G–BAKP	PA-E23 Aztec 250	Executair Ltd.	
G–BAKR	Jodel D.117	J. B. Fisher	
G–BAKS	A-B 206B JetRanger 2	G. M. H. Willis	
G–BAKT	A-B 206B JetRanger 2	Burnthills Plant Hire Ltd.	
G–BAKV	PA-18 Super Cub 150	Pounds Marine Shipping Ltd.	
G–BAKW	B.121 Pup 2	J. Trevor-Hicks Ltd.	
G–BAKY	Slingsby T.61C Falke	D. R. C. Reeves	
G–BAKZ	BN-2A Islander	Fairey Surveys Ltd.	
G–BALB	Air & Space Model 18A	Interflight Ltd.	
G–BALC	Bell 206B JetRanger 2	Park Farms Services Ltd.	
G–BALE	Enstrom F.28A	C.S.E. Aviation Ltd.	
G–BALF	Robin DR.400/140	F. A. Spear	
G–BALG	Robin DR.400/180	R. Jones	
G–BALH	Robin DR.400/140	Pennine Leisure Ltd.	
G–BALI	Robin DR.400 2+2	E. F. Rowe	
G–BALJ	Robin DR.400/180	Barlodz Ltd.	
G–BALK	SNCAN SV-4C	J. C. Brierley	
G–BALL	Bede BD-5	J. P. Turner	
G–BALM	Cessna 340	Patgrove Ltd.	
G–BALN	Cessna T.310Q	Wadkin (Aviation) Ltd.	
G–BALP	PA-39 Twin Comanche 160 C/R	Maynards (Heels) Ltd.	
G–BALR	Wittman W.8 Tailwind	D. O. A. Elmer	
G–BALS	Nipper T.66 Mk. 3	L. W. Shaw	
G–BALT	Enstrom F.28A	Franklin Aviation Ltd.	
G–BALU	PA-E23 Aztec 250C	Avio-Lec Ltd. (G–BADD)	
G–BALW	PA-28R Cherokee Arrow 200-II	H. A. W. Pilkington	
G–BALX	D.H.82A Tiger Moth (N6848)	C. P. B. Horsley & R. G. Annis	
G–BALY	Practavia Pilot Sprite 150	A. L. Young	
G–BAMB	Slingsby T.61C Falke	Universities of Glasgow & Strathclyde Gliding Club	
G–BAMC	Cessna F.150L	Sierra Aviation Services Ltd.	
G–BAME	Volmer VJ-22 Sportsman	V. H. Bellamy	
G–BAMF	MBB Bo 105D	Management Aviation Ltd.	

Notes	Reg.	Type	Owner or Operator
	G–BAMG	Avions Lobet Ganagobie	J. A. Brompton
	G–BAMI	Beech 95-B55 Baron	Ace Belmont International Ltd.
	G–BAMJ	Cessna 182P	Aylesbury Mushrooms Ltd.
	G–BAMK	Cameron D-96 hot-air airship	Cameron Balloons Ltd.
	G–BAML	Bell 206A JetRanger	Somerton-Rayner Helicopters Ltd.
	G–BAMM	PA-28 Cherokee 235	E. R. Walters
	G–BAMN	Cessna U.206C Super Skywagon	M. E. Robinson
	G–BAMR	PA-16 Clipper	H. Boyce
	G–BAMS	Robin DR.400/160	Bostock Aviation Ltd.
	G–BAMU	Robin DR.400/160	W. J. C. Scrope & ptnrs.
	G–BAMV	Robin DR.400/180	Craven Aviation Ltd.
	G–BAMY	PA-28R Cherokee Arrow 200D Mk. II	Briar Hire & Supply Ltd.
	G–BAMZ	PA-34-200-2 Seneca	G. R. Air Services Ltd.
	G–BANA	Robin DR.221	G. T. Pryor
	G–BANB	Robin DR.400/180	Time Electronics Ltd.
	G–BANC	GY-201 Minicab	C. D. B. Trollope
	G–BAND	Cameron O-84 balloon	Mid-Bucks Farmers Balloon Group *Clover*
	G–BANE	Cessna FRA.150L	Edinburgh Flying Services Ltd.
	G–BANF	Luton L.A.4A Minor	D. W. Bosworth
	G–BANG	Cameron O-84 balloon	R. Harrower *Salamander*
	G–BANK	PA-34-200-2 Seneca	Airde Ltd.
	G–BANL	BN-2A-8 Islander	Loganair Ltd.
	G–BANS	PA-34-200-2 Seneca	G. Knowles
	G–BANT	Cameron O-65 balloon	M. D. Tweedie & ptnrs. *Shades*
	G–BANU	Wassmer Jodel D.120	C. E. McKinney
	G–BANV	Phoenix Currie Wot	K. Knight
	G–BANW	CP.1330 Super Emeraude	J. D. McCracker & ptnrs.
	G–BANX	Cessna F.172M	A. J. Keen & ptnrs.
	G–BANY	Glos-AESL Airtourer 115	Bernell Aviation Ltd.
	G–BAOB	Cessna F.172M	Gordon King Aviation Ltd.
	G–BAOC	M.S.894E Rallye Minerva	P. V. & Mrs. E. M. Gilliar
	G–BAOD	M.S.880B Rallye Club	E. K. Chalke
	G–BAOF	M.S.880B Rallye Club	G. B. Instrument Panel Co. Ltd.
	G–BAOG	M.S.880B Rallye Club	P. L. M. Moss & S. W. Biroth
	G–BAOH	M.S.880B Rallye Club	S. P. Bryant & ptnrs.
	G–BAOJ	M.S.880B Rallye Club	John Finlay Concrete Pipes Ltd.
	G–BAOM	M.S.880B Rallye Club	J. A. Aldridge & ptnrs.
	G–BAOO	Cessna 421B	Ladies Pride Aviation Ltd.
	G–BAOP	Cessna FRA.150L	Rogers Aviation Sales Ltd.
	G–BAOS	Cessna F.172M	Allen Barker Photography Ltd.
	G–BAOT	M.S.880B Rallye Club	B. Findley
	G–BAOU	AA-5 Traveler	W. H. Ingram
	G–BAOV	AA-5 Traveler	Hornet Aviation Ltd.
	G–BAOW	Cameron O-65 balloon	P. A. White *Winslow Boy*
	G–BAOX	Cessna 310Q	Gempen Ltd.
	G–BAOY	Cameron S-31 balloon	Shell-Mex BP Ltd. *New Potato*
	G–BAPA	Fournier RF-5B Sperber	R. Pasold & D. Stuynor
	G–BAPB	DHC-1 Chipmunk 22	R. C. P. Brookhouse
	G–BAPC	Luton L.A.4A Minor	Midland Aircraft Preservation Soc.
	G–BAPD	V.814 Viscount	British Midland Airways Ltd.
	G–BAPE	V.814 Viscount	
	G–BAPF	V.814 Viscount	British Midland Airways Ltd.
	G–BAPG	V.814 Viscount	
	G–BAPH	Cessna FRA.150L	B. Flay & T. C. Hocking
	G–BAPI	Cessna FRA.150L	Industrial Supplies (Peterborough) Ltd.
	G–BAPJ	Cessna FRA.150L	P. H. Electronics Ltd.
	G–BAPK	Cessna F.150L	Ulster Flying Club (1961) Ltd.
	G–BAPL	PA-23 Aztec 250E	Scottish Malt Distillers Ltd.
	G–BAPM	Fuji FA.200-160	M. J. Flanagan
	G–BAPN	PA-28 Cherokee 180	Gilbert Business Management Ltd.
	G–BAPP	Evans VP-1	N. Crow
	G–BAPR	Jodel D.11	E. W. Osbourn & ptnrs.
	G–BAPS	*Campbell Cougar	British Rotorcraft Museum
	G–BAPT	Fuji FA.200-180	T. F. Turner
	G–BAPV	Robin DR.400/160	J. D. Millne & ptnrs.
	G–BAPW	PA-28R Cherokee Arrow 180	G. & R. Consultants Ltd.
	G–BAPX	Robin DR.400/160	A. J. Howes
	G–BAPY	Robin HR.100/210	Engineering Appliances Ltd.
	G–BARB	PA-34-200-2 Seneca	Maykind Ltd.
	G–BARC	Cessna FR.172J	Sir G. P. Grant-Suttie

Reg.	Type	Owner or Operator	Notes
G–BARD	Cessna 337C	Europa Aviation Ltd.	
G–BARF	Jodel D.112 Club	G. J. Anderson	
G–BARG	Cessna 310Q	G. E. Platt	
G–BARH	Beech C.23 Sundowner	D. Phillips	
G–BARN	Taylor JT.2 Titch	R. G. W. Newton	
G–BARP	Bell 206B JetRanger 2	Patgrove Ltd.	
G–BARR	H.S.125 Srs. 600B	Rolls-Royce Ltd.	
G–BARS	D.H.C.1. Chipmunk 22	T. I. Sutton	
G–BART	H.S.125 Srs. 600B	Green Shield Trading Stamp Co. Ltd.	
G–BARV	Cessna 310Q	Old England Watches	
G–BARW	Cessna 402B	C. Love & H. G. Woodsend	
G–BARX	Bell 206B JetRanger 2	Sutton & Son (St. Helens) Ltd.	
G–BARY	CP.301A Emeraude	W. C. C. Meyer	
G–BARZ	Scheibe SF.28A	RAFGSA	
G–BASB	Enstrom F-28A	Travel Centre (Norwich) Ltd.	
G–BASD	B.121 Pup 2	G. W. Archer	
G–BASE	Bell 206B JetRanger 2	Air Hanson	
G–BASG	AA-5 Traveler	Trinecare Ltd.	
G–BASH	AA-5 Traveler	Gilbert O'Sullivan Ltd.	
G–BASI	PA-28 Cherokee 140	CFS Aircraft Ltd.	
G–BASJ	PA-28 Cherokee 180	P. B. Lacey & M. A. Court	
G–BASL	PA-28 Cherokee 140	Air Navigation & Trading Ltd.	
G–BASM	PA-34-200-2 Seneca	A. Jacobs *Joie de Vivre*	
G–BASN	Beech C.23 Sundowner	M. F. Fisher	
G–BASO	Lake LA-4 Amphibian	P. R. M. Chapman	
G–BASP	B.121 Pup 1	R. Gilkes	
G–BASR	PA-25 Pawnee 235C	C. M. G. Ellis & M. M. James	
G–BASS	Cessna 421B	Bass Charrington Ltd.	
G–BAST	Cameron O-84 balloon	P. A. Salmons & D. A. Tinley *Honey*	
G–BASU	PA-31-350 Navajo Chieftain	Casair Aviation Services Ltd.	
G–BASV	Enstrom F-28A	C. S. E. Aviation Ltd.	
G–BASX	PA-34-200 Seneca	Trans Europe Air Charter Ltd.	
G–BASY	Jodel D.9 Bebe	R. L. Sambell	
G–BATA	H.S.125 Srs. 400B	Beecham-Imperial Aviation Ltd.	
G–BATB	MBB Bo 105C	British Caledonian Helicopters Ltd.	
G–BATC	MBB Bo 105C	Management Aviation Ltd.	
G–BATE	PA-23 Aztec 250E	Trehaven Trust Ltd.	
G–BATH	Cessna F.337G	S. White & L. H. Bedden	
G–BATJ	Jodel D.119	N. Shepherd & ptnrs.	
G–BATM	PA-32 Cherokee Six 300	Patgrove Ltd. & J. Wakeman & Co.	
G–BATN	PA-E23 Aztec 250	Marshall of Cambridge Ltd.	
G–BATR	PA-34-200 Seneca	Executive Aviation Ltd.	
G–BATS	Taylor JT.1 Monoplane	J. Jennings	
G–BATT	Hughes 269C	United Marine (1939) Ltd.	
G–BATU	Enstrom F-28A	Trent Park Stables	
G–BATV	PA-28 Cherokee 180D	The Scoresby Flying Group	
G–BATW	PA-28 Cherokee 140	Mooney Aviation Ltd.	
G–BATX	PA-23 Aztec 250E	Tayside Aviation Ltd.	
G–BAUA	PA-E23 Aztec 250	Alat Aviation Ltd.	
G–BAUC	PA-25 Pawnee 235C	P. M. Charles	
G–BAUD	Robin DR.400/160	Triton Air Travel Ltd.	
G–BAUE	Cessna 310Q	A. J. Dyer	
G–BAUF	Hughes 269C	Point to Point Helicopters Ltd.	
G–BAUH	Jodel D.112	I. G. Glenn & ptnrs.	
G–BAUI	PA-E23 Aztec 250	Simulated Flight Training Ltd.	
G–BAUJ	PA-E23 Aztec 250	Airde Ltd.	
G–BAUK	Hughes 269C	Curtis Engineering (Frome) Ltd.	
G–BAUM	Bell 206B JetRanger 2	PLM Helicopters Ltd.	
G–BAUN	Bell 206A JetRanger 2	Bristow Helicopters Ltd.	
G–BAUO	PA-E23 Aztec 250	Glencair (Aero Services) Ltd.	
G–BAUR	F.27 Friendship Mk. 200	Air UK	
G–BAUV	Cessna F.150L	J. Braithwaite (Aerial Photography) Ltd.	
G–BAUW	PA-E23 Aztec 250	Myson Group Ltd.	
G–BAUX	Limba Lapwing	B. J. Jacobson & R. M. Fisher	
G–BAUY	Cessna FRA.150L	Inverness Flying Services Ltd.	
G–BAUZ	Nord NC.854S	T. Atkinson	
G–BAVB	Cessna F.172M	Hudson Bell Aviation	
G–BAVC	Cessna F.150L	Elles Aviation	
G–BAVE	Beech A.100 King Air	Vernair Transport Services	
G–BAVF	Beech 58 Baron	Tarlworth Ltd.	
G–BAVG	Beech E.90 King Air	Allied Breweries (UK) Ltd.	
G–BAVH	D.H.C.1 Chipmunk 22	RAFGSA	

Notes	Reg.	Type	Owner or Operator
	G–BAVJ	PA-31-350 Navajo Chieftain	Vickers Ltd.
	G–BAVL	PA-E23 Aztec 250	Shipboard Maintenance Ltd.
	G–BAVM	PA-31-350 Navajo Chieftain	Air Commuter Ltd.
	G–BAVN	Boeing Stearman PT-17	Lake Green Garage
	G–BAVO	Boeing Stearman PT-17	Keenair Services Ltd.
	G–BAVP	Beech A.23-24 Musketeer	Wearside Flying Group
	G–BAVR	AA-5 Traveler	Rabhart Ltd.
	G–BAVS	AA-5 Traveler	Crystal Heart Salad Co. Ltd.
	G–BAVU	Cameron A-105 balloon	J. D. Michaelis
	G–BAVW	PA-E23 Aztec 250	Goldenbolt International Ltd.
	G–BAVX	HPR-7 Herald 214	British Air Ferries Ltd. *Timothy Keegan*
	G–BAVY	PA-E23 Aztec 250	Stellaris Ltd.
	G–BAVZ	PA-E23 Aztec 250	Dismore Aviation Ltd.
	G–BAWA	PA-28R-200-2 Cherokee Arrow	Airways Aero Associations Ltd.
	G–BAWB	PA-E23 Aztec 250	J. T. Tyer
	G–BAWG	PA-28R-200-2 Cherokee Arrow	Richard Flint & Co. Ltd.
	G–BAWI	Enstrom F-28A	Spooner Aviation Ltd.
	G–BAWK	PA-28 Cherokee 140	Newcastle-Upon-Tyne Aero Club Ltd.
	G–BAWL	Airborne Industries gas airship	A. F. J. Smith *The Santos-Dumont*
	G–BAWM	Jodel D.112	Wearside Flying Association
	G–BAWN	PA-30C Twin Comanche 160	J. & Y. Plastics (Aviation) Ltd.
	G–BAWO	Cessna 340	Etchborder Ltd.
	G–BAWR	Robin HR.100/210	Devonair Transport Ltd.
	G–BAWS	PA-25 Pawnee 235C	Farm Aviation Services Ltd.
	G–BAWT	M.S.894E Rallye Minerva	A. M. Collins
	G–BAWU	PA-30 Twin Comanche 160	C. P. Francis
	G–BAWV	PA-E23 Aztec 250	Golden Lion Plant Hire Ltd.
	G–BAWW	Thunder Ax7-77 balloon	Miss M. L. C. Hutchins *Taurus*
	G–BAWX	PA-28 Cherokee 180	Bawxair Ltd.
	G–BAWZ	Cessna 402B	E. M. Brain
	G–BAXE	Hughes 269A	Reethorpe Engineering Ltd.
	G–BAXF	Cameron O-77 balloon	I. R. Williams & ptnrs. *Granna*
	G–BAXH	Cessna 310Q	Fisher-Karpark Ltd.
	G–BAXI	PA-39 Twin Comanche C/R	C.S.E. (Aircraft Services) Ltd.
	G–BAXJ	PA-32 Cherokee Six 300	T. E. Elliott
	G–BAXK	Thunder Ax7-77 balloon	Newbury Balloon Group *Jack O'Newbury*
	G–BAXL	H.S.125 Srs. 3B	Dennis Vanguard International (Switchgear) Ltd.
	G–BAXM	Beech B.24R Sierra	Strangford Flying Group
	G–BAXN	PA-34-200-2 Seneca	Ards Aviation
	G–BAXP	PA-E23 Aztec 250	Peregrine Air Services Ltd.
	G–BAXR	Beech B.55 Baron	Rig Design Services Ltd.
	G–BAXS	Bell 47G-5	T. C. Barton
	G–BAXT	PA-28R-200 Cherokee Arrow	Williams & Griffin Ltd.
	G–BAXU	Cessna F.150L	W. Lancs Aero Club Ltd.
	G–BAXV	Cessna F.150L	J. H. A. Rogers
	G–BAXW	Cessna F.150L	Wycombe Air Centre Ltd.
	G–BAXX	Cessna F.150L	R. J. Forbes & ptnrs.
	G–BAXY	Cessna F.172M	Risk Management Group (London) Ltd.
	G–BAXZ	PA-28 Cherokee 140	A. J. Smith & D. N. Sharpe
	G–BAYC	Cameron O-65 balloon	A. T. Willmer *Viva Verdi II*
	G–BAYL	Nord 1203/111 Norecrin	D. M. Fincham
	G–BAYO	Cessna 150L	Cheshire Air Training School Ltd.
	G–BAYP	Cessna 150L	J. N. Collins
	G–BAYR	Robin HR.100/210	Thorney Machinery Co. Ltd.
	G–BAYT	H.S.125 Srs. 600B	Management Agency and Music Ltd.
	G–BAYX	Bell 47G-5	Helicopter Hire Ltd.
	G–BAYY	Cessna 310C	Tarlworth Ltd.
	G–BAYZ	Bellanca 7GC BC Citabria	Cambridge University Gliding Trust Ltd.
	G–BAZA	H.S.125 Srs. 400B	British Aerospace
	G–BAZB	H.S.125 Srs. 400B	Short Bros. Ltd.
	G–BAZC	Robin DR.400/160	Greataslot Ltd.
	G–BAZF	AA-5 Traveler	Carpfinch Ltd.
	G–BAZG	Boeing 737-204	Britannia Airways Ltd. *Florence Nightingale*

Reg.	Type	Owner or Operator	Notes
G–BAZH	Boeing 737-204	Britannia Airways Ltd. *Isambard Kingdom Brunel*	
G–BAZI	Boeing 737-204	Britannia Airways Ltd. *Sir Walter Rayleigh*	
G–BAZJ	HPR-7 Herald 209	Air UK	
G–BAZM	Jodel D.11	Bingley Flying Group	
G–BAZN	Bell 206 JetRanger	Somerton-Rayner Helicopters Ltd.	
G–BAZS	Cessna F.150L	Channel Island Aero Services Ltd.	
G–BAZT	Cessna F.172M	Murray Fraser (Aviation) Ltd.	
G–BAZU	PA-28R-200 Cherokee Arrow	R. Bolsover	
G–BAZV	PA-E23 Aztec 250	Casair Aviation Services Ltd.	
G–BBAB	M.S.894A Rallye Minerva	Stenloss Ltd.	
G–BBAE	L.1011–385 TriStar	British Airways	
G–BBAF	L.1011–385 TriStar	British Airways	
G–BBAG	L.1011–385 TriStar	British Airways	
G–BBAH	L.1011–385 TriStar	British Airways	
G–BBAI	L.1011–385 TriStar	British Airways	
G–BBAJ	L.1011–385 TriStar	British Airways	
G–BBAK	M.S.894A Rallye Minerva	B. G. J. Alimo	
G–BBAR	Jodel D.117	J. F. Wright	
G–BBAU	Enstrom F.28A	J. D. M. Roberts	
G–BBAV	PA-E23 Aztec 250	Pan Universal Aircraft Services (CI) Ltd.	
G–BBAW	Robin HR.100/210	Scoba Ltd.	
G–BBAX	Robin DR.400/140	S. R. Young	
G–BBAY	Robin DR.400/140	G. A. Pentelow & D. B. Roadnight	
G–BBAZ	Hiller UH-12E	Management Aviation Ltd.	
G–BBBA	Hiller UH-12E	Management Aviation Ltd.	
G–BBBB	Taylor JT.1 Monoplane	S. A. MacConnacher	
G–BBBC	Cessna F.150L	East Midland School of Flying	
G–BBBD	PA-E23 Aztec 250	Moseley Group (PSV) Ltd.	
G–BBBI	AA-5 Traveler	The Bravo India Group	
G–BBBJ	PA-E23 Aztec 250	Usoland Ltd.	
G–BBBK	PA-28 Cherokee 140	Bolton Air Training School Ltd.	
G–BBBL	Cessna 337B	Alderney Air Charter Ltd.	
G–BBBM	Bell 206B JetRanger 2	Keluma Ltd.	
G–BBBN	PA-28 Cherokee 180	G. Nicholson	
G–BBBO	SIPA 903	M. C. Wroe	
G–BBBP	Bell 212	Bristow Helicopters Ltd.	
G–BBBR	Enstrom F.28A	Spooner Aviation Ltd.	
G–BBBS	Cessna 182P Skylane	Hammond Aviation Services Ltd.	
G–BBBU	Pitts S-1D Special	R. N. Goode & N. A. Ball	
G–BBBW	FRED Series 2	D. L. Webster	
G–BBBX	Cessna E310L	Air Atlantique Ltd.	
G–BBBY	PA-28 Cherokee 140	Channel Aviation Ltd.	
G–BBBZ	Enstrom F-28A	Spooner Aviation (Enstrom Helicopters) Ltd.	
G–BBCA	Bell 206B JetRanger 2	Barkers Plant Hire (Alsager) Ltd.	
G–BBCB	Western O-65 balloon	M. Westwood *Cee Bee*	
G–BBCC	PA-E23 Aztec 250	Goodridge (UK) Ltd.	
G–BBCD	Beech 95 B.55 Baron	Jackson Aviation Ltd.	
G–BBCF	Cessna FRA.150L	Union Beech Ltd.	
G–BBCG	Robin DR.400/2+2	Headcorn Flying School Ltd.	
G–BBCH	Robin DR.400/2+2	Headcorn Flying School Ltd.	
G–BBCI	Cessna 150H	D. Machin	
G–BBCJ	Cessna 150J	Mattick Aviation Ltd.	
G–BBCK	Cameron O-77 balloon	R. J. Leathart *The Mary Gloster*	
G–BBCL	H.S.125 Srs. 600B	British Aerospace (G–BJCB)	
G–BBCM	PA-E23 Aztec 250	Keenair Services Ltd.	
G–BBCN	Robin HR.100/210	B. D. Steel Holdings Ltd.	
G–BBCP	Thunder Ax6-56 balloon	P. H. Keene *Jack Frost*	
G–BBCR	Z.326 Trener Master	Tiger Zlin Group	
G–BBCS	Robin DR.400/140	Albert J. Parsons & Sons Ltd.	
G–BBCT	PA-31-350 Navajo Chieftain	Express Aviation Services Ltd.	
G–BBCU	PA-E23 Aztec 250	Eastern-Air Executive Ltd.	
G–BBCV	Cessna A.188B Agtruck	Norfolk Aerial Spraying Ltd.	
G–BBCW	PA-E23 Aztec 250	Aerolease	
G–BBCX	Airship (hot-air) radio-controlled	E. A. Wills & G. W. Moger *Dew Drop*	
G–BBCY	Luton L.A.4A. Minor	C. H. Difford	
G–BBCZ	AA-5 Traveler	Stronghill Flying Group	
G–BBDA	AA-5 Traveler	Headcorn Flying School Ltd.	

Notes	Reg.	Type	Owner or Operator
	G–BBDB	PA-28 Cherokee 180	Eleuthero Ltd.
	G–BBDC	PA-28 Cherokee 140	Apache Aircraft Services Ltd.
	G–BBDD	PA-28 Cherokee 140	FR Aviation
	G–BBDE	PA-28R-200-2 Cherokee Arrow	Hardings of Winsford
	G–BBDG	Concorde 100	British Aerospace
	G–BBDH	Cessna F.172M	Rogers Aviation Sales Ltd.
	G–BBDI	PA-18-150 Super Cub	Scottish Gliding Union Ltd.
	G–BBDJ	Thunder Ax6-56 balloon	S. W. D. & H. B. Ashby *Jack Tar*
	G–BBDK	V.808F Viscount	Southern International Cargo
	G–BBDL	AA-5 Traveler	A. Howard
	G–BBDM	AA-5 Traveler	E. M. Pettit Construction Ltd.
	G–BBDN	Taylor JT.1 Monoplane	D. A. Nice
	G–BBDO	PA-E23 Aztec 250	R. Long
	G–BBDP	Robin DR.400/160	Braddon & Sons (Haulage) Ltd.
	G–BBDS	PA-31 Navajo	Broad Oak Air Services
	G–BBDT	Cessna 150H	Sherburn Aero Club
	G–BBDU	PA-31 Navajo	ITT Components Ltd.
	G–BBDV	SIPA S.903	A. W. Webster
	G–BBEA	Luton L.A.4A Minor	D. J. Wells & ptnrs.
	G–BBEB	PA-28R-200-2 Cherokee Arrow	Atkinson Aviation Ltd.
	G–BBEC	PA-28 Cherokee 180	J. H. Kimber
	G–BBED	M.S.894B Rallye Minerva	Trago Mills Ltd.
	G–BBEE	Learjet 25B	Pendleton Aviation Ltd.
	G–BBEF	PA-28 Cherokee 140	Air Navigation & Trading Co. Ltd.
	G–BBEI	PA-31 Navajo	Survey Flights Ltd.
	G–BBEJ	PA-31-350 Navajo Chieftain	Iceni Aviation Ltd.
	G–BBEL	PA-28R Cherokee Arrow 180	J. M. McRitchie
	G–BBEM	Beech B.55 Baron	M. J. Coburn & L. C. G. Hughes
	G–BBEN	Bellanca 7GC-BC Citabria	Ulster Gliding Club Ltd.
	G–BBEO	Cessna FRA.150L	Solitair Flight Management Ltd.
	G–BBEP	H.S.125 Srs. 600B	Albert Abela & Co. Ltd.
	G–BBEU	Bell 206B JetRanger 2	International Messengers Ltd.
	G–BBEV	PA-28 Cherokee 140	Spooner Aviation Ltd.
	G–BBEW	PA-E23 Aztec 250	Lease Air Ltd.
	G–BBEX	Cessna 185A Skywagon	R. G. Brooks & D. E. Wilson
	G–BBEY	PA-E23 Aztec 250	T. H. Knitwear Ltd.
	G–BBFB	Bell 206B JetRanger	Air Hanson Helicopters Ltd.
	G–BBFC	AA-1B Trainer	P. C. Blake
	G–BBFD	PA-28R-200-2 Cherokee Arrow	Truman Aviation Ltd.
	G–BBFE	Bell 206 JetRanger	W. Holmes
	G–BBFF	PA-34-200 Seneca	American Airspeed Ltd.
	G–BBFL	GY-201 Minicab	D. H. Greenwood
	G–BBFS	Van Den Bemden gas balloon	A. J. F. Smith *Le Tomate*
	G–BBFT	Cessna A.188B AgTruck	Mindacre Ltd.
	G–BBFU	PA-E23 Aztec 250	Southampton Airport Ltd.
	G–BBFV	PA-32 Cherokee Six 260	Southend Securities Ltd.
	G–BBFW	PA-E23 Aztec 250B	T. Bartlett
	G–BBFX	PA-34-200 Seneca	C. D. Weiswall
	G–BBFY	PA-31P Navajo	Peters Stores Ltd.
	G–BBFZ	PA-28R-200-2 Cherokee Arrow	Larkfield Garage (Chepstow) Ltd.
	G–BBGB	PA-E23 Aztec 250	K. J. F. Aircraft Ltd.
	G–BBGC	M.S.893E Rallye Commodore 180	Channel Island Mortgage Brokers (Alderney) Ltd.
	G–BBGE	PA-E23 Aztec 250	Dollar Air Services Ltd.
	G–BBGF	Cessna 340	Hillair Ltd & Lawrence Wilson & Son Ltd.
	G–BBGG	AA-5 Traveler	H. Richards & Sons
	G–BBGH	AA-5 Traveler	Cabair Ltd.
	G–BBGI	Fuji FA.200–160	Sureway Security Ltd.
	G–BBGJ	Cessna 180	Sussex Agricultural Aviation Services
	G–BBGK	Lake LA-4-200 Buccaneer	James Duncan (Plumbers) Ltd.
	G–BBGL	Baby Great Lakes	P. W. Thomas
	G–BBGN	Cameron A-375 balloon	J. P. R. Nott *Daffodil II*
	G–BBGO	Robin HR.100/210	Home & Overseas International Oil Co. Ltd.
	G–BBGR	Cameron O-65 balloon	Thames Valley Balloon Group
	G–BBGS	Sikorsky S-61N	Bristow Helicopters Ltd. *Indefatigable*
	G–BBGU	H.S.125 Srs. 400B	McAlpine Aviation Ltd.
	G–BBGX	Cessna 182P Skylane	Lockwoods Technical Services (Liverpool) Ltd.

Reg.	Type	Owner or Operator	Notes
G–BBGZ	CHABA 42 balloon	J. Haigh & ptnrs. *Phlogiston*	
G–BBHB	PA-31-300 Navajo	Distance No Object Ltd.	
G–BBHC	Enstrom F-28A	Dumbrill Plant & Engineering Co. Ltd.	
G–BBHD	Enstrom F-28A	Hyde Industrial Holdings Ltd.	
G–BBHE	Enstrom F-28A	Complant Engineering Services Ltd.	
G–BBHF	PA-23 Aztec 250E	Air Foyle Ltd.	
G–BBHG	Cessna E-310Q	Airwork Services Ltd.	
G–BBHI	Cessna 177RG	Hi-Lines Ltd.	
G–BBHJ	Piper J-3C-65 Cub	R. V. Miller & R. H. Heath	
G–BBHK	AT-16 Harvard IIB	Bob Warner Aviation	
G–BBHL	Sikorsky S-61N Mk II	Bristow Helicopters Ltd. *Glamis*	
G–BBHM	Sikorsky S-61N Mk II	Bristow Helicopters Ltd. *Braemar*	
G–BBHU	SA.341G Gazelle I	Solaria Investments Ltd.	
G–BBHW	SA.341G Gazelle I	McAlpine Aviation Ltd.	
G–BBHX	M.S.893E Rallye Commodore	P. J. Pitts	
G–BBIA	PA-28R-200 Cherokee Arrow	A. G. (Commodities) Ltd.	
G–BBIC	Cessna 310Q	ATA Grinding Processes	
G–BBID	PA-28 Cherokee 140	C.S.E. Aviation Ltd.	
G–BBIF	PA-E23 Aztec 250	Northern Executive Aviation Ltd.	
G–BBIH	Enstrom F-28A	Forecourt Services Ltd.	
G–BBII	Fiat G-46-3B	The Hon. Patrick Lindsay	
G–BBIJ	Cessna 421B	Owledge Ltd.	
G–BBIL	PA-28 Cherokee 140	Leisure Lease	
G–BBIM	Cessna E-310Q	Mann Aviation Sales Ltd.	
G–BBIN	Enstrom F-28A	Churnbury Ltd.	
G–BBIO	Robin HR.100/210	R. A. King	
G–BBIT	Hughes 269B	W. R. Finance Ltd.	
G–BBIU	Hughes 269C	W. R. Finance Ltd.	
G–BBIV	Hughes 269C	W. R. Finance Ltd.	
G–BBIW	Hughes 269C	W. R. Finance Ltd.	
G–BBIX	PA-28 Cherokee 140	M. Dukes & K. Learmonth	
G–BBJB	Thunder AX7-77 balloon	St. Crispin Balloon Group *Dick Darby*	
G–BBJD	Cessna 172M	A. F. Hall	
G–BBJH	Cessna A-188B Agtruck	W. R. C. Wauchope	
G–BBJI	Isaacs Spitfire	R. C. Tyler	
G–BBJT	Robin HR.200/100	M. J. McRobert	
G–BBJU	Robin DR.400/140	Tower Scaffolding (Bristol) Ltd.	
G–BBJV	Cessna F.177RG	Pilot Magazine	
G–BBJW	Cessna FRA.150L	Gordon King (Aviation) Ltd.	
G–BBJX	Cessna F.150L	Yorkshire Flying Services Ltd.	
G–BBJY	Cessna F.172M	J. Lucketti	
G–BBJZ	Cessna F.172M	Border Flying Group	
G–BBKA	Cessna F.150L	Executive Air	
G–BBKB	Cessna F.150L	Flightec Airline Support Ltd.	
G–BBKC	Cessna F.172M	D. E. H. Designs Ltd.	
G–BBKE	Cessna F.150L	Wickenby Aviation Ltd.	
G–BBKF	Cessna FRA.150L	George House Holdings Ltd.	
G–BBKG	Cessna FR.172J	S. Richman & M. J. H. Raymont	
G–BBKI	Cessna F.172M	G. Crawford	
G–BBKJ	Cessna FT.337G	Carters Gold Medal Soft Drinks Ltd.	
G–BBKL	CP.301A Emeraude	W. J. Walker	
G–BBKO	Thunder Ax7-77 balloon	J. A. Clarke & P. D. Morgan	
G–BBKP	Bell 47G-5A	Alan Mann Helicopters Ltd.	
G–BBKR	Scheibe SF.24A Motorspatz	D. W. Bridson	
G–BBKU	Cessna FRA.150L	Balgia Ltd.	
G–BBKV	Cessna FRA.150L	Laarbruch Powered Flying Club	
G–BBKX	PA-28 Cherokee 180	P. E. Eglington	
G–BBKY	Cessna F.150L	Channel Islands Aero Services Ltd.	
G–BBKZ	Cessna 172M	Exeter Flying Club Ltd.	
G–BBLA	PA-28 Cherokee 140	Southport Aviation Co. Ltd.	
G–BBLC	Hiller UH-12E	Agricopters Ltd.	
G–BBLD	Hiller UH-12E	Agricopters Ltd.	
G–BBLE	Hiller UH-12E	Agricopters Ltd.	
G–BBLG	Hiller UH-12E	Sloane Aviation Ltd.	
G–BBLH	Piper O-59A Grasshopper	P. A. Mann	
G–BBLI	R.500S Shrike Commander	Armstrong Aviation Ltd.	
G–BBLJ	Cessna 402B	Hallbret Aviation Ltd.	
G–BBLL	Cameron O-84 balloon	University of East Anglia Hot-Air Ballooning Club *Boadicea*	
G–BBLM	MS.880 Rallye 100 Sport	E. F. G. Flying Services Ltd.	
G–BBLP	PA-E23 Aztec 250D	Elecwind (Clay Cross) Ltd.	
G–BBLS	AA-5 Traveler	J. H. Johnstone	
G–BBLU	PA-34-200-2 Seneca	F. Tranter	

Notes	Reg.	Type	Owner or Operator
	G–BBMB	Robin DR.400/180	R. B. Tyler
	G–BBMD	H.S.125 Srs. 600B	McAlpine Aviation Ltd.
	G–BBME	BAC One-Eleven 401	British Airways (G–AZMI)
	G–BBMF	BAC One-Eleven 401	British Airways (G–ATVU)
	G–BBMG	BAC One-Eleven 408	British Airways (G–AWEJ)
	G–BBMH	E.A.A. Sports Biplane Model P.I.	K. Dawson
	G–BBMI	Dewoitine D.26 (282)	R. J. Willies
	G–BBMJ	PA-E23 Aztec 250	Northern Pig Development Co. Ltd.
	G–BBMK	PA-31-300 Navajo	Velcourt (East) Ltd.
	G–BBML	PA-31-300 Navajo	WF Aviation
	G–BBMN	D.H.C.I Chipmunk 22	R. Steiner
	G–BBMO	D.H.C.I Chipmunk 22	A. J. Hurst
	G–BBMP	D.H.C.I Chipmunk 22	M. J. Harvey & J. R. Rochester
	G–BBMT	D.H.C.I Chipmunk 22	A. T. Letts & ptnrs.
	G–BBMV	D.H.C.I Chipmunk 22 (WG348)	M. D. Payne & M. J. Mead
	G–BBMW	D.H.C.I Chipmunk 22	Cynthia I. Clarry
	G–BBMX	D.H.C.I Chipmunk 22	B. R. C. Wild
	G–BBMZ	D.H.C.I Chipmunk 22	A. J. Baggarley
	G–BBNA	D.H.C.I Chipmunk 22	Coventry Gliding Club Ltd.
	G–BBNB	D.H.C.I Chipmunk 22	A. J. Hurst
	G–BBND	D.H.C.I Chipmunk 22	West Johnson Property Holdings
	G–BBNE	*D.H.C.I Chipmunk T.10 (WZ873)	Aeroplane Collection Ltd.
	G–BBNF	D.H.C.I Chipmunk 22	A. Cullen
	G–BBNG	Bell 206B JetRanger	Bristow Helicopters Ltd.
	G–BBNH	PA-34-200-2 Seneca	Lawrence Goodwin Machine Tools Ltd.
	G–BBNI	PA-34-200-2 Seneca	Colneway Ltd.
	G–BBNJ	Cessna F.150L	K. D. Wickenden
	G–BBNK	PA-23 Aztec 250E	George House (Holdings) Ltd.
	G–BBNM	PA-23 Aztec 250	Fairflight Ltd.
	G–BBNN	PA-E23 Aztec 250D	British Caledonian Airways Ltd.
	G–BBNO	PA-E23 Aztec 250E	Distance No Object Ltd.
	G–BBNR	Cessna 340	J. Lipton
	G–BBNS	Cessna 310Q	Bridgeford Aviation Ltd.
	G–BBNT	PA-31-350 Navajo Chieftain	Simpson Ready Foods Ltd.
	G–BBNV	Fuji FA.200-160	C.S.E. (Aircraft Services) Ltd.
	G–BBNX	Cessna FRA.150L	Three Counties Aero Club Ltd.
	G–BBNY	Cessna FRA.150L	Fairoaks Aviation Services Ltd.
	G–BBNZ	Cessna F.172M	W. G. Fisher
	G–BBOA	Cessna F.172M	George House (Holdings) Ltd.
	G–BBOB	Cessna 421B	Barclays Export & Finances Co. Ltd.
	G–BBOC	Cameron O-77 balloon	A. G. Hopkins & ptnrs. *Bacchus*
	G–BBOD	Thunder 005 balloon	Thunder Balloons Ltd. *Eric the Lad*
	G–BBOE	Robin HR.200/100	Goodwood Terrena Ltd.
	G–BBOH	Pitts S-1S Special	P. Meeson
	G–BBOI	Bede BD-5B	Heather V. B. Wheeler
	G–BBOJ	PA-E23 Aztec 250	Kenton Utilities & Developments Ltd.
	G–BBOK	PA-E23 Aztec 250E	NEI Clarke Chapman Ltd.
	G–BBOL	PA-18 150 Super Cub	Lakes Gliding Club Ltd.
	G–BBOM	PA-E23 Aztec 250E	T. S. Grimshaw Ltd.
	G–BBOO	Thunder Ax6-56 balloon	K. Meehan *Tigerjack*
	G–BBOR	Bell 206B JetRanger 2	Cabair Ltd.
	G–BBOS	Bell 206B JetRanger 2	British Executive Air Services Ltd.
	G–BBOX	Thunder Ax7-77 balloon	R. C. Weyda *Rocinante*
	G–BBOY	Thunder Ax6-56A balloon	N. C. Faithfull *Eric of Titchfield*
	G–BBPJ	Cessna F.172M	Simmette Ltd.
	G–BBPK	Evans VP-1	P. D. Kelsey
	G–BBPM	Enstrom F-28A	P. A. Masters
	G–BBPN	Enstrom F-28A	C.S.E. Aviation Ltd.
	G–BBPO	Enstrom F-28A	Spooner Aviation Ltd.
	G–BBPP	PA-28 Cherokee 180	J. D. B. Hamilton
	G–BBPS	Jodel D.117	A. Appleby
	G–BBPU	Boeing 747-136	British Airways *Henry Hudson*
	G–BBPW	Robin HR.100–210	Parlway Ltd.
	G–BBPX	PA-34-200-2 Seneca	Richel Investments Ltd.
	G–BBPY	PA-28 Cherokee 180	J. Burgess
	G–BBPZ	PA-E23 Aztec 250D	Casair Aviation Services Ltd.
	G–BBRA	PA-E23 Aztec 250E	T. S. Grimshaw Ltd.
	G–BBRB	D.H.82A Tiger Moth (DF198)	R. Barham

Reg.	Type	Owner or Operator	Notes
G–BBRC	Fuji FA.200-180	W. & L. Installations & Co. Ltd.	
G–BBRE	Fuji FA.200-160	M. R. Howse	
G–BBRG	Bell 47G-5A	N. H. Andrew	
G–BBRH	Bell 47G-5A	Autair Helicopters Ltd.	
G–BBRI	Bell 47G-5A	Camlet Helicopters Ltd.	
G–BBRJ	PA-E23 Aztec 250E	Hellodair (Man) Ltd.	
G–BBRL	PA-31P Navajo	Brockhampton Plant Ltd.	
G–BBRN	Procter Kittiwake	RNGSA	
G–BBRO	H.S.125 Srs. 600B	McAlpine Aviation Ltd.	
G–BBRP	BN-2A-9 Islander	Army Parachute Association	
G–BBRS	Enstrom F-28A	A. G. Chrismas Ltd.	
G–BBRV	D.H.C.I Chipmunk 22	HSA (Chester) Sports & Social Club	
G–BBRW	PA-28 Cherokee 140	J. Martin	
G–BBRX	SIAI-Marchetti S.205-18F	W. Chrystal	
G–BBRY	Cessna 210	R. Q. & A. S. Bond	
G–BBRZ	AA-5 Traveler	Galaxy Publications Ltd.	
G–BBSA	AA-5 Traveler	Watsons Anodising Ltd.	
G–BBSB	Beech C23 Sundowner	Sundowner Group	
G–BBSC	Beech B24R Sierra	V. G. Orme	
G–BBSD	Beech 58 Baron	A.B.I. Caravans Ltd.	
G–BBSE	D.H.C.I Chipmunk 22	B. L. Sims & M. Jackson	
G–BBSF	Cessna 310Q	R. B. W. Enterprises Ltd.	
G–BBSL	PA-E23 Aztec 250E	C. Rowbotham & Sons (Management) Ltd.	
G–BBSM	PA-32 Cherokee Six 300	All Seasons Aviation Co. Ltd.	
G–BBSN	PA-E23 Aztec 250	Burnthills (Contractors) Ltd.	
G–BBSO	PA-28 Cherokee 140	C.S.E. (Aircraft Services) Ltd.	
G–BBSR	PA-E23 Aztec 250D	LDL Enterprises	
G–BBSS	D.H.C.IA Chipmunk	Northumbria Gliding Club	
G–BBST	PA-E23 Aztec 250	Thurston Aviation Ltd.	
G–BBSU	Cessna 421B	Westward Television Ltd.	
G–BBSV	Cessna 421B	Owledge Ltd.	
G–BBSW	Pietenpol Air Camper	J. K. S. Wills	
G–BBSZ	Douglas DC-10-10	Laker Airways *Canterbury Belle*	
G–BBTB	Cessna FRA.150L	George House (Holdings) Ltd.	
G–BBTG	Cessna F.172M	Light Planes (Lancashire) Ltd.	
G–BBTH	Cessna F.172M	H. J. C. Townley & ptnrs.	
G–BBTJ	PA-E23 Aztec 250E	Milford Haven Dry Dock Ltd.	
G–BBTK	Cessna FRA.150L	Airwork Services Ltd.	
G–BBTL	PA-E23 Aztec 250C	Air Navigation & Trading Co. Ltd.	
G–BBTS	Beech V35B Bonanza	Golden Sands Estates Ltd.	
G–BBTT	Cessna F.150L	Ulster Aviation Ltd.	
G–BBTU	ST-10 Diplomate	P. Campion	
G–BBTW	PA-31P Navajo	Burch (Engineering) Ltd.	
G–BBTX	Beech C23 Sundowner	J. P. Danton	
G–BBTY	Beech C23 Sundowner	Torlid Ltd.	
G–BBTZ	Cessna F.150L	Ulster Aviation Ltd.	
G–BBUD	Sikorsky S-61N Mk. II	British Airways Helicopters Ltd.	
G–BBUE	AA-5 Traveler	Jobar Properties Ltd.	
G–BBUF	AA-5 Traveler	Eglington Flying Club	
G–BBUG	PA-16 Clipper	C. M. L. Edwards	
G–BBUJ	Cessna 421B	The Automobile Association	
G–BBUK	Bell 47G-2	Bristow Helicopters Ltd.	
G–BBUL	Mitchell-Procter Kittiwake I	R. Bull	
G–BBUO	Cessna 150L	Exeter Flying Club Ltd.	
G–BBUP	B.121 Pup I	C. C. Brown	
G–BBUT	Western O-65 balloon	Wg. Cdr. G. F. Turnbull & Mrs. K. Turnbull *Christabelle II*	
G–BBUU	Piper L-4B Cub	Cooper Bros.	
G–BBUW	SA.102.5 Cavalier	P. C. H. Clarke	
G–BBUX	Bell 206B JetRanger 2	British Car Auctions Aviation Ltd.	
G–BBUY	Bell 206B JetRanger 2	N. Denes Aerodrome Ltd.	
G–BBVA	Sikorsky S-61N Mk. II	Bristow Helicopters Ltd. *Vega*	
G–BBVB	Sikorsky S-61N Mk. II	Bristow Helicopters Ltd. *Aboyne*	
G–BBVC	Slingsby T-59D Kestrel	Slingsby Engineering Ltd.	
G–BBVE	Cessna 340	R. M. Cox Ltd.	
G–BBVF	SA Twin Pioneer III	Flight One Ltd.	
G–BBVG	PA-E23 Aztec 250D	R. F. Wanbon & P. G. Warmerdan	
G–BBVH	V.807 Viscount	Gibraltar Airways Ltd.	
G–BBVI	Enstrom F-28A	C.S.E. Aviation Ltd.	
G–BBVJ	Beech B24R Sierra	Ian Anthony (Sales) Ltd.	
G–BBVM	Beech A.100 King Air	Laura Ashley Ltd.	
G–BBVO	Isaacs Fury II	D. B. Wilson	

Notes	Reg.	Type	Owner or Operator
	G–BBVP	Westland-Bell 47G-3B1	Freemans of Bewdley (Aviation) Ltd.
	G–BBVR	PA-31-350 Navajo Chieftain	Thurston Aviation Ltd.
	G–BBWM	PA-E23 Aztec 250E	Torvale Air Service Ltd.
	G–BBWN	D.H.C.1 Chipmunk 22	G. R. Tait & J. Ripley
	G–BBWZ	AA-1B Trainer	Warner Aviation
	G–BBXA	Beech B.55 Baron	La Villette Court Apartments Ltd.
	G–BBXB	Cessna FRA.150L	Cambridge Technical Developments Ltd.
	G–BBXD	Bensen B.8M-VW	J. Remington
	G–BBXE	PA-E23 Aztec 250	Sovereign Chemical Industries Ltd.
	G–BBXG	PA-34-200-11 Seneca	Lambourn Air
	G–BBXH	Cessna FR.172F	Vale Hire & Contracting Co. Ltd.
	G–BBXI	HPR-7 Herald 203	Air UK
	G–BBXK	PA-34-200 Seneca	Rebate Cleaning Organisation Ltd.
	G–BBXL	Cessna E310Q	Kearns-Barker Associates Ltd.
	G–BBXO	Enstrom F-28A	C.S.E. Aviation Ltd.
	G–BBXR	PA-31-350 Navajo Chieftain	Rolls-Royce Ltd.
	G–BBXS	Piper J-3C-65 Cub	N. Simpson (G–ALMA)
	G–BBXT	Cessna F.172M	Solent Flight Centre
	G–BBXU	Beech B24R Sierra	E. E. Taylor Air Services Ltd.
	G–BBXV	PA-28-151 Warrior	Beechwood Marine Ltd.
	G–BBXW	PA-28-151 Warrior	Hypozone Ltd.
	G–BBXX	PA-31-350 Navajo Chieftain	Rolls Royce Ltd.
	G–BBXY	Bellanca 7GC BC Citabria	Cambridge University Gliding Trust Ltd.
	G–BBXZ	Evans VP-1	N. M. Bloom
	G–BBYB	PA-18 Super Cub 95	E. C. Lee & ptnrs.
	G–BBYE	Cessna 195	Wilrow Products Ltd.
	G–BBYH	Cessna 182P	Sanderson (Forklifts) Ltd.
	G–BBYK	PA-E23 Aztec 250	Bristol Air Taxis Ltd.
	G–BBYL	Cameron O-77 balloon	Buckingham Balloon Club *Jammy*
	G–BBYM	H.P.137 Jetstream 200	The Morgan Crucible Co. Ltd. (G–AYWR)
	G–BBYN	PA-30 Twin Comanche 160	Express Aviation Services Ltd.
	G–BBYO	BN-2A Mk. III Trislander	Aurigny Air Services (G–BBWR)
	G–BBYP	PA-28 Cherokee 140	Channel Aviation Ltd.
	G–BBYR	Cameron O-65 balloon	D. M. Winder *Phoenix*
	G–BBYS	Cessna 182P Skylane	Forth Engineering Ltd.
	G–BBYT	Cessna 414	Ralli Bros. & Coney Ltd.
	G–BBYU	Cameron O-56 balloon	C. J. T. Davey *Chieftain*
	G–BBYW	PA-28 Cherokee 140	C.S.E. (Aircraft Services) Ltd.
	G–BBZB	PA-31-350 Navajo Chieftain	Fairflight Ltd.
	G–BBZE	PA-28 Cherokee 140	C.S.E. (Aircraft Services) Ltd.
	G–BBZF	PA-28 Cherokee 140	C.S.E. (Aircraft Services) Ltd.
	G–BBZH	PA-28R-200 Cherokee Arrow	George House (Holdings) Ltd.
	G–BBZI	PA-31-310 Navajo	Armstrong Aviation Ltd.
	G–BBZJ	PA-34-200 Seneca	C.S.E. (Aircraft Services) Ltd.
	G–BBZL	Westland-Bell 47G-3B1	Dollar Air Services Ltd.
	G–BBZN	Fuji FA.200-180	Jasper Carrott Ltd.
	G–BBZO	Fuji FA.200-160	D. Cowan & D. G. Foreman
	G–BBZP	PA-31-350 Navajo Chieftain	Fairflight Ltd.
	G–BBZR	Enstrom F-28A	C.S.E. Aviation Ltd.
	G–BBZS	Enstrom F-28A	Spooner Aviation (Enstrom Helicopters) Ltd.
	G–BBZV	PA-28R-200-2 Cherokee Arrow	Unicol Engineering
	G–BCAB	M.S.894A Rallye Minerva 220	P. V. & Mrs. E. M. Gilliar
	G–BCAC	M.S.894A Rallye Minerva 220	Velcourt (East) Ltd. & Hereford Farmplan Ltd.
	G–BCAD	M.S.894A Rallye Minerva 220	RMA Aviation
	G–BCAH	D.H.C.1 Chipmunk 22 (WG316)	D. E. Hughes & J. Powell
	G–BCAN	Thunder Ax7-77 balloon	J. P. Roche & ptnrs. *Billboard*
	G–BCAP	Cameron O-56 balloon	P. A. Salmons *Honey Child*
	G–BCAR	Thunder Ax7-77 balloon	T. J. Woodbridge *Marie Antoinette*
	G–BCAS	Thunder Ax7-77 balloon	Scout Association *Drifter*
	G–BCAT	PA-31-310 Turbo Navajo	C-Line Group (Aircraft Holdings) Ltd.
	G–BCAY	R. Commander 685	R. B. Tyler (Plant) Ltd.
	G–BCAZ	PA-12 Super Cruiser	Dr. C. J. Pennycuick
	G–BCBA	Boeing 720-023	Monarch Airlines Ltd.
	G–BCBB	Boeing 720-023	Monarch Airlines Ltd.

Reg.	Type	Owner or Operator	Notes
G-BCBD	Bede BD-5	Brockmore-Bede Aircraft (UK) Ltd.	
G-BCBG	PA-E23 Aztec 250	D. V. Beadon	
G-BCBH	Fairchild 24R-46A Argus III	A. Bowes	
G-BCBI	Cessna 402B	Systime Ltd.	
G-BCBJ	PA-25 Pawnee 235	Westwick Distributors Ltd.	
G-BCBK	Cessna 421B	Lloyds & Scottish Development Ltd.	
G-BCBL	Fairchild 24R-46A Argus III (HB751)	Battle of Britain Prints International Ltd.	
G-BCBM	PA-E23 Aztec 250	Edinburgh Flying Services Ltd.	
G-BCBN	Scheibe SF.27M-Ci	D. B. James	
G-BCBO	PA-31P Navajo	K. Gomm (Holdings) Ltd.	
G-BCBP	M.S.880B Rallye 100S Sport	B. W. J. Pring	
G-BCBR	AJEP/Wittman Tailwind	G. McMillan	
G-BCBV	PA-25 Pawnee 235	Farmair Ltd.	
G-BCBW	Cessna 182P	Brailsford Aviation Ltd.	
G-BCBX	Cessna F.150L	Ulster Aviation Ltd.	
G-BCBY	Cessna F.150L	Esquire Ltd.	
G-BCBZ	Cessna 337C	H. Tempest Ltd.	
G-BCCA	Cessna A.188B Agwagon	Mindacre Ltd.	
G-BCCB	Robin HR.200/100	Goodwood Terrena Ltd.	
G-BCCC	Cessna F.150L	Crescent Leasing Ltd.	
G-BCCD	Cessna F.172M	Clyde Developments Ltd.	
G-BCCE	PA-E23 Aztec 250	Barry Sheene Racing Ltd.	
G-BCCF	PA-28 Cherokee 180	J. T. Friskney Ltd.	
G-BCCG	Thunder Ax7-65 balloon	D. W. Claridge *Emitape*	
G-BCCH	Thunder Ax6-56A balloon	Wrangler, Bluebell Apparel Ltd. *Wrangler*	
G-BCCJ	AA-5 Traveler	T. Needham & ptnrs.	
G-BCCK	AA-5 Traveler	R. E. Carpenter	
G-BCCL	HS.125 Srs. 600B	McAlpine Aviation Ltd.	
G-BCCN	Robin HR.200/100	June Hall Services Ltd.	
G-BCCP	Robin HR.200/100	Northampton School of Flying Co. Ltd.	
G-BCCR	CP.301A Emeraude	A. B. Fisher	
G-BCCX	D.H.C.1 Chipmunk 22	RAFGSA	
G-BCCY	Robin HR.200/100	Goodwood Terrena Ltd.	
G-BCDA	Boeing 727–46	Dan-Air Services Ltd.	
G-BCDB	PA-34-200-11 Seneca	C.S.E. (Aircraft Services) Ltd.	
G-BCDC	PA-18 Super Cub 95	Aly Aviation Ltd.	
G-BCDH	MBB Bo 105D	North-Scottish Helicopters Ltd.	
G-BCDI	Cessna T.310Q-11	Leygatecourt Ltd.	
G-BCDJ	PA-28 Cherokee 140	Andrewsfield Flying Club	
G-BCDK	Partenavia P.68B	H. E. Peacock	
G-BCDL	Cameron O-42 balloon	D. P. & Mrs. B. O. Turner *Chums*	
G-BCDN	F.27 Friendship Mk. 200	Air UK	
G-BCDO	F.27 Friendship Mk. 200	Air UK	
G-BCDR	Thunder Ax7-77 balloon	W. G. Johnston & ptnrs. *Obelix*	
G-BCDS	PA-E23 Aztec 250	Hamilton Aviation Ltd.	
G-BCDU	Cessna 414	Kasugo Ltd.	
G-BCDV	Western O-65 balloon	T. A. Adams *Quo Vadis*	
G-BCDW	Hughes 269C	Point-to-Point Helicopters Ltd.	
G-BCDY	Cessna FRA. 150L	P. R. Pykett	
G-BCDZ	H.S.748 Srs. 2A	British Aerospace	
G-BCEA	Sikorsky S-61N Mk. II	British Airways Helicopters Ltd.	
G-BCEB	Sikorsky S-61N Mk. II	British Airways Helicopters Ltd.	
G-BCEC	Cessna F.172M	David Scott-Moncrieff & Son Ltd.	
G-BCEE	AA-5 Traveler	Stewarts Nurseries Ltd.	
G-BCEF	AA-5 Traveler	Echo Fox Ltd.	
G-BCEO	AA-5 Traveler	Amex Automobile Exports	
G-BCEP	AA-5 Traveler	Nottingham Industrial Cleaners Ltd.	
G-BCER	GY-201 Minicab	T. Worrall	
G-BCEU	Cameron O-42 balloon	Entertainment Services Ltd. *Harlequin*	
G-BCEX	PA-E23 Aztec 250	Weekes Bros. (Welling) Ltd.	
G-BCEY	D.H.C.1 Chipmunk 22	D. O. Wallis	
G-BCEZ	Cameron O-84 balloon	Anglia Aeronauts Ascension Association *Stars and Bars*	
G-BCFB	Cameron O-77 balloon	J. J. Harris & P. Pryce-Jones *Teutonic Turkey*	
G-BCFC	Cameron O-65 balloon	Miss D. I. Gray-Fisk & P. L. Gallagher *Candy Twist*	
G-BCFD	West balloon	E. D. & J. N. West *Hellfire*	
G-BCFE	Odyssey 4000 balloon	R. M. Glover *Odyssey*	

Notes	Reg.	Type	Owner or Operator
	G–BCFF	Fuji FA-200-160	C.S.E. (Aircraft Services) Ltd.
	G–BCFN	Cameron O-65 balloon	T. Aston & ptnrs. *Fireball*
	G–BCFO	PA-18-150 Super Cub	Bristol & Gloucestershire Gliding Club (Pty) Ltd.
	G–BCFP	Enstrom F-28A	Northfield Carpets Ltd.
	G–BCFR	Cessna FRA.150L	W. H. & J. Rogers Group Ltd.
	G–BCFV	Saab 91D Safir	J. Townhill
	G–BCFW	Saab 91D Safir	D. R. Williams
	G–BCFY	Luton L.A.4A Minor	G. F. M. Garner
	G–BCFZ	Cameron A-500 balloon	C. J. T. Davy & ptnrs. *Le Geant*
	G–BCGB	Bensen B8	A. Melody
	G–BCGC	D.H.C.1 Chipmunk 22	Culdrose Gliding Club
	G–BCGD	PA-28R-200 Cherokee Arrow	Joseph D. J. Jones & Co. Ltd.
	G–BCGE	Bell 212	British Executive Air Services Ltd.
	G–BCGG	AJEP/Wittman Tailwind	C. G. Gray
	G–BCGH	Nord NC.854S	R. A. Yates
	G–BCGI	PA-28 Cherokee 140	C.S.E.(Aircraft Services) Ltd.
	G–BCGJ	PA-28 Cherokee 140	C.S.E.(Aircraft Services) Ltd.
	G–BCGK	PA-28 Cherokee 140	C.S.E.(Aircraft Services) Ltd.
	G–BCGL	Jodel D.112	G. Hughes & R. E. Jones
	G–BCGM	Jodel D.120	I. E. Fisher
	G–BCGN	PA-28 Cherokee 140	C.S.E.(Aircraft Services) Ltd.
	G–BCGP	Gazebo Ax6 balloon	A. R. Wilson *Aries*
	G–BCGS	PA-28R-200 Cherokee Arrow	K. W. Norris
	G–BCGT	PA-28 Cherokee 140	C. Hewitt & ptnrs.
	G–BCGU	HP.137 Jetstream 200	Terravia (Aircraft) Ltd. (G–AXRI)
	G–BCGW	Jodel D.11	G. H. & M. D. Chittenden
	G–BCGX	Bede BD-5A/B	R. Hodgson
	G–BCHJ	Cessna F.172H	Toon Ghose Aviation Ltd.
	G–BCHK	Cessna F.172H	Hornet Aviation Ltd.
	G–BCHL	D.H.C.1 Chipmunk 22A	B. D. Bate & R. Rutherford
	G–BCHM	SA.341G Gazelle	Westland Helicopters Ltd.
	G–BCHP	CP.1310-C3 Super Emeraude	B. V. Mayo
	G–BCHT	Schleicher ASK.16	K. M. Barton & ptnrs.
	G–BCHU	Dawes VP-2	G. Dawes
	G–BCHV	D.H.C.1 Chipmunk 22	N. F. Charles
	G–BCHX	SF.23A Sperling	T. Davies & P. H. Chamberlain
	G–BCID	PA-34-200-11 Seneca	Comanche Air Services Ltd.
	G–BCIE	PA-28-151 Warrior	Channel Aviation Ltd.
	G–BCIF	PA-28 Cherokee 140	Cotgrave Service Station Ltd.
	G–BCIH	D.H.C.1 Chipmunk 22	J. M. Hosey & R. A. Schofield
	G–BCII	Cessna 500 Citation	IDS Aircraft Ltd.
	G–BCIJ	AA-5 Traveler	C. Wilkinson & ptnrs.
	G–BCIK	AA-5 Traveler	W. Nutt & Son Ltd.
	G–BCIL	AA-1B Trainer	D. C. Cumberland
	G–BCIM	AA-1B Trainer	Special Alloys (Northern) Ltd.
	G–BCIN	Thunder Ax7-77 balloon	Isambard Kingdom Brunel Balloon Group *Isambard Kingdom Brunel*
	G–BCIO	PA-39 Twin Comanche C/R	C.S.E. (Aircraft Services) Ltd.
	G–BCIP	PA-39 Twin Comanche C/R	C.S.E. (Aircraft Services) Ltd.
	G–BCIR	PA-28-151 Warrior	W. M. F. Taylor
	G–BCIS	Beagle B.206 Srs. 1	R. Nathan
	G–BCIT	CIT/AI Srs. 1	Cranfield Institute of Technology
	G–BCIU	Beagle B.206 Srs. 1	Bradburn & Wedge (Garages) Ltd.
	G–BCIV	Beagle B.206 Srs. 1	Bradburn & Wedge Ltd.
	G–BCIW	D.H.C.1 Chipmunk 22 (WZ868)	B. J. Clack
	G–BCJF	Beagle B.206 Srs. 1	A. A. Mattacks
	G–BCJH	Mooney M.20F	V. J. Cousin
	G–BCJI	PA-31-350 Navajo Chieftain	Hay & Co. (Lerwick) Ltd.
	G–BCJL	PA-28 Cherokee 140	C.S.E. (Aircraft Services) Ltd.
	G–BCJM	PA-28 Cherokee 140	MB Aviation Ltd.
	G–BCJN	PA-28 Cherokee 140	C. P. B. Horsley
	G–BCJO	PA-28R-200 Cherokee Arrow	Express Aviation Services Ltd.
	G–BCJP	PA-28 Cherokee 140	C. C. Smith
	G–BCJR	PA-E23 Aztec 250	Anglo-Normandy Aviation Ltd.
	G–BCJS	PA-E23 Aztec 250	Woodgate Aviation Ltd.
	G–BCJU	HS.125 Srs. 600B	McAlpine Aviation Ltd.
	G–BCKB	R. Thrush Commander 600	G. W. Peek
	G–BCKC	R. Thrush Commander 600	ADS (Aerial) Ltd.
	G–BCKD	PA-28R-200-2 Cherokee Arrow	A. B. Plant (Aviation) Ltd.
	G–BCKF	SA.102.5 Cavalier	K. Fairness
	G–BCKJ	PA-23 Aztec 250	Renco (Aviation) Ltd.

Reg.	Type	Owner or Operator	Notes
G–BCKM	Cessna 500 Citation	IDS Fanjets Ltd.	
G–BCKN	D.H.C.1A Chipmunk 22	RAFGSA	
G–BCKO	PA-E23 Aztec 250	NPD Aviation Ltd.	
G–BCKP	Luton L.A.5A Major	J. R. Callow	
G–BCKS	Fuji FA.200-180	J. T. Hicks	
G–BCKT	Fuji FA.200-180	Littlewick Green Service Station Ltd.	
G–BCKU	Cessna FRA.150L	Airwork Services Ltd.	
G–BCKV	Cessna FRA.150L	Airwork Services Ltd.	
G–BCLA	Sikorsky S-61N	Bristow Helicopters Ltd.	
G–BCLC	Sikorsky S-61N	Bristow Helicopters Ltd. *Craigievar*	
G–BCLD	Sikorsky S-61N	Bristow Helicopters Ltd. *Slains*	
G–BCLI	AA-5 Traveler	C. G. Whittaker (Commercials) Ltd.	
G–BCLJ	AA-5 Traveler	M. A. Lenihan	
G–BCLK	R. 500S Shrike Commander	International Systems & Applications Ltd.	
G–BCLL	PA-28 Cherokee 180	Agricultural & Industrial Services (Yorkshire) Ltd.	
G–BCLM	GY-201 Minicab	J. H. Hill	
G–BCLS	Cessna 170B	C. W. Proffitt-White	
G–BCLU	Jodel D.117	Cotswold Flying Group	
G–BCLV	Bede BD-5A	R. A. Gardiner	
G–BCLW	AA-1B Trainer	Whisky Flying Club Ltd.	
G–BCMB	Partenavia P.68B	DK Aviation Ltd.	
G–BCMC	Bell 212	Bristow Helicopters Ltd.	
G–BCMD	PA-19 Super Cub 95	R. W. M. & B. N. C. Mogg	
G–BCMF	Levi Go-Plane RL-6 Srs. 1	R. Levi	
G–BCMJ	SA.102.5 Cavalier	M. Johnson	
G–BCML	SIPA S.903	Re-registered G–BGME	
G–BCMR	Robin HR.100/285	Home & Overseas International Oil Co. Ltd.	
G–BCMT	Isaacs Fury II	M. H. Turner	
G–BCNC	GY.201 Minicab	J. R. Wraight	
G–BCNO	BN-2A Mk. III-1 Trislander	Aurigny Air Services	
G–BCNP	Cameron O-77 balloon	B. A. Nathan & J. E. Quilliam	
G–BCNR	Thunder Ax7-77A balloon	S. J. Milliken & ptnrs. *Howdy*	
G–BCNS	Cameron O-84 balloon	Cathay Pacific Airways Ltd. *Cathay I*	
G–BCNT	Partenavia P.68B	DK Aviation Ltd.	
G–BCNX	Piper J-3C-65 Cub	N. J. R. Empson & ptnrs.	
G–BCNZ	Fuji FA.200-160	Condor Flying Club	
G–BCOA	Cameron O-65 balloon	E. E. J. & Mrs. E. Sutton *Bunny*	
G–BCOB	Piper J-3C-65 Cub	R. W. & Mrs. J. W. Marjoram	
G–BCOD	PA-31-350 Navajo Chieftain	C. F. Doyle Ltd.	
G–BCOE	HS.748 Srs. 2A	British Airways	
G–BCOF	HS.748 Srs. 2A	British Airways	
G–BCOG	Jodel D.112	B. A. Bower & ptnrs.	
G–BCOH	Avro 683 Lancaster 10	Strathallan Aircraft Collection	
G–BCOI	D.H.C.1 Chipmunk 22	D. S. McGregor & A. T. Letham	
G–BCOJ	Cameron O-56 balloon	Anglian Industrial Gases Ltd. *Vulcan*	
G–BCOL	Cessna F.172M	J. Birkett	
G–BCOM	Piper J-3C-65 Cub	R. H. B. Cox & ptnrs.	
G–BCOO	D.H.C.1 Chipmunk 22	T. G. Fielding & M. S. Morton	
G–BCOP	PA-28R-200 Cherokee Arrow	J. B. Lowe	
G–BCOR	SOCATA Rallye 100ST	H. J. Pincombe	
G–BCOS	D.H.C.1 Chipmunk 22	Chitair Ltd.	
G–BCOT	Enstrom F-28C	Spooner Aviation Ltd.	
G–BCOU	D.H.C.1 Chipmunk 22	P. J. Loweth	
G–BCOV	Hawker Sea Fury TT.20 (VX302)	D. W. Arnold	
G–BCOX	Bede BD-5A	H. J. Cox	
G–BCOY	D.H.C.1 Chipmunk 22	Coventry Gliding Club Ltd.	
G–BCPA	SA.315B Alouette II	Dollar Air Services Ltd.	
G–BCPB	Howes radio-controlled model free balloon	R. B. & Mrs. C. Howes *Posbee I*	
G–BCPD	GY.201 Minicab	A. H. K. Denniss	
G–BCPE	Cessna F.150M	Channel Islands Aero Holdings Ltd.	
G–BCPF	PA-23 Aztec 250	Keenair Services Ltd.	
G–BCPG	PA-28R-200 Cherokee Arrow	Zipline Engineering Co.	
G–BCPH	Piper J-3C-65 Cub	J. A. Chandler	
G–BCPI	PA-18 Super Cub 135	Kent Gliding Club Ltd.	
G–BCPJ	Piper J-3C-65 Cub	M. C. Barraclough & T. M. Storey	
G–BCPK	Cessna F.172M	Skegness Air Taxi Services Ltd.	
G–BCPL	AA-5 Traveler	T. Neville & Co. (Engineering) Ltd.	
G–BCPN	AA-5 Traveler	B.W. Agricultural Equipments Ltd.	

Notes	Reg.	Type	Owner or Operator
	G–BCPO	Partenavia P.68B	Astra Aviation Ltd.
	G–BCPU	D.H.C.I Chipmunk T.10	P. Waller
	G–BCPX	SZEP HFC.125	A Szep
	G–BCPZ	R.500S Shrike Commander	Cleanacres Ltd.
	G–BCRA	Cessna F.150M	Three Counties Aero Club
	G–BCRB	Cessna F.172M	F. D. & M. D. Forbes
	G–BCRC	D.31 Turbulent	R. A. G. Chapman
	G–BCRE	Cameron O-77 balloon	A. R. Langston & R. J. Fuller *Snapdragon*
	G–BCRF	PA-23 Aztec 250	Linco Poultry Machinery Co. Ltd.
	G–BCRG	MBB Bo 105D	GKN Group Services Ltd.
	G–BCRH	Alaparma Baldo B.75	A. L. Scadding
	G–BCRI	Cameron O-65 balloon	V. J. Thorne *Joseph*
	G–BCRJ	Taylor JT.I Monoplane	Canary Flying Group
	G–BCRK	SA.102.5 Cavalier	R. Y. Kendal
	G–BCRL	PA-28-151 Warrior	Nordec (Builders) Ltd.
	G–BCRN	Cessna FRA.150L	Airwork Services Ltd.
	G–BCRP	PA-E23 Aztec 250	LEC Refrigeration Ltd.
	G–BCRR	AA-5B Tiger	Travelworth Ltd.
	G–BCRT	Cessna F.150M	Executive Air
	G–BCRX	D.H.C.I Chipmunk 22	J. P. V. Hunt & P. G. H. Tony
	G–BCSA	D.H.C.I Chipmunk 22	RAFGSA
	G–BCSB	D.H.C.I Chipmunk 22	RAFGSA
	G–BCSL	D.H.C.I Chipmunk 22	Jalawain Ltd.
	G–BCSM	Bellanca 8GC BC Scout	Hendon Air Services Ltd.
	G–BCSR	Bellanca 7ECA Citabria	Hendon Air Services Ltd.
	G–BCSS	M.S.892A Rallye Commodore 150	R. E. Ward & ptnrs.
	G–BCST	M.S.893A Rallye Commodore 180	P. J. Wilcox
	G–BCSV	Cessna 421B	Northair Aviation Ltd.
	G–BCSX	Thunder Ax7-77 balloon	Whoopski Balloon Group *Whoopski*
	G–BCSY	Taylor JT.2 Titch	P. L. Mines
	G–BCSZ	PA-28R-200 Cherokee Arrow	Marlow Chemical Co. Ltd.
	G–BCTA	PA-28-151 Warrior	G. Capes
	G–BCTB	Cameron O-65 balloon	G. W. Reader *Selebian*
	G–BCTD	Scheibe SF.25B Falke	E. R. Boyle
	G–BCTF	PA-28-151 Warrior	Seasons & Systems Ltd.
	G–BCTH	PA-28 Cherokee 140	John V. White Ltd.
	G–BCTI	Schleicher ASK.16	R. J. Steward
	G–BCTJ	Cessna 310Q	Airwork Services Ltd.
	G–BCTK	Cessna FR.172J	Shobdon Aviation Co. Ltd.
	G–BCTR	Taylor JT.2 Titch	T. Reagan
	G–BCTT	Evans VP-2	B. J. Boughton
	G–BCTU	Cessna FRA.150M	Mona Aviation Ltd.
	G–BCTV	Cessna F.150M	Hartmann Ltd.
	G–BCTW	Cessna F.150M	Ulster Aviation Ltd.
	G–BCUB	Piper J-3C-65 Cub	M. J. Mead
	G–BCUF	Cessna F.172M	G. H. C. B. Kirke
	G–BCUH	Cessna F.150M	Gordon King (Aviation) Ltd.
	G–BCUI	Cessna F.172M	Hillhouse Estates Ltd.
	G–BCUJ	Cessna F.150M	S.B. International Aviation Ltd.
	G–BCUK	Cessna F.172M	Q. Quitmann Ltd.
	G–BCUL	SOCATA Rallye 100ST	Gwynedd Air Services Ltd.
	G–BCUM	Stinson HW-75	P. J. Sellar
	G–BCUW	Cessna F.177RG Cardinal	A. J. Wheeler
	G–BCUY	Cessna FRA.150M	Vectair Aviation Ltd.
	G–BCUZ	Beech A200 Super King Air	Allied Breweries (UK) Ltd.
	G–BCVA	Cameron O-65 balloon	J. C. Bass & ptnrs. *Crepe Suzette*
	G–BCVB	PA-17 Vagabond	A. T. Nowak & B. Holland
	G–BCVC	SOCATA Rallye 100ST	Brettshire Ltd.
	G–BCVE	Evans VP-2	G. A. Bentley
	G–BCVF	Practavia Pilot Sprite	G. B. Castle
	G–BCVG	Cessna FRA.150L	Westair Flying Services Ltd.
	G–BCVH	Cessna FRA.150L	W. Lancs Aero Club Ltd.
	G–BCVI	Cessna FR.172J	B. D. & Mrs. W. Phillips
	G–BCVJ	Cessna F.172M	J. Males
	G–BCVV	PA-28-151 Warrior	Channel Islands Aero Holdings Ltd.
	G–BCVW	GY-80 Horizon 180	Horizon Flyers Ltd.
	G–BCVY	PA-34-200T Seneca	Cega Aviation Ltd.
	G–BCWA	BAC One-Eleven 518	Dan-Air Services Ltd. (G–AXMK)
	G–BCWB	Cessna 182P	Gaymark Trustees Ltd.
	G–BCWE	HPR-7 Herald 206	Air UK

Reg.	Type	Owner or Operator	Notes
G-BCWF	Scottish Aviation Twin Pioneer	Flight One Ltd. (G-APRS)	
G-BÇWH	Practavia Pilot Sprite	K. B. Parkinson & R. Tasker	
G-BCWI	Bensen B.8M	A. Barley	
G-BCWK	Alpavia Fournier RF-3	D. I. Nickolls & ptnrs.	
G-BCWL	Westland Lysander III (V9281)	D. W. Arnold	
G-BCWM	AB-206B JetRanger 2	N. R. Foster	
G-BCWU	BN-2A-21 Islander	Marchwiel Plant & Engineering Co. Ltd.	
G-BCWW	HP.137 Jetstream 200	The Distillers Co. Ltd. (G-AXUN)	
G-BCXB	SOCATA Rallye 100ST	Inchberry Ltd.	
G-BCXD	Pitts S-2A Special	Atalona Ltd.	
G-BCXE	Robin DR.400 '2+2'	Headcorn Flying School Ltd.	
G-BCXH	PA-28 Cherokee 140F	Mann Aviation Sales Ltd.	
G-BCXI	G-164A Ag-Cat	W. P. Miller	
G-BCXJ	Piper J-3C-65 Cub	W. F. Stockdale	
G-BCXN	D.H.C.1 Chipmunk 22 (WP800)	J. D. Scott	
G-BCXO	MBB Bo 105D	Management Aviation Ltd.	
G-BCXR	BAC One-Eleven 517	Monarch Airlines Ltd. (G-BCCV)	
G-BCXS	R. Turbo Commander 690A	Ferranti Ltd.	
G-BCXT	Cessna F.150M	Peterborough Aero Club Ltd.	
G-BCXV	BN-2A Mk. III-1 Trislander	Aurigny Air Services	
G-BCXZ	Cameron O-56 balloon	Olives from Spain Ltd. *Olives from Spain*	
G-BCYC	BN-2A Mk. III-2 Trislander	Loganair Ltd.	
G-BCYE	D.H.C.1 Chipmunk 22	L. T. Mersh	
G-BCYF	Dassault Mystère 20	Falcon Jet Centre Ltd.	
G-BCYH	Privateer Mk.2 Motor Glider	D. B. Limbert	
G-BCYI	Schleicher ASK-16	D. J. Pearce & ptnrs.	
G-BCYJ	D.H.C.1 Chipmunk 22 (WG307)	G. A. Warner	
G-BCYK	Avro CF.100 Mk. 4 Canuck (18393)	—	
G-BCYL	D.H.C.1 Chipmunk 22	P. C. Henry	
G-BCYM	D.H.C.1 Chipmunk 22	C. R. R. Eagleton	
G-BCYP	AB-206B JetRanger 2	Alan Mann Helicopters Ltd.	
G-BCYR	Cessna F.172M	Laxtonbridge Ltd.	
G-BCYS	Piper J-3C-65 Cub	R. P. M. Dawson	
G-BCYY	Westland-Bell 47G-3B1	Minster Helicopters Ltd.	
G-BCYZ	Westland-Bell 47G-3B1	Fife & Kinross Motor Auctions Ltd.	
G-BCZF	PA-28 Cherokee 180	Proto Kote Ltd.	
G-BCZG	HPR-7 Herald 202	Air UK	
G-BCZH	D.C.H.1 Chipmunk 22	The London Gliding Club (Pty.) Ltd.	
G-BCZI	Thunder Ax7-77 balloon	Motor Tyres & Accessories *Motorway for Tyres*	
G-BCZL	Westland-Bell 47G-3B1	British Executive Air Services Ltd.	
G-BCZM	Cessna F.172M	Air Transport Ltd.	
G-BCZN	Cessna F.150M	T. M. G. Hanley	
G-BCZO	Cameron O-77 balloon	W. O. T. Holmes *Leo*	
G-BDAB	SA.102.5 Cavalier	A. H. Brown	
G-BDAC	Cameron O-77 balloon	Cameron Balloons Ltd. *Chocolate Ripple*	
G-BDAD	Taylor JT.1 Monoplane	J. F. Bakewell	
G-BDAE	BAC One-Eleven 518	Dan-Air Services Ltd. (G-AXMI)	
G-BDAG	Taylor JT.1 Monoplane	R. S. Basinger	
G-BDAH	Evans VP-1	R. W. Lowe	
G-BDAI	Cessna FRA.150M	Seair Design Services Ltd.	
G-BDAJ	R. Commander 112A	Macro 4 Ltd.	
G-BDAK	R. Commander 112A	Chalke Valley Constructing Ltd.	
G-BDAL	R. 500S Shrike Commander	Micro Consultants Ltd.	
G-BDAM	AT-16 Harvard IIB (FE992)	R. H. Reeves	
G-BDAO	SIPA 91	A. L. Rose	
G-BDAP	AJEP Tailwind	J. Whiting	
G-BDAR	Evans VP-1	S. C. Foggin & M. T. Dugmore	
G-BDAS	BAC One-Eleven 518	Dan-Air Services Ltd. (G-AXMH)	
G-BDAT	BAC One-Eleven 518	Dan-Air Services Ltd. (G-AYOR)	
G-BDAU	Cessna FRA.150M	Airwork Services Ltd.	
G-BDAV	PA-23 Aztec 250	Sound Powered Telephone Co. Ltd.	
G-BDAW	Enstrom F-28A	RHM Investments Ltd.	
G-BDAX	PA-E23 Aztec 250	A. M. Collins	
G-BDAY	Thunder Ax5-42A balloon	Thunder Balloons Ltd. *Tom's Balloon*	
G-BDBA	Thunder Ax7-77 balloon	P. Porati *Briaza*	
G-BDBB	Cessna F.150M	Northamptonshire School of Flying Ltd	

Notes	Reg.	Type	Owner or Operator
	G–BDBD	Wittman W.8 Tailwind	H. Best-Devereux
	G–BDBE	Thunder Ax7-77A balloon	St. Ivel Yoghurt *Prize Guy*
	G–BDBF	FRED Srs. 2	W. T. Morrell
	G–BDBH	Bellanca 7GCBC Citabria	Mastergear Co. Ltd.
	G–BDBI	Cameron O-77 balloon	Royston Cooper Design Consultants *Sunny Money*
	G–BDBJ	Cessna 182P	Cosworth Engineering Ltd.
	G–BDBK	Cameron O-56 balloon	J. G. Green *Square Baby*
	G–BDBL	D.H.C.1 Chipmunk 22	Wycombe Gliding School Syndicate
	G–BDBM	Cameron O-56 balloon	A. H. K. Olphin *Scorpio*
	G–BDBP	D.H.C.1 Chipmunk 22	Sherwood Flying Club Ltd.
	G–BDBR	AB-206B JetRanger 2	P.L.M. Helicopters Ltd.
	G–BDBS	Short SD3-30	Short Bros. Ltd.
	G–BDBU	Cessna F.150M	Channel Islands Aero Holdings Ltd.
	G–BDBV	Jodel D.11A	J. P. de Hevingham
	G–BDBW	Heintz Zenith 100 A18	D. B. Winstanley
	G–BDBX	Evans VP-1	Montgomeryshire Ultra-Light Flying Club
	G–BDBZ	WS.55 Whirlwind Srs. 2	Autair Ltd.
	G–BDCA	SOCATA Rallye 150ST	B. W. J. Pring & ptnrs.
	G–BDCB	D.H.C.1 Chipmunk 22	D. R. Hodgson
	G–BDCC	D.H.C.1 Chipmunk 22	Coventry Gliding Club Ltd.
	G–BDCD	Piper J-3C-65 Cub (480308)	Suzanne C. Brooks
	G–BDCE	Cessna F.172H	H. A. Baillie
	G–BDCI	CP.301A Emeraude	H. A. R. Horesign
	G–BDCK	AA-5 Traveler	Sequoia Air Ltd.
	G–BDCM	Cessna F.177RG	Park Plant Ltd.
	G–BDCO	B.121 Pup 1	Dr. R. D. H. & Mrs. K. N. Maxwell
	G–BDCP	AB-206B JetRanger 2	Ben Turner & Sons (Helicopters) Ltd.
	G–BDCS	Cessna 421B	Marchwiel Aviation Ltd.
	G–BDCT	PA-25 Pawnee 235C	Sprayfields (Scothern) Ltd.
	G–BDCU	Cameron O-77 balloon	York Hot-Air Balloon Club Ltd.
	G–BDDA	Sikorsky S-61N Mk. II	British Airways Helicopters Ltd.
	G–BDDD	D.H.C.1 Chipmunk 22	RAE Aero Club Ltd.
	G–BDDE	Douglas DC-8-54F	British Cargo Airlines Ltd.
	G–BDDF	Jodel D.120	C. A. Parker
	G–BDDG	Jodel D.112	A. C. Watts & I. C. Greig
	G–BDDH	F.27 Friendship Mk. 200	Air UK
	G–BDDJ	Luton LA.4A Minor	D. D. Johnson
	G–BDDS	PA-25 Pawnee 260	Farmair Ltd.
	G–BDDT	PA-25 Pawnee 260	Farm Aviation Services Ltd.
	G–BDDV	BN-2A-8 Islander	Loganair Ltd.
	G–BDDX	Whittaker MW.2B Excalibur	Trago Mills Ltd.
	G–BDDZ	CP.301A Emeraude	M. Jones
	G–BDEA	Boeing 707-338C	British Caledonian Airways *Loch Thom*
	G–BDEB	SOCATA Rallye 100ST	W. G. Dunn & ptnrs.
	G–BDEC	SOCATA Rallye 100ST	Cambridge Chemical Co. Ltd.
	G–BDED	SOCATA Rallye 100ST	J. Scott
	G–BDEF	PA-34-200T-11 Seneca	European Paper Sales Ltd.
	G–BDEH	Jodel D.120A	L. I. Lee
	G–BDEI	Jodel D.9 Bebe	P. J. Griggs
	G–BDEJ	R. Commander 112	F. K. Greensmith (Long Eaton) Ltd.
	G–BDEN	SIAI-Marchetti SF.260	Micro Consultants Ltd.
	G–BDER	Auster AOP.9 (WZ672)	F. & H. (Aircraft) Ltd.
	G–BDES	Sikorsky S-61N Mk. II	British Airways Helicopters Ltd.
	G–BDET	D.H.C.1 Chipmunk 22	R. C. Lock & ptnrs.
	G–BDEU	D. H. C.1 Chipmunk 22 (WP808)	A. Taylor
	G–BDEV	Taylor JT.1 Monoplane	P. J. Houston
	G–BDEW	Cessna FRA.150M	George House (Holdings) Ltd.
	G–BDEX	Cessna FRA.150M	George House (Holdings) Ltd.
	G–BDEY	Piper J-3C-65 Cub	W. J. & J. Morecraft
	G–BDEZ	Piper J-3C-65 Cub	P. Elliot
	G–BDFB	Currie Wot	D. F. Faulkner-Bryant
	G–BDFC	R. Commander 112A	Zaar International Cinema & TV Programmes Ltd.
	G–BDFE	HPR-7 Herald 206	British Air Ferries Ltd. *Rory Keegan*
	G–BDFF	Supermarine S.5 Replica (N220)	Leisure Sport Ltd.
	G–BDFG	Cameron O-65 balloon	N. A. Robertson *Golly II*
	G–BDFH	Auster AOP.9 (XR240)	F. & H. (Aircraft) Ltd.

Reg.	Type	Owner or Operator	Notes
G–BDFI	Cessna F.150M	Coventry Civil Aviation Ltd.	
G–BDFJ	Cessna F.150M	American Airspeed Inc. Ltd.	
G–BDFK	Cessna 414	Northair Aviation Ltd.	
G–BDFL	PA-28R-200-2 Cherokee Arrow	Tiger Airways Ltd.	
G–BDFM	Caudron C.270 Luciole	G. V. Gower	
G–BDFN	PA-31-350 Navajo Chieftain	Fairflight Charters Ltd.	
G–BDFO	Hiller UH-12E	Heliscot Ltd. & F. F. Chamberlain	
G–BDFP	Hughes 369HS	John E. Clarke & Co. (Bournemouth) Ltd.	
G–BDFR	Fuji FA.200-160	C.S.E. (Aircraft Services) Ltd.	
G–BDFS	Fuji FA.200-160	D. J. Carding	
G–BDFT	V.668 Varsity T.1 (WJ897)	D. W. Mickleburgh	
G–BDFU	Dragonfly MPA Mk. 1	R. J. A. Hardy & R. Churcher	
G–BDFW	R. Commander 112A	Ace Aviation	
G–BDFX	Auster 5	I. A. Haddon & M. J. Kirk	
G–BDFY	AA-5 Traveler	E. O. Liebert	
G–BDFZ	Cessna F.150M	Skyviews & General Ltd.	
G–BDGA	Bushby-Long Midget Mustang	J. R. Owen	
G–BDGB	GY-20 Minicab	D. G. Burden	
G–BDGG	BN-2A Mk. III-2 Trislander	Aurigny Air Services	
G–BDGH	Thunder Ax6-56 balloon	The London Balloon Club Ltd. *London Pride III*	
G–BDGJ	Cameron O-56 balloon	Cathay Pacific Airways Ltd. *Cathy II*	
G–BDGK	Beechcraft D.17S	Customline Ltd.	
G–BDGL	Cessna U.206 Super Skywagon	C. Wren	
G–BDGM	PA-28-151 Warrior	Channel Islands Aero Services Ltd.	
G–BDGN	AA-5B Tiger	Hamblin & Glover Oil Field (Services) Ltd.	
G–BDGO	Thunder Ax7-77 balloon	International Distillers & Vintners Ltd. *J. & B. Rare*	
G–BDGP	Cameron V-65 balloon	P. G. Dunnington	
G–BDGX	Scheibe SF.25E Super Falke	British Gliding Association Ltd.	
G–BDGY	PA-28 Cherokee 140	R. E. Woolridge	
G–BDHA	Douglas DC-8-54F	*For Sale*	
G–BDHB	Isaacs Fury II	D. H. Berry	
G–BDHC	D.H.C.6 Twin Otter 310	Chubb Security Services Ltd.	
G–BDHJ	Pazmany PL.1	H. James	
G–BDHK	Piper J-3C-65 Cub (329417)	H. Knight	
G–BDHL	PA-E23 Aztec 250E	Granpack Ltd.	
G–BDHM	SA.102.5 Cavalier	D. H. Mitchell	
G–BDHU	BN-2A-26 Islander	BN (Bembridge) Ltd.	
G–BDIB	Enstrom F-280 Shark	Warner Holidays Ltd.	
G–BDIC	D.H.C.1 Chipmunk 22	R. Merry	
G–BDID	D.H.C.1 Chipmunk 22	Coventry Gliding Club Ltd.	
G–BDIE	R. Commander 112A	T. Saveker Ltd.	
G–BDIF	D.H.106 Comet 4C	Dan-Air Services Ltd. (*withdrawn*)	
G–BDIG	Cessna 182P	Roger Clarke (Air Transport) Ltd.	
G–BDIH	Jodel D.117	J. Chisholm	
G–BDII	Sikorsky S-61N Mk. II	Bristow Helicopters Ltd. *Drum*	
G–BDIJ	Sikorsky S-61N Mk. II	Bristow Helicopters Ltd. *Crathes*	
G–BDIK	V.708 Viscount	Alidair Ltd.	
G–BDIM	D.H.C.1 Chipmunk 22	R. A. F. Dimblebee	
G–BDIT	D.H.106 Comet 4C	Dan-Air Services Ltd. (*withdrawn*)	
G–BDIU	D.H.106 Comet 4C	Dan-Air Services Ltd. (*withdrawn*)	
G–BDIV	D.H.106 Comet 4C	Dan-Air Services Ltd. (*withdrawn*)	
G–BDIW	D.H.106 Comet 4C	Dan-Air Services Ltd. (*withdrawn*)	
G–BDIX	D.H.106 Comet 4C	Dan-Air Services Ltd. (*withdrawn*)	
G–BDIY	Luton L.A.4A Minor	M. A. Musselwhite	
G–BDJB	Taylor JT.1 Monoplane	J. F. Barber	
G–BDJC	AJEP Tailwind	A. Whiting	
G–BDJD	Jodel D.112	J. V. Derrick	
G–BDJE	H.S.125 Srs. 600B	McAlpine Aviation Ltd.	
G–BDJF	Bensen B.8MV	R. P. White	
G–BDJG	Luton L.A.4A Minor	D. J. Gaskin	
G–BDJN	Robin HR.200/100	Northampton School of Flying Co. Ltd.	
G–BDJP	Piper J-3C-65 Cub	Mrs. J. M. Pothecary	
G–BDJR	Nord NC.858	V. Panteli & J. Spanton	
G–BDJY	BN-2A-27 Islander	J. Cunningham	
G–BDKB	SOCATA Rallye 150ST	R. G. Moffatt	
G–BDKC	Cessna A185F	Bridge of Tilt Co. Ltd.	

Notes	Reg.	Type	Owner or Operator
	G–BDKD	Enstrom F-28A	Renco (Aviation) Ltd.
	G–BDKG	Beech 65-A80 Queen Air	I. R. Martin
	G–BDKH	CP.301A Emeraude	R. F. Bridge
	G–BDKI	Sikorsky S-61N Mk. II	British Airways Helicopters Ltd.
	G–BDKJ	SA.102.5 Cavalier	H. B. Yardley
	G–BDKK	Bede BD-5B	A. W. Odell
	G–BDKL	Hughes 369HM	A & B Cars (Distributors) Ltd.
	G–BDKM	SIPA 903	S. W. Markham
	G–BDKR	BN-2A Mk. III-2 Trislander	Loganair Ltd.
	G–BDKS	Pitts S-2A Special	Rothmans International Ltd.
	G–BDKT	Cameron O-84 balloon	A. R. M. Fraser & R. Fermor-Hesketh
	G–BDKU	Taylor JT.1 Monoplane	N. F. Whistler & ptnrs.
	G–BDKV	PA-28R-200-2 Cherokee Arrow	Patricia Swanson
	G–BDKW	R. Commander 112A	Link Systems Ltd.
	G–BDLM	Boeing 707-338C	British Caledonian Airways *Loch Fyne*
	G–BDLO	AA-5A Cheetah	T. R. Bamber
	G–BDLR	AA-5B Tiger	G. Morton-Smith
	G–BDLS	AA-1B Trainer	Warner Aviation
	G–BDLT	R. Commander 112A	Wintergrain Ltd.
	G–BDLV	Chilton DW.1A	R. E. Nerou
	G–BDLX	SA Twin Pioneer 3	Transportation Ltd.
	G–BDLY	SA.102.5 Cavalier	B. S. Reeve
	G–BDLZ	B.175 Britannia 253F	Air Faisel *Al Mubarak*
	G–BDMB	Robin HR.100/210	R. J. Hitchman & Son
	G–BDMC	MBB Bo 105D	Management Aviation Ltd.
	G–BDME	Robin DR.400/140B	Miss S. A. Pound
	G–BDMM	Jodel D.11	D. M. Metcalf
	G–BDMO	Thunder Ax7-77A balloon	Villiers Insurance Consultants *Flash Harry*
	G–BDMS	Piper J-3C-65 Cub	D. M. Squires
	G–BDMW	Jodel DR.100	J. T. Nixon
	G–BDNC	Taylor JT.1 Monoplane	N. J. Cole
	G–BDNE	Harker Hawk	D. Harker
	G–BDNF	Bensen B.8M	W. F. O'Brien
	G–BDNG	Taylor JT.1 Monoplane	D. J. Phillips
	G–BDNO	Taylor JT.1 Monoplane	R. A. Bragger
	G–BDNP	BN-2A Islander	Jersey European Airways
	G–BDNR	Cessna FRA.150M	Solent Flight Centre
	G–BDNS	Saro Skeeter 12 (XM529)	M. Eastman & M. Ridley
	G–BDNT	Jodel D.92	J. S. Barker
	G–BDNU	Cessna F.172M	Skyline Aviation Ltd.
	G–BDNW	AA-1B Trainer	J. G. Hill
	G–BDNX	AA-1B Trainer	J. A. Nott
	G–BDNY	AA-1B Trainer	M. R. Langford
	G–BDNZ	Cameron O-77 balloon	N. R. Page *Winston Churchill*
	G–BDOA	H.S.125 Srs. 600B	McAlpine Aviation Ltd.
	G–BDOC	Sikorsky S-61N Mk. II	Bristow Helicopters Ltd. *Tolquhoun*
	G–BDOD	Cessna F.150M	Pegasus Aviation Ltd.
	G–BDOE	Cessna FR.172J	P.A.V.H. (International) Ltd.
	G–BDOF	Cameron O-56 balloon	New Holker Estates Co. *Fred Cavendish*
	G–BDOG	SA Bullfinch	British Aerospace
	G–BDOH	Hiller UH-12E	Bridge Helicopters Ltd.
	G–BDOI	Hiller UH-12E	Heli-Spray Ltd.
	G–BDOL	Piper J-3C-65 Cub	Allman & Son (Redhill)
	G–BDOM	BN-2A Mk. III-2 Trislander	Loganair Ltd.
	G–BDON	Thunder Ax7-77A balloon	York Hot-Air Balloon Club *Fred*
	G–BDOO	Thunder Ax7-77 balloon	Thunder Balloons Ltd.
	G–BDOR	Thunder Ax6-56A balloon	Outspan Sales Ltd. *Peeler*
	G–BDOS	BN-2A Mk. III-2 Trislander	Loganair Ltd.
	G–BDOU	Cessna FRA.150M	Cleansing Service (Southern Counties) Ltd.
	G–BDOW	Cessna FRA.150M	Hartmann Ltd.
	G–BDOY	Hughes 369HS	Cosworth Engineering Ltd.
	G–BDOZ	Sportavia-Fournier RF-5	B. A. Allsopp
	G–BDPA	PA-28-151 Warrior	Noblair Ltd.
	G–BDPB	Falconar F-11-3	A. E. Pritchard
	G–BDPC	Bede BD-5A	P. R. Cremer
	G–BDPF	Cessna F.172M	Westair Flying Services Ltd.
	G–BDPG	Cessna F.150M	Flightec Airlines Support Ltd.
	G–BDPH	Cessna F.172M	Astro Data Ltd.

Reg.	Type	Owner or Operator	Notes
G–BDPI	PA-25 Pawnee 235B	Farmair Ltd.	
G–BDPJ	PA-25 Pawnee 235B	Farmair Ltd.	
G–BDPK	Cameron O-56 balloon	Manpower Ltd. *Manpower*	
G–BDPL	Falconar F-11	A. J. Watson	
G–BDPR	BN-2A Islander	BN (Bembridge) Ltd.	
G–BDPV	Boeing 747-136	British Airways	
G–BDPZ	Boeing 747-148	British Airways	
G–BDRB	AA-5B Tiger	Northern Strip Mining Ltd.	
G–BDRC	V.724 Viscount	Guernsey Airlines Ltd. *Sarnia II*	
G–BDRD	Cessna FRA.150M	Airwork Services Ltd.	
G–BDRE	AA-1B Trainer	S. J. Hammond & ptnrs.	
G–BDRF	Taylor JT.1 Monoplane	R. A. Codling	
G–BDRI	PA-34-200T-11 Seneca	Aylesbury Mushrooms Ltd.	
G–BDRJ	D.H.C.1 Chipmunk 22 (WP857)	J. K. Avis	
G–BDRK	Cameron O-65 balloon	D. L. Smith *Smirk*	
G–BDRL	Stitts SA-3 Playboy	D. L. MacLean	
G–BDRY	Hiller UH-12E	G. & S. G. Neal (Helicopters) Ltd.	
G–BDSB	PA-28-181 Archer II	MTS Electronic Holdings Ltd.	
G–BDSC	Cessna F.150M	Lyndair (Aviation) Ltd.	
G–BDSD	Evans VP-1	J. E. Worthington	
G–BDSE	Cameron O-77 balloon	British Airways *Concorde*	
G–BDSF	Cameron O-56 balloon	J. R. Joiner *Itzuma*	
G–BDSG	Woody Pusher	D. S. Gingell	
G–BDSH	PA-28 Cherokee 140	Bamberhurst Ltd.	
G–BDSJ	Boeing 707-338C	British Caledonian Airways *Loch Tay*	
G–BDSK	Cameron O-65 balloon	Southern Balloon Group *Carousel II*	
G–BDSL	Cessna F.150M	Cleveland Flying School Ltd.	
G–BDSM	Slingsby/Kirby Cadet Mk. 3	D. W. Savage	
G–BDSN	Wassmer WA.52 Europa	Rollason Aircraft & Engines Ltd. (G–BADN)	
G–BDSO	Cameron O-31 balloon	Budget Rent-a-Car *Baby Budget*	
G–BDSP	Cessna U.206F Stationair	F. M. Usher-Smith	
G–BDSR	PA-25 Pawnee 235	Norfolk Aerial Spraying Ltd.	
G–BDSZ	BN-2A Islander	Fletcher & Stewart Ltd.	
G–BDTB	Evans VP-1	T. E. Boyes	
G–BDTL	Evans VP-1	A. K. Lang	
G–BDTT	Bede BD-5	The TT Group	
G–BDTU	Omega III gas balloon	Mrs. K. E. Turnbull *Omega III*	
G–BDTV	Mooney M.20F	J. A. Holgate Ltd.	
G–BDTW	Cassutt Racer	B. E. Smith & C. S. Thompson	
G–BDTX	Cessna F.150M	A. A. & R. N. Croxford	
G–BDUI	Cameron V-56 balloon	G. H. Dorrell *Pig Bucket*	
G–BDUJ	PA-31-310 Navajo	Vickers Shipbuilding Group Ltd.	
G–BDUK	R. Commander 685	Air Service Training Ltd.	
G–BDUL	Evans VP-1	C. Goodman	
G–BDUM	Cessna F.150M	Lattey Aviation Ltd.	
G–BDUN	PA-34-200T-11 Seneca	Coriolis Ltd.	
G–BDUO	Cessna F.150M	Clark Masts Ltd.	
G–BDUP	B.175 Britannia 253	Afrek Ltd.	
G–BDUR	B.175 Britannia 253	Afrek Ltd.	
G–BDUX	Slingsby T.31B motor glider	R. H. Hearn	
G–BDUY	Robin DR.400/160B	F. J. Franklin	
G–BDUZ	Cameron V-56 balloon	Balloon Stable Ltd. *Hot Lips*	
G–BDVA	PA-17 Vagabond	Mrs. H. S. & I. M. Callier	
G–BDVB	PA-15 (PA-17) Vagabond	B. P. Gardner	
G–BDVC	PA-17 Vagabond	A. T. Christian	
G–BDVG	Thunder Ax6-56A balloon	R. F. Pollard *Argonaut*	
G–BDVH	H.S.748 Srs. 2A	British Aerospace	
G–BDVI	D.H.82A Tiger Moth	K. D. J. Ecclestone	
G–BDVJ	Westland-Bell 47G-3B1	Hawkspare Ltd.	
G–BDVR	Cessna 180F	The Missionary Aviation Fellowship	
G–BDVS	F.27 Friendship 200	Air UK	
G–BDVT	F.27 Friendship 200	Air UK	
G–BDVU	Mooney M.20F	Freecomp Ltd.	
G–BDVW	BN-2A Islander	Loganair Ltd.	
G–BDWA	SOCATA Rallye 150ST	Air Touring Services Ltd.	
G–BDWB	SOCATA Rallye 150ST	M. Hazzeldine	
G–BDWE	Flaglor Scooter	D. W. Evernden	
G–BDWG	BN-2A Islander	Euroair Transport Ltd.	
G–BDWH	SOCATA Rallye 150ST	O. G. Owen	
G–BDWI	PA-34-200T-11 Seneca	T. W. Walker	
G–BDWJ	SE-5A Replica (F8010)	M. L. Beach	

Notes	Reg.	Type	Owner or Operator
	G–BDWK	Beech 95-B58 Baron	David Huggett Motor Factors Ltd.
	G–BDWL	PA-25 Pawnee 235	Summerhouse Farm Ltd.
	G–BDWM	Mustang replica	D. C. Bonsall
	G–BDWN	SA.318C Alouette II	AB Aviation Holdings Ltd.
	G–BDWO	Howes Ax6-6 balloon	R. B. & Mrs. C. Howes *Griffin*
	G–BDWP	PA-32R-300 Cherokee Lance	Starling (Sales Ideas) Ltd.
	G–BDWU	BN-2A Mk.III-2 Trislander	Anglo-Thai Corporation Ltd.
	G–BDWW	Cameron O-77 balloon	Dan-Air Hot-Air Balloon Group *Dan-Air*
	G–BDWX	Jodel D.120A	J. P. Lassey
	G–BDWY	PA-28 Cherokee 140	Cormack Aircraft Services Ltd.
	G–BDXA	Boeing 747-236B	British Airways
	G–BDXB	Boeing 747-236B	British Airways
	G–BDXC	Boeing 747-236B	British Airways
	G–BDXD	Boeing 747-236B	British Airways
	G–BDXE	Boeing 747-236B	British Airways
	G–BDXF	Boeing 747-236B	British Airways
	G–BDXG	Boeing 747-236B	British Airways
	G–BDXH	Boeing 747-236B	British Airways
	G–BDXI	Boeing 747-236B	British Airways
	G–BDXJ	Boeing 747-236B	British Airways
	G–BDXK	Boeing 747-236B	British Airways
	G–BDXL	Boeing 747-236B	British Airways
	G–BDXW	PA-28R-200 Cherokee Arrow	Camlet Helicopters Ltd.
	G–BDXX	Nord NC.858S	V. Pantelli
	G–BDXY	Auster AOP.9 (XR269)	Tyre & Tyne Transport Ltd.
	G–BDXZ	Pitts S-1S Special	P. Meeson
	G–BDYB	AA-5B Tiger	M. D. Joy
	G–BDYC	AA-1B Trainer	Cabair Ltd.
	G–BDYD	R. Commander 114	Glos-Air Ltd.
	G–BDYF	Cessna 421C	Mining Supplies Ltd.
	G–BDYG	Percival Provost T.1 (WV493)	Strathallan Aircraft Collection
	G–BDYH	Cameron V-56 balloon	Aeolus Balloon Company *Novacastrian*
	G–BDYL	Beech C23 Sundowner	J. C. & Mrs. V. Hale
	G–BDYM	Skysales S-31 balloon	Miss A. I. Smith & M. J. Moore *Cheeky Devil*
	G–BDYY	Hiller UH-12E	A. G. Norrie & Sons
	G–BDYZ	MBB Bo 105D	Management Aviation Ltd.
	G–BDZA	Scheibe SF.25E Super Falke	Norfolk Gliding Club Ltd.
	G–BDZB	Cameron S-31 balloon	Kenning Motor Group Ltd. *Kenning*
	G–BDZC	Cessna F.150M	Air Tows Ltd.
	G–BDZD	Cessna F.172M	Three Counties Aero Club Ltd.
	G–BDZF	G.164 Ag-Cat B	Miller Aerial Spraying Ltd.
	G–BDZS	Scheibe SF.25E Super Falke	A. D. Gubbay
	G–BDZU	Cessna 421C	Owledge Ltd.
	G–BDZV	HPR-7 Herald 214	British Air Ferries
	G–BDZW	PA-28 Cherokee 140	Fairoaks Flight Centre
	G–BDZX	PA-28-151 Warrior	Antlink Ltd.
	G–BDZY	Phoenix LA-4A Minor	P. J. Dalby
	G–BEAA	Taylor JT.1 Monoplane	R. C. Hobbs
	G–BEAB	Jodel DR.1051	Laxtonbridge Ltd.
	G–BEAC	PA-28 Cherokee 140	Eileen R. Purfield
	G–BEAD	Westland WG.13 Lynx	Westland Helicopters Ltd.
	G–BEAE	PA-25 Pawnee 235	Farm Aviation Services Ltd.
	G–BEAG	PA-34-200T-II Seneca	R. P. Yeoward
	G–BEAH	J/2 Arrow	W. J. & Mrs. M. D. Horler
	G–BEAK	L-1011-385 TriStar	British Airways
	G–BEAL	L-1011-385 TriStar	British Airways
	G–BEAM	L-1011-385 TriStar	British Airways
	G–BEAN	L-1011-385 TriStar	British Airways
	G–BEAO	L-1011-385 TriStar	British Airways
	G–BEAP	L-1011-385 TriStar	British Airways
	G–BEBA	H.S.748 Srs. 2	Dan-Air Services Ltd.
	G–BEBB	HPR-7 Herald 214	British Air Ferries
	G–BEBC	WS.55 Whirlwind Srs. 3	Milford Car Services Ltd.
	G–BEBE	AA-5A Cheetah	Special Alloys (Northern) Ltd.
	G–BEBF	Auster AOP.9	M. D. N. & Mrs. A. C. Fisher
	G–BEBG	WSK-PZL SDZ-45A Ogar	Anglo-Polish Sailplanes Ltd.
	G–BEBH	Cessna FR.172J	D. L. B. Upjohn
	G–BEBI	Cessna F.172M	Calder Equipment Ltd.
	G–BEBJ	PA-E23 Aztec 250	Burnthills Aviation Ltd.

Reg.	Type	Owner or Operator	Notes
G–BEBK	PA-31-300 Turbo Navajo	Dubilier Ltd.	
G–BEBL	Douglas DC-10-30	British Caledonian Airways *Sir Alexander Fleming—The Scottish Challenger*	
G–BEBM	Douglas DC-10-30	British Caledonian Airways *Robert Burns—The Scottish Bard*	
G–BEBN	Cessna 177B	J. Parry	
G–BEBO	Phoenix Currie Wot	C. Turner	
G–BEBR	GY-201 Minicab	A. S. Jones & D. R. Upton	
G–BEBS	Andreasson BA-4B	D. M. Fenton	
G–BEBT	Andreasson BA-4B	D. M. Fenton	
G–BEBU	R. Commander 112A	Judd Studios Ltd.	
G–BEBZ	PA-28-151 Warrior	Goodwood Terrena Ltd.	
G–BECA	SOCATA Rallye 100ST	P. D. Radcliffe	
G–BECB	SOCATA Rallye 100ST	Air Touring Services Ltd.	
G–BECC	SOCATA Rallye 150ST	Air Touring Services Ltd.	
G–BECD	SOCATA Rallye 150ST	J. B. Roberts	
G–BECE	Aerospace Developments Srs. B1 airship	Aerospace Developments (London) Ltd.	
G–BECF	Scheibe SF.25A Falke	D. A. Wilson & ptnrs.	
G–BECG	Boeing 737-204ADV	Britannia Airways Ltd. *Amy Johnson*	
G–BECH	Boeing 737-204ADV	Britannia Airways Ltd. *Viscount Montgomery of Alamein*	
G–BECJ	Partenavia P.68B	Ryburn Air Ltd.	
G–BECK	Cameron V-56 balloon	Miss D. M. Bragg *Noddy*	
G–BECL	C.A.S.A. C.352L	War Birds of Great Britain Ltd.	
G–BECM	Aerotec Pitts S-2A Special	Rothmans International Ltd.	
G–BECN	Piper J-3C-65 Cub	E. J. Cottle & Sons	
G–BECO	Beech A.36 Bonanza	Thorney Machinery Co. Ltd.	
G–BECP	PA-31-310 Turbo Navajo	Kenton Utilities & Developments Ltd.	
G–BECR	Cessna 182B	Midland Parachute Centre	
G–BECS	Thunder Ax6-56A balloon	Thunder Balloons Ltd. *Sine Nomine*	
G–BECT	C.A.S.A. 1.131 Jungmann	D. C. Flavell	
G–BECW	C.A.S.A. 1.131 Jungmann	N. C. Jensen	
G–BECX	C.A.S.A. 1.131 Jungmann	R. A. Seeley	
G–BECZ	CAARP CAP.10B	Aerobatic Associates Ltd.	
G–BEDA	C.A.S.A. 1.131 Jungmann	R. P. Lewis	
G–BEDB	Nord 1203 Norecrin	B. F. G. Lister	
G–BEDD	Jodel D.117A	B. R. Vickers	
G–BEDE	Bede BD-5A	Biggin Hill BD5 Syndicate	
G–BEDF	Boeing B-17G-105-VE (485784)	M. H. Campbell	
G–BEDG	R. Commander 112A	N. J. Orr	
G–BEDH	R. Commander 114	Ian Anthony (Sales) Ltd.	
G–BEDI	Sikorsky S-61N Mk. II	British Airways Helicopters Ltd.	
G–BEDJ	Piper J-3C-65 Cub	D. J. Elliott & A. E. Molton	
G–BEDK	Hiller UH-12E	T. C. Jay	
G–BEDL	Cessna T337D	R. M. Cox Ltd.	
G–BEDO	BN-2A Mk. III-2 Trislander	BN (Bembridge) Ltd.	
G–BEDR	BN-2A Mk. III-2 Trislander	Fairflight Ltd.	
G–BEDS	Thunder Ax7-77A balloon	Thunder Balloons Ltd.	
G–BEDU	Scheibe SF.23C Sperling	Doncaster & District Gliding Club Ltd.	
G–BEDV	V.668 Varsity T.1 (WJ945)	D. S. Selway & A. G. Battersby	
G–BEDZ	BN-2A Islander	Loganair Ltd.	
G–BEEA	SOCATA Rallye 235GT	Air Sinclair	
G–BEEE	Thunder Ax6-56A balloon	Thunder Balloons Ltd. *Avia*	
G–BEEF	Thunder Ax6-56A balloon	V. Trimble & P. D. Morgan *Beefeater*	
G–BEEG	BN-2A Islander	Loganair Ltd.	
G–BEEH	Cameron V-56 balloon	J. M. Langley *Kaleidoscope*	
G–BEEI	Cameron N-77 balloon	D. W. A. Legg *Master McGrath*	
G–BEEJ	Cameron O-77 balloon	DAL (Builders Merchants) Ltd. *Dal's Pal*	
G–BEEK	Enstrom F-280C Shark	Uniweld Ltd.	
G–BEEL	Enstrom F-280C Shark	Slea Aviation Ltd.	
G–BEEN	Cameron O-56 balloon	Swire Bottlers Ltd. *Coke*	
G–BEEP	Thunder Ax5-42 balloon	Mrs. B. C. Faithful *Also Kenneth*	
G–BEER	Isaacs Fury II	M. J. Clark	
G–BEEU	PA-28 Cherokee 140E	Berkshire Aviation Services Ltd.	
G–BEEV	PA-28 Cherokee 140E	Tayside Aviation Ltd.	
G–BEEW	Taylor JT.1 Monoplane	K. Wigglesworth	
G–BEFA	PA-28-151 Warrior	B. J. Whitemore	

Notes	Reg.	Type	Owner or Operator
	G–BEFC	AA-5B Tiger	T. Nevill & Co. (Engineers) Ltd.
	G–BEFD	Jodel D.112	R. N. Taylor
	G–BEFE	Cameron N-77 balloon	S. R. & I. S. Bridge *Atlasgas*
	G–BEFF	PA-28 Cherokee 140	Sherwood Flying Club Ltd.
	G–BEFH	—	—
	G–BEFM	BN-2A Islander	George Wimpey & Co. Ltd.
	G–BEFR	Fokker DR.1 Repilca (F1425)	Leisure Sport Ltd.
	G–BEFT	Cessna 421C	Lucas Industries Ltd.
	G–BEFU	Sturgeonair MJ.7 Mustang	W. E. Wilks
	G–BEFV	Evans VP-2	Yeadon Aeroplane Group
	G–BEFW	PA-39 Twin Comanche C/R	Warnell Motors Ltd. (G–AYZP)
	G–BEFX	Hiller UH-12E	Agricopters Ltd.
	G–BEFY	Hiller UH-12E	Heliscot Ltd.
	G–BEGA	Westland Bell 47G-3B1	Harold Poupart Ltd.
	G–BEGG	Scheibe SF.25E Super Falke	RAFGSA
	G–BEGV	PA-23 Aztec 250F	Stoitwell Ltd.
	G–BEGX	BN-2A Mk. III-2 Trislander	BN (Bembridge) Ltd.
	G–BEGZ	Boeing 727-193	Dan-Air Services Ltd.
	G–BEHC	BN-2A Mk. III-2 Trislander	BN (Bembridge) Ltd.
	G–BEHD	BN-2A Mk. III-2 Trislander	BN (Bembridge) Ltd.
	G–BEHE	BN-2A Mk. III-2 Trislander	BN (Bembridge) Ltd.
	G–BEHF	BN-2A Mk. III-2 Trislander	BN (Bembridge) Ltd.
	G–BEHG	AB-206B JetRanger 2	Willowbrook International Ltd.
	G–BEHH	PA-32R-300 Cherokee Lance	SMK Engineering Ltd.
	G–BEHJ	Evans VP-1	K. Heath
	G–BEHL	Agusta-Bell 47G-3B1	Dollar Air Services Ltd.
	G–BEHM	Taylor JT.1 Monoplane	H. McGovern
	G–BEHN	Westland-Bell 47G-3B1	Dollar Air Services Ltd.
	G–BEHR	Beech A200 Super King Air	Eagle Aircraft Services Ltd.
	G–BEHS	PA-25 Pawnee 260C	Farm Aviation Services Ltd.
	G–BEHT	PA-25 Pawnee 260C	Farm Aviation Services Ltd.
	G–BEHU	PA-34-200T-11 Seneca	Nottingham Self-Fly & Aircraft Hire
	G–BEHV	Cessna F.172N	Surgeoners
	G–BEHW	Cessna F.150M	J. P. Hudson & S. R. Taylor
	G–BEHX	Evans VP-2	G. S. Adams
	G–BEHY	PA-28-181 Archer II	Aerogulf Services Co. Ltd.
	G–BEIA	Cessna FRA.150M	Airwork Services Ltd.
	G–BEIB	Cessna F.172N	R. L. Orsborn & Son Ltd.
	G–BEIC	Sikorsky S-61N	British Airways Helicopters Ltd.
	G–BEID	Sikorsky S-61N	British Airways Helicopters Ltd.
	G–BEIE	Evans VP-2	F. G. Morris
	G–BEIF	Cameron O-65 balloon	R. N. Sprackling *Windfall*
	G–BEIG	Cessna F.150M	Gordon King (Aviation) Ltd.
	G–BEIH	PA-25 Pawnee 235D	Farmair Ltd.
	G–BEII	PA-25 Pawnee 235D	Miller Aerial Spraying Ltd.
	G–BEIJ	G.164B Ag-Cat	Miller Aircraft Hire Ltd.
	G–BEIK	Beech A.36 Bonanza	Scot-Stock Ltd.
	G–BEIL	SOCATA Rallye 150T	D. H. Tonkin
	G–BEIP	PA-28-181 Archer II	S. Webb
	G–BEIS	Evans VP-1	D. J. Park
	G–BEIT	G.164B Ag-Cat	Miller Aerial Spraying Ltd.
	G–BEIZ	Cessna 500 Citation	Air Commuter Ltd.
	G–BEJA	Thunder Ax6-56A balloon	P. A. Hutchins *Jackson*
	G–BEJB	Thunder Ax6-56A balloon	International Distillers & Vintners Ltd. *Baby J. & B.*
	G–BEJD	H.S.748 Srs. 1	Dan-Air Services Ltd.
	G–BEJE	H.S.748 Srs. 1	Dan-Air Services Ltd.
	G–BEJK	Cameron S-31 balloon	Esso Petroleum Ltd.
	G–BEJL	Sikorsky S-61N	British Airways Helicopters Ltd.
	G–BEJM	BAC One-Eleven 423	Ford Motor Co. Ltd.
	G–BEJN	R. Turbo Commander 690A	Leopard Investments Ltd.
	G–BEJO	Saffery S.250 free balloon	Cupro Sapphire Ltd. *Firefly*
	G–BEJP	D.H.C.-6 Twin Otter 310	Loganair Ltd.
	G–BEJT	PA-23 Aztec 250F	Aviation Beauport (Finance) Ltd.
	G–BEJV	PA-34-200T-II Seneca	C.S.E. Aviation Ltd.
	G–BEJW	BAC One-Eleven 423	Ford Motor Co. Ltd.
	G–BEJX	Partenavia P.68B	DK Aviation Ltd.
	G–BEJY	Hughes 369D	Auto-Alloys Foundries Ltd.
	G–BEJZ	Aerostar 601PE	Mann Aviation Sales Ltd.
	G–BEKA	BAC One-Eleven 520	Dan-Air Services Ltd.
	G–BEKB	PA-23 Aztec 250	Alidair Ltd.
	G–BEKC	H.S.748 Srs. 1	Dan-Air Services Ltd.
	G–BEKD	H.S.748 Srs. 1	Dan-Air Services Ltd.

Reg.	Type	Owner or Operator	Notes
G–BEKE	H.S.748 Srs. I	Dan-Air Services Ltd.	
G–BEKG	H.S.748 Srs. I	Dan-Air Services Ltd.	
G–BEKH	AB-206B JetRanger 2	Barratt Developments Ltd.	
G–BEKK	Scheibe SF.25E Super Falke	Polish Flying Club Ltd.	
G–BEKL	Bede BD-4	Brockmore-Bede Aircraft (UK) Ltd.	
G–BEKM	Evans VP-I	G. J. McDill	
G–BEKN	Cessna FRA.150M	Balgin Ltd.	
G–BEKO	Cessna F.182Q	C. J. Leonard	
G–BEKR	Rand KR-2	K. B. Raven	
G–BEKS	PA-25 Pawnee 235D	Farmair Ltd.	
G–BEKT	PA-25 Pawnee 235D	Peter Charles (Air Farmers) Ltd.	
G–BELK	BN-2A Islander	Brown & Root (UK) Ltd.	
G–BELN	BN-2A Islander	BN (Bembridge) Ltd.	
G–BELO	Douglas DC-10-10	Laker Airways *Southern Belle*	
G–BELP	PA-28-151 Warrior	Channel Aviation Ltd.	
G–BELR	PA-28 Cherokee 140	D. McSorley	
G–BELS	D.H.C.-6 Twin Otter 310	Loganair Ltd.	
G–BELT	Cessna F.150J	Yorkshire Light Aircraft Ltd. (G–AWUV)	
G–BELV	Cessna 404 Titan	Executive Express Ltd.	
G–BELX	Cameron V-56 balloon	Cameron Balloons Ltd. *Topsy*	
G–BEMA	Cessna 310R II	Air Charter & Travel Ltd.	
G–BEMB	Cessna F.172M	Northair Aviation Ltd.	
G–BEMC	Cessna F.172M	Aberdeen Aero Club	
G–BEMD	Beech 95-B55 Baron	Vaux (Aviation) Ltd.	
G–BEMF	Taylor JT.I Monoplane	D. Magill	
G–BEMI	Thunder Ax6-56A balloon	Thunder Balloons Ltd.	
G–BEMM	Slingsby T.31B Motor Cadet	M. N. Martin	
G–BEMT	Bede BD-5G	G. Smith	
G–BEMU	Thunder Ax5-42 balloon	P. J. Langford	
G–BEMV	PA-28 Cherokee 140	J. Traynor	
G–BEMW	PA-28-181 Archer II	J. Traynor	
G–BEMX	Cessna 404 Titan	Owledge Ltd.	
G–BEMY	Cessna FRA.150M	Te-Elf Ltd.	
G–BENC	Cessna 402B	Medburn Air Services Ltd.	
G–BEND	Cameron V-56 balloon	Dante Balloon Group *Le Billet*	
G–BENE	Cessna 402B	Greenline Refrigerated Transport Ltd.	
G–BENF	Cessna T.210L	Loughton Aviation Ltd.	
G–BENG	Westland Bell 47G-3B1	GSM Helicopters Ltd.	
G–BENH	Phoenix LA.5A Major	C. D. Macartney	
G–BENI	Bell 47G-4A	R. J. A. Brown	
G–BENJ	R. Commander 112B	F. T. Arnold	
G–BENK	Cessna F.172M	Capeston Asbestos Ltd.	
G–BENL	PA-25 Pawnee 235D	Farm Aviation Services	
G–BENM	PA-31-325 Navajo	Gull Air Ltd.	
G–BENN	Cameron V-56 balloon	English Rose Kitchens Ltd.	
G–BENO	Enstrom F-280C Shark	Spooner Aviation Ltd.	
G–BENP	D.31A Turbulent	N. H. Ponsford	
G–BENS	Saffery S.330 balloon	D. Whitlock *Hot Plastic*	
G–BENT	Cameron N-77 balloon	N. Tasker	
G–BEOC	BN-2A Islander	Alderney Air Ferries Ltd.	
G–BEOD	Cessna 180	Farm Supply (Thirsk) Ltd.	
G–BEOE	Cessna FRA.150M	Shirlstar Container Transport Ltd.	
G–BEOF	Cessna F.150M	Aircraft Mart Ltd.	
G–BEOG	Cessna F.182Q	Balgone Aviation	
G–BEOH	PA-28R-201T Arrow III	Flockvale Ltd.	
G–BEOI	PA-18-150 Super Cub	S. Down Gliding Club Ltd.	
G–BEOK	Cessna F.150M	Gordon King Aviation Ltd.	
G–BEON	Sikorsky S-61N Mk. II	British Airways Helicopters Ltd.	
G–BEOO	Sikorsky S-61N Mk. II	British Airways Helicopters Ltd.	
G–BEOT	PA-25 Pawnee 235D	C.S.E. Aviation Ltd.	
G–BEOV	PA-28 Cherokee 140	Aerogulf Services Co. Ltd.	
G–BEOW	PA-28 Cherokee 140	Aerogulf Services Co. Ltd.	
G–BEOX	*L-414 Hudson IIIA (A16-199)	Strathallan Aircraft Collection	
G–BEOY	Cessna FRA.150L	Wycombe Air Centre Ltd.	
G–BEOZ	A.W.650 Argosy 101	Air Bridge Carriers Ltd.	
G–BEPB	Pereira Osprey II	J. J. & A. J. C. Zwetsloot	
G–BEPC	SNCAN SV-4C	M. Harbron	
G–BEPD	SA.102.5 Cavalier	P. & Mrs. E. A. Donaldson	
G–BEPE	SC.5 Belfast	HeavyLift Cargo Airlines Ltd. (G–ASKE)	
G–BEPF	SNCAN SV-4A	M. Stelfox & R. G. Harrington	

Notes	Reg.	Type	Owner or Operator
	G–BEPI	BN-2A Mk. III-2 Trislander	BN (Bembridge) Ltd.
	G–BEPJ	BN-2A Mk. III-2 Trislander	BN (Bembridge) Ltd.
	G–BEPK	BN-2A Mk. III-2 Trislander	BN (Bembridge) Ltd.
	G–BEPO	Cameron N-77 balloon	Sungas Ltd. *Sungas*
	G–BEPP	AB-206B JetRanger 2	T. W. Walker Ltd.
	G–BEPS	SC.5 Belfast	HeavyLift Cargo Airlines Ltd.
	G–BEPV	Fokker S.11-1 Instructor	Strathallan Aircraft Collection
	G–BEPY	R. Commander 112B	Astrapalm Ltd.
	G–BEPZ	Cameron D-96 hot-air airship	IAZ International (UK) Ltd.
	G–BERA	SOCATA Rallye 150ST	I. M. White
	G–BERB	SOCATA Rallye 150ST	Air Touring Services Ltd.
	G–BERC	SOCATA Rallye 150ST	Velcourt (East) Ltd.
	G–BERD	Thunder Ax6-56A balloon	Thunder Balloons Ltd. *Goldfinger*
	G–BERE	Thunder Ax6-56A balloon	Fisons Ltd. *Fisons*
	G–BERG	SA 330J Puma	Bristow Helicopters Ltd.
	G–BERH	SA 330J Puma	Bristow Helicopters Ltd.
	G–BERI	R. Commander 114	Rockville Motors Ltd.
	G–BERJ	Bell 47G-4A	Hon. G. M. H. Wills
	G–BERL	AA-5B Tiger	Scotia Safari Ltd.
	G–BERM	AA-5A Cheetah	John Farbon & Co. Ltd.
	G–BERN	Saffery S-330 balloon	B. Martin *Beeze*
	G–BERR	Thunder Ax7-77A balloon	M. Kendrick *Woolworths*
	G–BERT	Cameron V-56 balloon	Southern Balloon Group *Bert*
	G–BERW	R. Commander 114	Allison (Contractors) Ltd.
	G–BERY	AA-1B Trainer	North West Flying Services Ltd.
	G–BESA	Beech 95-58 Baron	Oldham Aviation
	G–BESE	BN-2A Islander	BN (Bembridge) Ltd.
	G–BESG	BN-2A Islander	BN (Bembridge) Ltd.
	G–BESJ	BN-2A Islander	BN (Bembridge) Ltd.
	G–BESL	BN-2A Islander	BN (Bembridge) Ltd.
	G–BESO	BN-2A Islander	Jersey European Airways
	G–BESP	BN-2A Islander	Rothmans International Ltd.
	G–BESR	BN-2A Islander	T. E. Oxley
	G–BESS	Hughes 369D	Micro Consultants Ltd.
	G–BESW	BN-2A Islander	Alderney Air Ferries Ltd.
	G–BESX	BN-2A Islander	Fairoaks Aviation Services Ltd.
	G–BETA	Rollason B.2A Beta	J. L. Kinch
	G–BETC	Cameron V-56 balloon	P. G. Dunnington *Etcetera*
	G–BETD	Robin HR.200/100	D. W. Jackson
	G–BETE	Rollason B.2 Beta	T. M. Jones
	G–BETF	Cameron 'Champion' balloon	Balloon Stable Ltd. *Champion*
	G–BETG	Cessna 180K Skywagon	E. H. S. Warner
	G–BETH	Thunder Ax6-56A balloon	Debenhams Ltd. *Debenhams I*
	G–BETI	Pitts S-1D Special	E. B. Bray
	G–BETJ	Douglas DC-8-33	Transmeridian Air Cargo Ltd.
	G–BETL	PA-25 Pawnee 235D	Lincs Aerial Spraying Co.
	G–BETM	PA-25 Pawnee 235D	Skegness Air Taxi Services Ltd.
	G–BETO	MS.885 Super Rallye	A. Somerville
	G–BETP	Cameron O-65 balloon	J. R. Rix & Sons Ltd.
	G–BETR	Cessna A.188B AgTruck	Voygate Ltd.
	G–BETS	Cessna A.188B AgTruck	Mindacre Ltd.
	G–BETT	PA-34-200 II Seneca	M.S.G. Aviation Ltd.
	G–BETU	Piper J-3C-65 Cub	M. J. Curran & R. G. Fitton
	G–BETV	HS.125 Srs. 600B	Tenneco Aviation Ltd.
	G–BETW	Rand KR-2	T. A. Wiffen
	G–BEUA	PA-18-150 Super Cub	London Gliding Club (Pty.) Ltd.
	G–BEUC	PA-28-181 Archer II	P. C. T. Warner
	G–BEUD	Robin HR.100/285R	Home & Overseas International Oil Co. Ltd.
	G–BEUE	Bell 47G-2	Bristow Helicopters Ltd.
	G–BEUF	Bell 47G-2	Bristow Helicopters Ltd.
	G–BEUG	Bell 47G-2	Bristow Helicopters Ltd.
	G–BEUH	Bell 47G-2	Bristow Helicopters Ltd.
	G–BEUI	Piper J-3C-65 Cub	B. P. Gardner
	G–BEUK	Fuji FA-200-160	C.S.E. Aviation Ltd.
	G–BEUL	Beech 95-58 Baron	Basic Metal Co. Ltd.
	G–BEUM	Taylor JT.1 Monoplane	M. T. Taylor
	G–BEUN	Cassutt Racer 111m	M. S. Crossley
	G–BEUP	Robin DR.400/180	Steelfields Ltd.
	G–BEUS	SNCAN SV-4C	G. V. Gower
	G–BEUU	PA-19 Super Cub 95	E. D. Burke & M. R. Smith
	G–BEUV	Thunder Ax6-56A balloon	Thunder Balloons Ltd. *Debenhams II*
	G–BEUW	AA-5A Cheetah	Mission Control Music Centre Ltd.

Reg.	Type	Owner or Operator	Notes
G–BEUX	Cessna F.172N	Light Planes (Lancashire) Ltd.	
G–BEUY	Cameron N-31 balloon	Southern Balloon Group	
G–BEUZ	Beech A200 Super King Air	Anglian Double Glazing Ltd.	
G–BEVA	SOCATA Rallye 150ST	J. G. Wilson	
G–BEVB	SOCATA Rallye 150ST	Wallis & Son Ltd.	
G–BEVC	SOCATA Rallye 150ST	B. W. Walpole	
G–BEVE	Thunder Ax7-77A	Thunder Balloons Ltd.	
G–BEVG	PA-34-200T II Seneca	C. G. Strasser	
G–BEVH	Holland D.700 balloon	D. I. Holland *Sally*	
G–BEVI	Thunder Ax7-77A balloon	The Painted Clouds Balloon Co. Ltd.	
G–BEVJ	Cessna U-206F Stationair	Twinguard Leasing Ltd.	
G–BEVK	PA-31-350 Navajo Chieftain	Vickers Ltd. (G–BCSE)	
G–BEVL	Cessna 421C	Vickers Ltd.	
G–BEVM	BN-2A Islander	BN (Bembridge) Ltd.	
G–BEVN	Boeing 707-321C	Dan-Air Services Ltd.	
G–BEVO	Sportavia-Pützer RF-5	D. Lister & R. F. Bradshaw	
G–BEVP	Evans VP-2	I. S. Walsh	
G–BEVR	BN-2A Mk. III-2 Trislander	BN (Bembridge) Ltd.	
G–BEVS	Taylor JT.1 Monoplane	D. Hunter	
G–BEVT	BN-2A Mk. III-2 Trislander	BN (Bembridge) Ltd.	
G–BEVV	BN-2A Mk. III-2 Trislander	BN (Bembridge) Ltd.	
G–BEVW	SOCATA Rallye 150ST	W. D. Hall	
G–BEVY	BN-2A Mk. III-2 Trislander	BN (Bembridge) Ltd.	
G–BEWJ	Westland-Bell 47G-3B-1	Trent Helicopters Ltd.	
G–BEWL	Sikorsky S-61N Mk. II	British Airways Helicopters Ltd.	
G–BEWM	Sikorsky S-61N Mk. II	British Airways Helicopters Ltd.	
G–BEWN	D.H.82A Tiger Moth	H. D. Labouchere	
G–BEWO	Zlin 326 Trener Master	H. D. Labouchere	
G–BEWP	Cessna F.150M	Pegasus Aviation Ltd.	
G–BEWR	Cessna F.172N	Mersey Flying Club Ltd.	
G–BEWS	Cameron 56 'Lamp Bulb' hot-air balloon	Osram (GEC) Ltd. *Osram*	
G–BEWW	HS.125 Srs. 600B	McAlpine Aviation Ltd. (G–AZUF)	
G–BEWX	PA-28R-201 Arrow III	Gidney & Kirby Holdings Ltd.	
G–BEWY	Bell 206B JetRanger II	Bristow Helicopters Ltd.	
G–BEWZ	PA-32-300 Cherokee Six C	B. G. Havenhand	
G–BEXK	PA-25 Pawnee 235D	Howard Avis (Aviation) Ltd.	
G–BEXL	PA-25 Pawnee 235D	Plains Aerial Spraying Ltd.	
G–BEXN	AA-1C Lynx	Scotia Safari Ltd.	
G–BEXO	PA-23 Apache 160	B. Burton	
G–BEXR	Mudry/CAARP CAP-10B	R. P. Lewis	
G–BEXS	Cessna F.150M	Hartmann Ltd.	
G–BEXV	Cameron O-56 balloon	Revlon (Hong Kong) Ltd. *Charlie*	
G–BEXW	PA-28-181 Archer II	Dickson Bros. Ltd.	
G–BEXX	Cameron V-56 balloon	A. Tyler & ptnrs. *Rupert of Rutland*	
G–BEXY	PA-28 Cherokee 140	D. H. L. Wigan	
G–BEXZ	Cameron N-56 balloon	M. J. Moroney *Valor*	
G–BEYA	Enstrom F-280C Shark	Guy Morton & Sons Ltd.	
G–BEYB	Fairey Flycatcher (replica) (S1287)	John S. Fairey	
G–BEYD	HPR-7 Herald 401	British Air Ferries	
G–BEYE	HPR-7 Herald 401	British Air Ferries *Rupert Keegan*	
G–BEYF	HPR-7 Herald 401	British Air Ferries	
G–BEYG	HPR-7 Herald 401	British Air Ferries *Jeremy Keegan*	
G–BEYH	HPR-7 Herald 401	British Air Ferries	
G–BEYJ	HPR-7 Herald 401	British Air Ferries	
G–BEYK	HPR-7 Herald 401	Air UK	
G–BEYL	PA-28 Cherokee 180	B. G. & G. Airlines Ltd.	
G–BEYM	Cessna F.150M	Anglian Flight Training Ltd.	
G–BEYN	Evans VP-2	C. D. Denham	
G–BEYO	PA-28 Cherokee 140	J. Traynor	
G–BEYP	Fuji FA-200-180AO	A. C. Pritchard	
G–BEYR	Enstrom F-280C Shark	W. Shyvers Ltd.	
G–BEYS	PA-28-181 Archer II	Scotia Safari Ltd.	
G–BEYU	Fuji FA-200-160	C.S.E. Aviation Ltd.	
G–BEYV	Cessna T.210L	Valley Motors	
G–BEYW	Taylor JT.1 Monoplane	R. J. Smythe	
G–BEYX	PA-31P Navajo	Air Commuter Ltd.	
G–BEYY	PA-31-310 Turbo Navajo	Trans Europe Air Charters Ltd.	
G–BEYZ	Jodel DR.1051/MI	M. J. McCarthy & S. Aarons	
G–BEZA	Zlin 226T Trener	L. Bezak	
G–BEZB	HPR-7 Herald 209	Express Air Services Ltd.	
G–BEZC	AA-5 Traveler	Newbranch Ltd.	

Notes	Reg.	Type	Owner or Operator
	G–BEZE	Rutan VariEze	M. F. Sharples & ptnrs.
	G–BEZF	AA-5 Traveler	W. Wheatley
	G–BEZG	AA-5 Traveler	R. J. Stevens
	G–BEZH	AA-5 Traveler	R. J. Herbert
	G–BEZI	AA-5 Traveler	Capital Aviation Sales (UK) Ltd.
	G–BEZJ	MBB Bo 105D	Management Aviation Ltd.
	G–BEZK	Cessna F.172H	R. J. Lock & ptnrs.
	G–BEZL	PA-31-310 Navajo	Barratt Developments (Northern) Ltd.
	G–BEZM	Cessna F.182Q	Northair Aviation Ltd.
	G–BEZO	Cessna F.172M	Rogers Aviation Ltd.
	G–BEZP	PA-32-300D Cherokee Six	Advanced Marketing Management Ltd.
	G–BEZR	Cessna F.172M	Kirmington Aviation Ltd.
	G–BEZS	Cessna FR.172J	Rogers Aviation Sales Ltd.
	G–BEZU	PA-31-350 Navajo Chieftain	Fairflight Charters Ltd.
	G–BEZV	Cessna F.172M	J. E. Bundy
	G–BEZW	Practavia Pilot Sprite	T. S. Wilkins & ptnrs.
	G–BEZY	Rutan VariEze	R. J. Jones
	G–BEZZ	Jodel D.112	A. J. Stevens & ptnrs.
	G–BFAA	GY-80 Horizon 160	Dustalong Ltd.
	G–BFAB	Cameron N-56 balloon	Phonogram Ltd. *Phonogram*
	G–BFAC	Cessna F.177 RG	Brendan Butler (London) Ltd.
	G–BFAD	PA-28-161 Warrior II	J. G. Hogg & Broad Oak Chartering Ltd.
	G–BFAF	Aeronca 7BCM	D. C. W. Harper
	G–BFAH	Phoenix Currie Wot	T. G. Thomson
	G–BFAI	R. Commander 114	Biograft Ltd.
	G–BFAK	MS.892A Rallye Commodore 150	R. Jennings & ptnrs.
	G–BFAM	PA-31P Navajo	International Air Charter Ltd.
	G–BFAN	H.S.125 Srs. 600F	British Aerospace (G–AZHS)
	G–BFAO	PA-20 Pacer 135	C. A. G. Schofield
	G–BFAP	SIAI-Marchetti S.205-20R	Miss M. A. Eccles
	G–BFAR	Cessna 500-1 Citation	Yewlands Engineering Co. Ltd.
	G–BFAS	Evans VP-1	A. I. Sutherland
	G–BFAV	Orion model free balloon	D. C. Boxall
	G–BFAW	D.H.C.-1 Chipmunk 22	R. H. Reeves
	G–BFAX	D.H.C.-1 Chipmunk 22	Leo Designs
	G–BFAY	Hughes 369D	Sloane Aviation Ltd.
	G–BFBA	Jodel DR.100A	Wasp Flying Group
	G–BFBB	PA-23 Aztec 250E	Worldair Sales
	G–BFBC	Taylor JT.1 Monoplane	A. Brooks
	G–BFBD	Partenavia P.68B	Pegasus Aviation Ltd.
	G–BFBE	Robin HR 200/100	Charles Major Ltd.
	G–BFBF	PA-28 Cherokee 140	P. H. de Havilland
	G–BFBG	PA-28-161 Warrior II	J. G. Hogg & Broad Oak Chartering Ltd.
	G–BFBH	PA-31-325 Turbo Navajo	Aviation Beauport (Finance) Ltd.
	G–BFBJ	PA-34-200T II Seneca	Shorgard Ltd.
	G–BFBK	Agusta-Bell 47G-3B1	Dollar Air Services Ltd.
	G–BFBM	Saffery S.330 balloon	B. Martin *Beeze II*
	G–BFBN	PA-25 Pawnee 235D	J. E. F. Aviation Ltd.
	G–BFBR	PA-28-161 Warrior II	J. Hogg & Broad Oak Chartering Ltd.
	G–BFBS	Boeing 707-351B	Laker Airways Ltd.
	G–BFBU	Partenavia P-68B	Clyde Developments Ltd.
	G–BFBV	Brügger Colibri M.B.2	J. D. Hutton & ptnrs.
	G–BFBW	PA-25 Pawnee 235D	Lawrence Agricultural Aviation Services Ltd.
	G–BFBX	PA-25 Pawnee 235D	Bowker Aircraft Services Ltd.
	G–BFBY	Piper J-3C-65 Cub	D. S. Quirk
	G–BFCA	L-1011-385 TriStar 500	British Airways
	G–BFCB	L-1011-385 TriStar 500	British Airways
	G–BFCC	L-1011-385 TriStar 500	British Airways
	G–BFCD	L-1011-385 TriStar 500	British Airways
	G–BFCE	L-1011-385 TriStar 500	British Airways
	G–BFCF	L-1011-385 TriStar 500	British Airways *Elizabeth of Glamis*
	G–BFCG	—	British Airways
	G–BFCH	—	British Airways
	G–BFCI	—	British Airways
	G–BFCJ	—	British Airways
	G–BFCK	—	British Airways
	G–BFCL	—	British Airways

Reg.	Type	Owner or Operator	Notes
G–BFCR	BN-2A Islander	BN (Bembridge) Ltd.	
G–BFCT	Cessna TU-206F	Cecil Aviation Ltd.	
G–BFCV	BN-2A Islander	BN (Bembridge) Ltd.	
G–BFCW	BN-2A Islander	BN (Bembridge) Ltd.	
G–BFCX	BN-2A Islander	Loganair Ltd.	
G–BFCY	AB-206B JetRanger 2	Window Boxes Ltd.	
G–BFCZ	Sopwith Camel (B7270)	Leisure-Sport Ltd.	
G–BFDA	PA-31-350 Navajo Chieftain	Air Charter (Scotland) Ltd	
G–BFDB	PA-31-350 Navajo Chieftain	Fairflight Ltd.	
G–BFDC	D.H.C.-1 Chipmunk 22	N. F. O'Neill	
G–BFDD	BN-2A Mk III-2 Trislander	BN (Bembridge) Ltd.	
G–BFDE	Sopwith Tabloid (replica)	D. M. Cashmore	
G–BFDF	SOCATA Rallye 235GT	S. G. Lawrence	
G–BFDG	PA-28R-201T Turbo Arrow III	The Electric Rainbow Co. Ltd.	
G–BFDI	PA-28-181 Archer II	A. P. Grant & J. R. Sunderland	
G–BFDK	PA-28-161 Warrior II	C.S.E. Aviation Ltd.	
G–BFDL	Piper L-4J Cub	R. Windley	
G–BFDM	Jodel D.120	Worcestershire Gliding Ltd.	
G–BFDN	PA-31-350 Navajo Chieftain	Lease Air Ltd.	
G–BFDO	PA-28R-201T Turbo-Arrow III	American Airspeed Inc.	
G–BFDP	CP.301A Emeraude	J. P. McGrath & J. Robson	
G–BFDR	Beech 60 Duke	Penarth Commercial Properties Ltd.	
G–BFDZ	Taylor JT.1 Monoplane	D. C. Barber	
G–BFEB	Jodel D.150	P. A. Ellway & ptnrs.	
G–BFEC	PA-23 Aztec 250F	Midas Airways	
G–BFED	PA-31-350 Navajo Chieftain	Woodgate Aviation Ltd.	
G–BFEE	Beech 95-E55 Baron	J. F. Pugh	
G–BFEF	Agusta-Bell 47G-3B1	Trent Helicopters Ltd.	
G–BFEG	Westland-Bell 47G-3B1	Trent Helicopters Ltd.	
G–BFEH	Jodel D.117A	E. J. Horsfall	
G–BFEI	Westland-Bell 47G-3B1	Trent Helicopters Ltd.	
G–BFEJ	Agusta-Bell 47G-3B1	Trent Helicopters Ltd.	
G–BFEK	Cessna F.152	P. R. J. Stone & P. B. Cantley	
G–BFEL	Cessna F.150M	A. L. Strongman	
G–BFEM	Cessna 421C	Brush Electrical Machines Ltd.	
G–BFEO	Boeing 707-323C	Tradewinds Airways Ltd.	
G–BFET	Partenavia P.68B	DK Aviation Ltd.	
G–BFEU	SA.330J Puma	Bristow Helicopters Ltd.	
G–BFEV	PA-25 Pawnee 235	Bowker Aircraft Services Ltd.	
G–BFEW	PA-25 Pawnee 235	Agricola Aerial Work Ltd.	
G–BFEX	PA-25 Pawnee 235	Harvest Air Ltd.	
G–BFEY	PA-25 Pawnee 235	Harvest Air Ltd.	
G–BFEZ	Beech B60 Duke	Turner Engineering Co. Ltd.	
G–BFFB	Evans VP-2	D. Bradley	
G–BFFC	Cessna F.152-II	Yorkshire Flying Services Ltd.	
G–BFFD	Cessna F.152-II	Skegness Air Taxi Service Co. Ltd.	
G–BFFE	Cessna F.152-II	Veetstor Ltd.	
G–BFFF	Cessna 188B Ag Truck	Shoreham Flight Simulation	
G–BFFG	Beech 95-B55 Baron	Basic Metal Co. Ltd.	
G–BFFI	PA-31-350 Navajo Chieftain	Apache Air Taxis	
G–BFFJ	Sikorsky S-61N Mk. II	British Airways Helicopters Ltd.	
G–BFFK	Sikorsky S-61N Mk. II	British Airways Helicopters Ltd.	
G–BFFM	Cessna 421C	Northair Aviation Ltd.	
G–BFFN	Enstrom F.28A	Spooner Aviation Ltd.	
G–BFFP	PA-18-150 Super Cub	Airways Aero Associations Ltd.	
G–BFFR	PA-31-350 Navajo Chieftain	Vickers Shipbuilding Group	
G–BFFT	Cameron V-56 balloon	J. M. Stables *Stables of York*	
G–BFFV	Agusta Bell 47G-3B1	Dollar Air Services Ltd.	
G–BFFW	Cessna F.152	W. A. Hanton	
G–BFFX	Cessna F.172-II	Northair Aviation Ltd.	
G–BFFY	Cessna F.150M	Edinburgh Flying Club	
G–BFFZ	Cessna FR.172 Hawk XP	Colton Aviation Services Ltd.	
G–BFGA	MS.880B Rallye 150ST	Selles Dispensing Chemists Ltd.	
G–BFGB	R. Commander 685	British Airports Authority	
G–BFGC	Cessna F.150M	Edinburgh Flying Club	
G–BFGD	Cessna F.172N-II	Northair Aviation Ltd.	
G–BFGE	Cessna F.172N-II	Northair Aviation Ltd.	
G–BFGF	Cessna F.177RG	Landlink Two Ltd.	
G–BFGG	Cessna FRA.150M	Airwork Services Ltd.	
G–BFGH	Cessna F.337G	Barfax Distributing Co. Ltd.	
G–BFGI	Douglas DC-10-30	British Caledonian Airways. *David Livingstone— The Scottish Explorer*	

Notes	Reg.	Type	Owner or Operator
	G–BFGJ	Cameron D-96 hot-air airship	Cameron Balloons Ltd. *Sultan*
	G–BFGK	Jodel D.117	R. G. M. Bissett
	G–BFGL	Cessna FA.152	Yorkshire Flying Services Ltd.
	G–BFGM	Boeing 727-095	Dan-Air Services Ltd.
	G–BFGN	Boeing 727-095	Dan-Air Services Ltd.
	G–BFGO	Fuji FA.200-160	Hillcrest Garage (Heseldon) Ltd.
	G–BFGP	D.H.C.-6 Twin Otter 310	Aurigny Air Services
	G–BFGR	R. Commander 685	Hamdani Investments Ltd.
	G–BFGS	SOCATA Rallye 180GT	F. G. & F. J. Storey
	G–BFGW	Cessna F.150H	South Humberside Flying Group Ltd.
	G–BFGX	Cessna FRA.150M	Airwork Services Ltd.
	G–BFGY	Cessna F.182Q	Oxford Controls Co. Ltd.
	G–BFGZ	Cessna FRA.150M	Airwork Services Ltd.
	G–BFHD	C.A.S.A. C.352L	Warbirds of Great Britain Ltd.
	G–BFHE	C.A.S.A. C.352L	Warbirds of Great Britain Ltd.
	G–BFHF	C.A.S.A. C.352L	Warbirds of Great Britain Ltd.
	G–BFHG	C.A.S.A. C.352L	Warbirds of Great Britain Ltd.
	G–BFHH	D.H.82A Tiger Moth	P. Harrison & B. Haddican
	G–BFHI	Piper J-3C-65 Cub	J. M. Robinson
	G–BFHJ	Cessna FRA.150M	Airwork Services Ltd.
	G–BFHK	Cessna F.177RG-11	A. F. Hall
	G–BFHL	Cessna FRA.150M	Denham Flying Training School Ltd.
	G–BFHM	Steen Skybolt	W. R. Penaluna
	G–BFHN	Scheibe SF.25E Super Falke	W. J. Dyer & R. D. C. Hart
	G–BFHO	PA-31 Navajo	Willow Vale Electronics Ltd.
	G–BFHP	Champion 7GCAA Citabria	Buckminster Gliding Club Ltd.
	G–BFHR	Jodel DR.220 2+2	J. B. Keith & J. Bugg
	G–BFHS	AA-5B Tiger	J. E. Fazackerley
	G–BFHT	Cessna F.152	Solitair Flight Management Ltd.
	G–BFHU	Cessna F.152	Solitair Flight Management Ltd.
	G–BFHV	Cessna F.152	P. J. Brown
	G–BFHX	Evans VP-1	D. F. Gibson
	G–BFIB	PA-31-310 Turbo Navajo	Mann Aviation Ltd.
	G–BFIC	H.S.125 Srs. 600B	McAlpine Aviation Ltd.
	G–BFID	Taylor JT.2 Titch Mk. III	W. F. Adams
	G–BFIE	Cessna FRA.150M	Northair Aviation Ltd.
	G–BFIF	Cessna FR.172XP	Air Navigation & Trading Co. Ltd.
	G–BFIG	Cessna FR.172XP	D. M. Balfour
	G–BFII	PA-23 Aztec 250E	European Ferries Ltd.
	G–BFIJ	AA-5A Cheetah	Cabair Ltd.
	G–BFIK	AA-5A Cheetah	Cabair Ltd.
	G–BFIL	AA-5A Cheetah	Cabair Ltd.
	G–BFIM	AA-5A Cheetah	D. E. Nixon & C. Heathcote
	G–BFIN	AA-5A Cheetah	Cabair Ltd.
	G–BFIP	Wallbro Monoplane 1909 Replica	K. H. Wallis
	G–BFIR	Avro 652A Anson 21 (WD413)	G. M. K. Fraser
	G–BFIT	Thunder Ax6-56Z balloon	R. J. Nicholson & C. MacKinnon
	G–BFIU	Cessna FR.172XP	Fletcher Bakeries Ltd.
	G–BFIV	Cessna F.177RG	Murr International Constructions Ltd.
	G–BFIX	Thunder Ax7-77A balloon	Thunder Balloons Ltd.
	G–BFIY	Cessna F.150M	R. J. Ford-Sagers
	G–BFIZ	Cessna FT.337GP	Gold Star Publications Ltd.
	G–BFJA	AA-5B Tiger	G. W. Hind
	G–BFJB	Bell 212	British Airways Helicopters Ltd.
	G–BFJC	—	—
	G–BFJD	—	—
	G–BFJE	—	—
	G–BFJF	—	—
	G–BFJH	SA.102-5 Cavalier	B. F. J. Hope
	G–BFJI	Robin HR.100/250	H. Deville & C. J. Lear
	G–BFJJ	Evans VP-1	P. R. Pykett & B. J. Dyke
	G–BFJK	PA-23 Aztec 250E	Air London (Executive Travel) Ltd.
	G–BFJM	Cessna F.152	Pegasus Aviation Ltd.
	G–BFJN	Westland-Bell 47G-3B1	A & B Agricultural Aviation Ltd.
	G–BFJO	G.164B Ag-Cat 450	Miller Aerial Spraying Ltd.
	G–BFJP	G.164B Ag-Cat 450	Lawrence Agricultural Aviation Services Ltd.
	G–BFJR	Cessna F.337G	John Roberts Services Ltd.
	G–BFJS	Cessna 340A	Northair Aviation Ltd.
	G–BFJT	Westland-Bell 47G-3B1	GSM Helicopters Ltd.
	G–BFJU	Westland-Bell 47G-3B1	GSM Helicopters Ltd.

Reg.	Type	Owner or Operator	Notes
G–BFJV	Cessna F.172H	N. R. Fineman	
G–BFJW	AB-206B JetRanger 2	Shawline Helicopters Ltd.	
G–BFJZ	Robin DR.400/140B	Alexander Howden Group	
G–BFKA	Cessna F.172N	C. Blackburn	
G–BFKB	Cessna F.172N	Air UK	
G–BFKC	Rand KR.2	K. K. Cutt	
G–BFKD	R. Commander 114B	K. W. Ford	
G–BFKF	Cessna FA.152	Trehaven Trust Ltd.	
G–BFKG	Cessna F.152	Western Air Training Ltd.	
G–BFKH	Cessna F.152	Western Air Training Ltd.	
G–BFKJ	PA-31-310 Navajo	Kent Messenger Ltd.	
G–BFKK	P.66 Pembroke (XF796)	G. M. Fraser	
G–BFKL	Cameron N-56 balloon	Merrythought Toys Ltd. *Merrythought*	
G–BFKM	Westland-Bell 47G-3B1	A. & B Agricultural Aviation Ltd.	
G–BFKN	PA-23 Aztec 250F	Anglo-African Machinery Co. Ltd.	
G–BFKO	PA-34 200T-II Seneca	Skycabs	
G–BFKP	Partenavia P.68B	Twin Flight Ltd.	
G–BFKR	PA-24 Comanche 250	Conforex Ltd.	
G–BFKS	Cessna F.172L	Wycombe Air Centre Ltd.	
G–BFKT	Cessna F.172M	Wycombe Air Centre Ltd.	
G–BFKU	PA-25 Pawnee 235D	Farmair Ltd.	
G–BFKV	PA-25 Pawnee 235D	Miller Aircraft Hire Ltd.	
G–BFKW	Concorde 102	British Aerospace	
G–BFKY	PA-34-200 Seneca	Northair Aviation Ltd.	
G–BFKZ	SA.330J Puma	Bristow Helicopters Ltd.	
G–BFLA	Cessna 414	Swaziland Portland Cement (Pty) Ltd.	
G–BFLB	Beech V35B Bonanza	K. A. Forbes	
G–BFLC	Cessna 210L	Jaserve Ltd.	
G–BFLD	Boeing 707-338C	British Midland Airways Ltd.	
G–BFLE	Boeing 707-338C	British Midland Airways Ltd.	
G–BFLH	PA-34-200T-II Seneca	C.S.E. (Aircraft Services) Ltd.	
G–BFLI	PA-28R-201T Turbo Arrow III	Europalite Ltd.	
G–BFLK	Cessna F.152	Gordon King Aviation Ltd.	
G–BFLL	H.S.748 Srs. 2A	Dan-Air Services Ltd.	
G–BFLM	Cessna 150M	Cornwall Flying Club Ltd.	
G–BFLN	Cessna 150M	Appaglow Ltd.	
G–BFLO	Cessna F.172M	W. A. Cook & H. Preston	
G–BFLP	Amethyst Ax6 balloon	K. J. Hendry *Amethyst*	
G–BFLR	Hiller UH-12E	Heliscot Ltd. & Major F. F. Chamberlain	
G–BFLS	Westland-Bell 47G-3B1	Hon. M. H. Wills	
G–BFLT	—	—	
G–BFLU	Cessna F.152	Inverness Flying Services Ltd.	
G–BFLV	Cessna F.172N	Rogers Aviation Ltd.	
G–BFLW	PA-39 Twin Comanche 160CR	Harris Aviation Services Ltd.	
G–BFLX	AA-5A Cheetah	Sheffield Auto Hire	
G–BFLZ	Beech 95-A55 Baron	K. K. Demel Ltd.	
G–BFMC	BAC One-Eleven 414	Ford Motor Co. Ltd.	
G–BFME	Cameron V-56 balloon	R. B. & Mrs. G. Girling	
G–BFMF	Cassutt Racer Mk. III	P. H. Lewis	
G–BFMG	PA-28-161 Warrior II	Segdale Ltd.	
G–BFMH	Cessna 177B	Span Aviation	
G–BFMJ	AA-5B Tiger	John Farbon & Co. Ltd.	
G–BFMK	Cessna FA.152	Denham Flying Training School Ltd.	
G–BFMM	PA-28-181 Archer II	Bristol & Wessex Aeroplane Club Ltd.	
G–BFMN	PA-28R-201T Turbo Arrow III	Adrian Alan Ltd.	
G–BFMR	PA-20 Pacer 135	C. C. & Mrs. J. M. Lovell	
G–BFMS	MS.893E Rallye 180GT	Mills Marketing Services Ltd.	
G–BFMT	Robin HR.200/100	Eagle Forms (GB) Ltd.	
G–BFMU	AA-5A Cheetah	Tentergate Trading	
G–BFMV	PA-28 Cherokee 160C	Overlantic Ltd.	
G–BFMW	V.735 Viscount	Alidair Ltd.	
G–BFMX	Cessna F.172N	Ackroyd Westwood & Associates Ltd.	
G–BFMY	Sikorsky S-61N	Bristow Helicopters Ltd.	
G–BFMZ	Payne Ax6 balloon	G. F. Payne	
G–BFNA	PA-25 Pawnee 235D	Miller Aircraft Hire Ltd.	
G–BFNB	PA-25 Pawnee 235D	A. G. Edwards	
G–BFNC	AS.350B Ecureuil	McAlpine Helicopters Ltd.	
G–BFND	AB-206B JetRanger 2	Warner Holidays	
G–BFNE	C.A.S.A. I.131 Jungmann	J. Lipton	
G–BFNG	Jodel D.112	K. D. Bass	

Notes	Reg.	Type	Owner or Operator
	G–BFNH	Cameron V-77 balloon	C. R. Kirby *Red Pepper*
	G–BFNI	PA-28-161 Warrior II	C.S.E. (Aircraft Services) Ltd.
	G–BFNJ	PA-28-161 Warrior II	C.S.E. (Aircraft Services) Ltd.
	G–BFNK	PA-28-161 Warrior II	C.S.E. (Aircraft Services) Ltd.
	G–BFNL	BN-2A Islander	BN (Bembridge) Ltd.
	G–BFNM	Globe GC.1 Swift	Nottingham Flying Group
	G–BFNO	BN-2A Islander	BN (Bembridge) Ltd.
	G–BFNP	Saffery S.330 balloon	N. H. Ponsford *Sidney*
	G–BFNS	BN-2A Islander	BN (Bembridge) Ltd.
	G–BFNU	BN-2A Islander	BN (Bembridge) Ltd.
	G–BFNV	BN-2A Islander	Loganair Ltd.
	G–BFNX	BN-2A Islander	BN (Bembridge) Ltd.
	G–BFOD	Cessna F.182Q	Staverton Flying School Ltd.
	G–BFOE	Cessna F.152	Northair Aviation Ltd.
	G–BFOF	Cessna F.152	Staverton Flying School Ltd.
	G–BFOG	Cessna 150M	Sierra Aviation Services Ltd.
	G–BFOH	Westland-Bell 47G-3B1	Helicopter Hire Ltd.
	G–BFOI	Westland-Bell 47G-3B1	Helicopter Hire Ltd.
	G–BFOJ	AA-1 Yankee	P. A. MacCarty Ltd.
	G–BFOL	Beech A.200 Super King Air	A. Ogden & Sons Ltd.
	G–BFOM	PA-31-325 Navajo	Matthew Royce Ltd.
	G–BFON	PA-31-310 Navajo	Air Kilroe
	G–BFOP	Jodel D.120	H. Cope
	G–BFOR	Thunder Ax6-56Z balloon	Thunder Balloons Ltd.
	G–BFOS	Thunder Ax6-56A balloon	Thunder Balloons Ltd. *Milton Keynes*
	G–BFOT	Thunder Ax6-56A balloon	Thunder Balloons Ltd.
	G–BFOU	Taylor JT.1 Monoplane	I. N. M. Cameron
	G–BFOV	Cessna F.172N	Gooda Walker Ltd.
	G–BFOW	Cessna F.172N	Sajan Ltd.
	G–BFOX	D.H.83 Fox Moth	M. C. Russell & ptnrs.
	G–BFOY	Cessna A.188B AgTruck	Plains Aerial Spraying Ltd.
	G–BFOZ	Thunder Ax6-56 balloon	Thunder Balloons Ltd. *Motorway II*
	G–BFPA	Scheibe SF.25B Super Falke	Yorkshire Gliding Club (Pty) Ltd.
	G–BFPB	AA-5B Tiger	Simon Deverell Print Ltd.
	G–BFPC	AA-5B Tiger	K.K. Aviation
	G–BFPD	AA-5A Cheetah	Scotia Safari Ltd.
	G–BFPE	PA-28 Cherokee 140	Clacton Light Aviation Co.
	G–BFPF	Sikorsky S-61N	British Airways Helicopters Ltd.
	G–BFPH	Cessna F.172K	R. W. Dorch
	G–BFPJ	Procter Petrel	S. G. Craggs
	G–BFPK	Beech A.23 Musketeer	Medway Flying Group
	G–BFPL	Fokker D.VII Replica	Leisure Sport Ltd.
	G–BFPM	Cessna F.172M	Abbey Windows Ltd.
	G–BFPO	R. Commander 112B	Millet Shipping Ltd.
	G–BFPP	Bell 47J-2	Shawline Helicopters Ltd.
	G–BFPR	PA-25 Pawnee 235D	Miller Aircraft Hire Ltd.
	G–BFPS	PA-25 Pawnee 235D	Bowker Aircraft Services Ltd.
	G–BFPX	Taylor JT.1 Monoplane	E. A. Taylor
	G–BFPY	PA-32 Cherokee Six 260	Colton Aviation Services Ltd.
	G–BFPZ	Cessna F.177RG	Aspenwell Ltd.
	G–BFRA	R. Commander 114	Glos-Air Ltd.
	G–BFRB	Cessna F.152	W. Shyvers Ltd.
	G–BFRC	AA-5A Cheetah	North-West Flying Services Ltd.
	G–BFRD	Bowers Flybaby 1A	F. R. Donaldson
	G–BFRE	AA-5B Tiger	Harlands of Hull Ltd.
	G–BFRF	Taylor JT.1 Monoplane	E. R. Bailey
	G–BFRH	PA-28 Cherokee 140	E. Ruscoe
	G–BFRI	Sikorsky S-61N	Bristow Helicopters Ltd.
	G–BFRJ	HPR-7 Herald 209	Express Air Services Ltd.
	G–BFRK	HPR-7 Herald 209	Express Air Services Ltd.
	G–BFRL	Cessna F.152	Light Planes (Lancashire) Ltd.
	G–BFRM	Cessna 550 Citation II	Marshall of Cambridge (Engineering) Ltd.
	G–BFRN	Cessna F.152	Rogers Aviation Sales Ltd.
	G–BFRO	Cessna F.150M	Jack Braithwaite (Aerial Photography) Ltd.
	G–BFRP	Cessna F.150M	Transgap Ltd.
	G–BFRR	Cessna FRA.150M	Pegasus Aviation Ltd.
	G–BFRS	Cessna F.172N	Poplar Toys Ltd.
	G–BFRT	Cessna FR.172XP	English Country Magazines Ltd.
	G–BFRU	Cessna F.152	A. G. Chrismas Ltd.
	G–BFRV	Cessna FA.152	Rogers Aviation Ltd.
	G–BFRW	Agusta-Bell 47G-3B1	Rotair Ltd.

Reg.	Type	Owner or Operator	Notes
G–BFRX	PA-25 Pawnee 260	Aero Cropcare Ltd.	
G–BFRY	PA-25 Pawnee 260	Aero Cropcare Ltd.	
G–BFSA	Cessna F.182Q	Clark Masts Ltd.	
G–BFSB	Cessna F.152	Rogers Aviation Ltd.	
G–BFSC	PA-25 Pawnee 235D	Farm Aviation Services Ltd.	
G–BFSD	PA-25 Pawnee 235D	Farm Aviation Services Ltd.	
G–BFSJ	Westland-Bell 47G-3B1	R.K.B. Leasing Services	
G–BFSK	PA-23 Apache 160	A.F. Aviation Ltd.	
G–BFSL	Cessna U.206F Stationair	Range Air	
G–BFSM	Hughes 500D	Sloane Aviation Ltd.	
G–BFSN	Agusta-Bell 47G-2	Minster Helicopters Ltd.	
G–BFSO	H.S.125 Srs. 700B	De Beers International Air Services Ltd.	
G–BFSP	H.S.125 Srs. 700B	De Beers International Air Services Ltd.	
G–BFSR	Cessna F.150J	Aero Group 78	
G–BFSS	Cessna FR.172G	Minerva Services	
G–BFST	Partenavia P.68B	Air Oxford	
G–BFSU	Partenavia P.68B	John Holman & Sons Ltd.	
G–BFSY	PA-28-181 Archer II	Arun Aviation Ltd.	
G–BFSZ	PA-28-161 Warrior II	Four-by-Four (Aviation) Ltd.	
G–BFTA	PA-28-161 Warrior II	N. Clayton	
G–BFTB	PA-28R-201 Arrow III	First Class Furniture Ltd.	
G–BFTC	PA-28R-201T Turbo Arrow III	Newpet Ltd.	
G–BFTE	AA-5A Cheetah	B. Refson	
G–BFTF	AA-5B Tiger	F. C. Burrow Ltd.	
G–BFTG	AA-5B Tiger	Corrugated Service Engineers Ltd.	
G–BFTH	Cessna F.172N	The Hammer People Ltd.	
G–BFTI	BN-2A Islander	BN (Bembridge) Ltd.	
G–BFTJ	BN-2A Islander	BN (Bembridge) Ltd.	
G–BFTK	BN-2A Islander	BN (Bembridge) Ltd.	
G–BFTL	BN-2A Islander	BN (Bembridge) Ltd.	
G–BFTM	BN-2A Islander	BN (Bembridge) Ltd.	
G–BFTN	G.164B Ag-Cat 450	Miller Aerial Spraying Ltd.	
G–BFTO	Rotorway Scorpion 133	Rotorway Helicopters Ltd.	
G–BFTR	Bell 206L Long Ranger	Air Hanson Ltd.	
G–BFTS	Westland-Bell 47G-3B1	Rotair Ltd.	
G–BFTT	Cessna 421C	Comben Group Services Ltd.	
G–BFTU	Cessna FA.152	Toon Ghose Aviation Ltd.	
G–BFTW	PA-23 Aztec 250F	Southampton Airport Ltd.	
G–BFTX	Cessna F.172N	Westair Flying Services Ltd.	
G–BFTY	Cameron V-77 balloon	P.S.H. Frewer *Regal Motors*	
G–BFTZ	MS.880B Rallye Club	R. & B. Legge Ltd.	
G–BFUB	PA-32RT-300 Turbo Lance II	Bumbles Ltd.	
G–BFUD	Scheibe SF.25E Super Falke	R. C. Bull	
G–BFUG	Cameron N-77 balloon	R. Sisterson *Blue Boar*	
G–BFUN	Hughes 269C	H. Downs & Sons (Huddersfield) Ltd	
G–BFUO	PA-23 Aztec 250F	Portland Motor Supplies	
G–BFUS	Cessna 404 Titan	Euroair Transport Ltd.	
G–BFUZ	Cameron V-77 balloon	Skysales Ltd.	
G–BFVA	Boeing 737-204ADV	Britannia Airways Ltd. *Sir John Alcock*	
G–BFVB	Boeing 737-204ADV	Britannia Airways Ltd. *Sir Thomas Sopwith*	
G–BFVF	PA-38-112 Tomahawk	C.S.E. (Aircraft Services) Ltd.	
G–BFVG	PA-28-181 Archer II	J. M. Henderson	
G–BFVH	D.H.2 Replica (5964)	Leisure Sport Ltd.	
G–BFVI	H.S.125 Srs. 700B	Bristow Helicopters Ltd.	
G–BFVM	Westland-Bell 47G-3B1	Heliwork Ltd.	
G–BFVO	Partenavia P.68B	Rothmans of Pall Mall Ltd.	
G–BFVP	PA-23 Aztec 250	Dukes Transport (Craigavon) Ltd.	
G–BFVS	AA-5B Tiger	J. J. Mudford	
G–BFVU	Cessna 150L	IIP Associates	
G–BFVV	SA.365 Dauphin 2	Management Aviation Ltd.	
G–BFVW	SA.365 Dauphin 2	Management Aviation Ltd.	
G–BFVX	Beech C90 King Air	Vernair Transport Services	
G–BFVY	Beech C90 King Air	Vernair Transport Services	
G–BFVZ	Beech A200 Super King Air	Eagle Aircraft Services Ltd.	
G–BFWA	BN-2A Islander	BN (Bembridge) Ltd.	
G–BFWB	PA-28-161 Warrior II	C.S.E. (Aircraft Services) Ltd.	
G–BFWD	Currie Wot	F. E. Nuthall	
G–BFWE	PA-23 Aztec 250	Air Navigation & Trading Co. Ltd.	
G–BFWF	Cessna 421B	Alcon Oil Ltd.	
G–BFWG	R. Commander 112A	Alcon Oil Ltd.	

Notes	Reg.	Type	Owner or Operator
	G–BFWJ	Slingsby T.65A Vega	S. Y. Duxbury
	G–BFWK	PA-28-161 Warrior II	Woodgate Aviation Ltd.
	G–BFWL	Cessna F.150L	Eglinton Flying Club
	G–BFWM	Hiller UH-12D	Management Aviation Ltd.
	G–BFWV	BN-2A Islander	BN (Bembridge) Ltd.
	G–BFWW	Robin HR.100/210	Willingair Ltd.
	G–BFWX	BN-2A Islander	BN (Bembridge) Ltd.
	G–BFWY	BN-2A Islander	BN (Bembridge) Ltd.
	G–BFWZ	BN-2A Islander	BN (Bembridge) Ltd.
	G–BFXC	Mooney M.20C	E. W. Passmore
	G–BFXD	PA-28-161 Warrior II	C.S.E. (Aircraft Services) Ltd.
	G–BFXE	PA-28-161 Warrior II	C.S.E. (Aircraft Services) Ltd.
	G–BFXF	Andreasson BA.4B	A. Brown
	G–BFXG	D.31 Turbulent	S. Griffin
	G–BFXH	Cessna F.152	Dayton Aerohire
	G–BFXI	Cessna F.172M	D. J. Stockley & P. S. Nicholson
	G–BFXJ	Cessna 210	E. Harper
	G–BFXK	PA-28 Cherokee 140	G. S. & Mrs. M. T. Pritchard
	G–BFXL	Albatross D.5A	Leisure Sport Ltd.
	G–BFXM	Jurca MJ.5 Sirocco	D. I. & W. A. Barker
	G–BFXN	PA-36-375 Brave	Miller Aerial Spraying Ltd.
	G–BFXO	Taylor JT.1 Monoplane	A. S. Nixon
	G–BFXR	Jodel D.112	D. A. Porter
	G–BFXS	R. Commander 114	Glos-Air Ltd.
	G–BFXT	H.S.125 Srs. 700B	Coca Cola Export Corporation
	G–BFXU	American Beta Z Airship	G. Turnbull
	G–BFXV	Air Tractor AT-302	ADS Aerial Ltd.
	G–BFXW	AA-5B Tiger	Cabair Ltd.
	G–BFXX	AA-5B Tiger	M. A. Lenihan
	G–BFXY	AA-5A Cheetah	Cabair Ltd.
	G–BFXZ	PA-28-181 Archer II	C. Lohr
	G–BFYA	MBB Bo 105D	British Caledonian Helicopters Ltd.
	G–BFYB	PA-28-161 Warrior II	C.S.E. (Aircraft Services) Ltd.
	G–BFYC	PA-32RT-300 Lance II	Avingate Ltd.
	G–BFYD	PA-32RT-300T Turbo Lance II	C.S.E. Aviation Ltd.
	G–BFYE	Robin HR.100/285	Tollbridge Machine Tool Co. Ltd.
	G–BFYF	Westland-Bell 47G-3B1	H. J. Hofman
	G–BFYG	BN-2A Islander	BN (Bembridge) Ltd.
	G–BFYI	Westland-Bell 47G-3B1	Dollar Air Services Ltd.
	G–BFYJ	Hughes 369HE	Wilford Aviation Ltd.
	G–BFYL	Evans VP.2	A. G. Wilford
	G–BFYM	PA-28-161 Warrior II	C.S.E. (Aircraft Services) Ltd.
	G–BFYN	Cessna FA.152	Denham Flying Training School Ltd.
	G–BFYO	Spad XIII (replica)	Leisure Sport Ltd.
	G–BFYP	Benson B.7	A. J. Philpotts
	G–BFYR	BN-2A Islander	BN (Bembridge) Ltd.
	G–BFYS	BN-2A Islander	BN (Bembridge) Ltd.
	G–BFYT	BN-2A Islander	BN (Bembridge) Ltd.
	G–BFYU	SC.5 Belfast	HeavyLift Cargo Airlines Ltd.
	G–BFYW	Slingsby T.65A Vega	F. J. Sheppard
	G–BFZA	Alpavia Fournier RF-3	T. J. Hartwell & D. R. Wilkinson
	G–BFZB	Piper J-3C-65 Cub	P. F. Ansell & ptnrs.
	G–BFZC	Sikorsky S-61N	Management Aviation Ltd.
	G–BFZD	Cessna FR.182Q	J. & R. Mac (Machine Tools) Ltd.
	G–BFZE	AS.350B Ecureuil	TBF Transport Ltd.
	G–BFZF	Boeing 707-321C	Scimitar Airlines Ltd.
	G–BFZG	PA-28-161 Warrior II	C.S.E. (Aircraft Services) Ltd.
	G–BFZH	PA-28R-200 Cherokee Arrow	R. R. Henderson
	G–BFZK	EMB-110P2 Bandeirante	Fairflight Ltd.
	G–BFZL	V.836 Viscount	British Midland Airways Ltd.
	G–BFZM	R. Commander 112TC	Glos-Air Ltd.
	G–BFZN	Cessna FA.152	Leicestershire Aero Club Ltd.
	G–BFZO	AA-5A Cheetah	North-West Flying Services Ltd.
	G–BFZP	AA-5B Tiger	Scotia Safari Ltd.
	G–BFZS	Cessna F.152	R. T. Haddow
	G–BFZT	Cessna FA.152	Denham Flying Training School Ltd.
	G–BFZU	Cessna FA.152	Rogers Aviation Sales Ltd.
	G–BFZV	Cessna F.172M	S. M. Rourke
	G–BFZZ	BN-2A Islander	BN (Bembridge) Ltd.
	G–BGAA	Cessna 152 II	Solent Flight Centre
	G–BGAB	Cessna F.152 II	Humberside Aero Club
	G–BGAD	Cessna F.152 II	Humberside Aero Club

Reg.	Type	Owner or Operator	Notes
G–BGAE	Cessna F.152 II	Tayside Aviation Ltd.	
G–BGAF	Cessna FA.152	E. P. Collier	
G–BGAG	Cessna F.172N	Medway Flying Group Ltd.	
G–BGAH	FRED Srs. 2	G. A. Harris	
G–BGAI	Westland-Bell 47G-3B1	Dollar Air Services Ltd.	
G–BGAJ	Cessna F.182Q II	Channel Islands Aero Holdings Ltd.	
G–BGAK	Cessna F.182Q II	Northair Aviation Ltd.	
G–BGAL	Saffery S.330 balloon	A. M. Lindsay *English Lady*	
G–BGAN	BN-2A Islander	BN (Bembridge) Ltd.	
G–BGAO	Cessna 401A	D. Grepne	
G–BGAP	Cessna F.182RG	B. D. Harris	
G–BGAS	Colting Ax8-105A balloon	British Gas Corporation	
G–BGAT	Douglas DC-10-30	British Caledonian Airways *James Watt – The Scottish Engineer*	
G–BGAU	Rearwin 9000L	Shipping & Airlines Ltd.	
G–BGAV	Rearwin 8135T	Shipping & Airlines Ltd.	
G–BGAW	Rotorway Scorpion 133	Rotorway Helicopters Ltd.	
G–BGAX	PA-28 Cherokee 140	D. E. Nicholl	
G–BGAY	Cameron O-77 balloon	Dante Balloon Group *Antonia*	
G–BGAZ	Cameron V-77 balloon	Cameron Balloons Ltd. *Silicon Chip*	
G–BGBA	Robin R.2100A	D. Faulkner	
G–BGBB	L.1011-385 TriStar 200	British Airways	
G–BGBC	L.1011-385 TriStar 200	British Airways	
G–BGBD	Evans VP-1	C. Watson	
G–BGBE	Jodel DR.1050	D. G. Perry	
G–BGBF	D.31A Turbulent	R. Clark	
G–BGBG	PA-28-181 Archer II	Truman Aviation Ltd.	
G–BGBH	PA-23 Aztec 250F	Bruce Harrison Ltd.	
G–BGBI	Cessna F.150L	Colon Aviation Services Ltd.	
G–BGBK	PA-38-112 Tomahawk	C.S.E. (Aircraft Services) Ltd.	
G–BGBM	SIPA 903	P. G. Phelps	
G–BGBN	PA-38-112 Tomahawk	R. T. Dickson & I. Conway	
G–BGBO	Cessna F.172N	Renco (Aviation) Ltd.	
G–BGBP	Cessna F.152	Brailsford Aviation Ltd.	
G–BGBR	Cessna F.172N	Steer Aviation Ltd.	
G–BGBS	PA-23 Aztec 250	Penarth Commercial Properties Ltd.	
G–BGBT	Partenavia P.68B	DK Aviation Ltd.	
G–BGBU	Auster AOP.9	K. J. Garrett	
G–BGBV	Slingsby T.65A Vega	R. Richards	
G–BGBW	PA-38-112 Tomahawk	C.S.E. (Aircraft Services) Ltd.	
G–BGBX	PA-38-112 Tomahawk	C.S.E. (Aircraft Services) Ltd.	
G–BGBY	PA-38-112 Tomahawk	Express Aviation Services Ltd.	
G–BGBZ	R. Commander 114	R. S. Fenwick	
G–BGCA	Slingsby T.65A Vega	E. C. Neighbour	
G–BGCB	Slingsby T.65A Vega	D. J. Dawson	
G–BGCC	PA-31-325 Navajo	Noble Aviation Ltd.	
G–BGCD	Cameron A-140 balloon	Skysales Ltd. *Maxim*	
G–BGCH	PA-38-112 Tomahawk	D. & J. Fountain-Barber	
G–BGCK	AA-5A Cheetah	E. C. Heathcote	
G–BGCL	AA-5A Cheetah	Goddard Aviation	
G–BGCM	AA-5A Cheetah	Publicity Innovations Leasing Ltd.	
G–BGCN	AA-5A Cheetah	Goddard Aviation	
G–BGCO	PA-44-180 Seminole	Aviation Beauport Ltd.	
G–BGCR	EMB-110P1 Bandeirante	Ashbon Associates Ltd.	
G–BGCS	EMB-110P1 Bandeirante	Ashbon Associates Ltd.	
G–BGCU	Slingsby T.65A Vega	D. Tucker	
G–BGCV	AS.350B Ecureuil	Valley of Gleneagles Helicopters Ltd.	
G–BGCX	Taylor JT.2 Titch	G. M. R. Walters	
G–BGCY	Taylor JT.1 Monoplane	D. V. Reeves	
G–BGDA	Boeing 737-236	British Airways	
G–BGDB	Boeing 737-236	British Airways	
G–BGDC	Boeing 737-236	British Airways	
G–BGDD	Boeing 737-236	British Airways	
G–BGDE	Boeing 737-236	British Airways	
G–BGDF	Boeing 737-236	British Airways	
G–BGDG	Boeing 737-236	British Airways	
G–BGDH	Boeing 737-236	British Airways	
G–BGDI	Boeing 737-236	British Airways	
G–BGDJ	Boeing 737-236	British Airways	
G–BGDK	Boeing 737-236	British Airways	
G–BGDL	Boeing 737-236	British Airways	
G–BGDN	Boeing 737-236	British Airways	
G–BGDO	Boeing 737-236	British Airways	

Notes	Reg.	Type	Owner or Operator
	G–BGDP	Boeing 737-236	British Airways
	G–BGDR	Boeing 737-236	British Airways
	G–BGDS	Boeing 737-236	British Airways
	G–BGDT	Boeing 737-236	British Airways
	G–BGDU	Boeing 737-236	British Airways
	G–BGDV	—	British Airways
	G–BGDW	—	British Airways
	G–BGDX	—	British Airways
	G–BGDY	—	British Airways
	G–BGDZ	—	British Airways
	G–BGEA	Cessna F.150M	R. L. Beverley
	G–BGEB	AB-206A JetRanger	Heliwork Ltd.
	G–BGEC	Cameron V-77 balloon	Central Garage (Southborough) Ltd. *Crikey*
	G–BGED	Cessna U.206F	Lubair (Transport Services) Ltd.
	G–BGEE	Evans VP-1	A. Morris
	G–BGEF	Jodel D.112	F. W. J. Ellis
	G–BGEH	Monnet Sonerai II	G. K. Penson
	G–BGEI	Baby Great Lakes	D. H. Greenwood
	G–BGEJ	BN-2A Islander	BN (Bembridge) Ltd.
	G–BGEK	PA-38-112 Tomahawk	H. Snelson Engineers Ltd.
	G–BGEL	PA-38-112 Tomahawk	H. Snelson Engineers Ltd.
	G–BGEM	Partenavia P.68B	Link-Hampson Ltd.
	G–BGEN	D.H.C.-6 Twin Otter 310	Loganair Ltd.
	G–BGEO	PA-31-350 Navajo Chieftain	Christian Salveson (Cold Storage) Ltd.
	G–BGEP	Cameron C-38 balloon	Cameron Balloons Ltd.
	G–BGES	Currie Wot	J. Roberts
	G–BGET	PA-38-112 Tomahawk	J. V. & J. Leasing
	G–BGEV	PA-38-112 Tomahawk	J. Apthorp Aviation & Leasing Ltd.
	G–BGEW	Nord NC.854S	R. A. Yates
	G–BGEX	Brookland Mosquito 2	J. G. Beesley
	G–BGFA	—	—
	G–BGFB	—	—
	G–BGFC	Evans VP-2	R. W. Eastman & J. R. Jones
	G–BGFD	PA-32-300 Cherokee Six	Fairoaks Aviation Services Ltd.
	G–BGFE	—	—
	G–BGFF	FRED Srs. 2	G. R. G. Smith
	G–BGFG	AA-5A Cheetah	Cabair Ltd.
	G–BGFH	Cessna F.182Q	TBF (Transport) Ltd.
	G–BGFI	AA-5A Cheetah	J. W. Hannay & Co. Ltd.
	G–BGFJ	Jodel D.9 Bebe	C. M. Fitton
	G–BGFK	Evans VP-1	D. Beaumont
	G–BGFM	Rollason-Luton Beta 4	G. H. C. Jiggins
	G–BGFN	PA-25 Pawnee 235	Miller Aircraft Hire Ltd.
	G–BGFS	Westland-Bell 47G-3B1	G. S. Mason
	G–BGFT	PA-34-200T-II Seneca	C.S.E. (Aircraft Services) Ltd.
	G–BGFU	BN-2A Islander	BN (Bembridge) Ltd.
	G–BGFV	BN-2A Islander	BN (Bembridge) Ltd.
	G–BGFW	BN-2A Islander	BN (Bembridge) Ltd.
	G–BGFX	Cessna F.152	Shirlstar Container Transport Ltd.
	G–BGFY	PA-31-350 Navajo Chieftain	Airmore International Ltd.
	G–BGGA	Bellanca 7GCBC Citabria	Hendon Air Services Ltd.
	G–BGGB	Bellanca 7GCBC Citabria	Hendon Air Services Ltd.
	G–BGGC	Bellanca 7GCBC Citabria	Embassy Air Services
	G–BGGD	Bellanca 8GCBC Scout	Bristol & Gloucestershire Gliding Club
	G–BGGE	PA-38-112 Tomahawk	C.S.E. (Aircraft Services) Ltd.
	G–BGGF	PA-38-112 Tomahawk	C.S.E. (Aircraft Services) Ltd.
	G–BGGG	PA-38-112 Tomahawk	C.S.E. (Aircraft Services) Ltd.
	G–BGGH	PA-38-112 Tomahawk	C.S.E. (Aircraft Services) Ltd.
	G–BGGI	PA-38-112 Tomahawk	C.S.E. (Aircraft Services) Ltd.
	G–BGGJ	PA-38-112 Tomahawk	C.S.E. (Aircraft Services) Ltd.
	G–BGGK	PA-38-112 Tomahawk	C.S.E. (Aircraft Services) Ltd.
	G–BGGL	PA-38-112 Tomahawk	C.S.E. (Aircraft Services) Ltd.
	G–BGGM	PA-38-112 Tomahawk	C.S.E. (Aircraft Services) Ltd.
	G–BGGN	PA-38-112 Tomahawk	C.S.E. (Aircraft Services) Ltd.
	G–BGGO	Cessna F.152	Rogers Aviation Sales Ltd.
	G–BGGP	Cessna F.152	E. Midlands School of Flying Ltd.
	G–BGGS	H.S.125 Srs. 700B	British Aerospace
	G–BGGT	Zenith CH.200	P. R. M. Nind
	G–BGGU	Wallis WA–116R-R	K. H. Wallis
	G–BGGV	Wallis WA–120 Srs. 2	K. H. Wallis
	G–BGGW	Wallis WA–122	K. H. Wallis

Reg.	Type	Owner or Operator	Notes
G–BGGX	AB-206B JetRanger 3	Willowbrook International Ltd.	
G–BGGY	AB-206B JetRanger 3	Willowbrook International Ltd.	
G–BGGZ	—	—	
G–BGHA	Cessna F.152	M. D. Ward	
G–BGHC	Saffery Hot Pants Firefly balloon	H. C. Saffery *Petunia*	
G–BGHD	Saffery Helios Blister balloon	H. C. Saffery	
G–BGHE	Convair L-13A	J. Davis	
G–BGHF	Westland WG.30	Westland Helicopters Ltd.	
G–BGHG	AS.350B Ecureuil	McAlpine Helicopters Ltd.	
G–BGHI	Cessna F.152	Taxon Ltd.	
G–BGHJ	Cessna F.172N	P.A.C.K. Enterprises (Sussex) Ltd.	
G–BGHK	Cessna F.152	Solent Flight Centre	
G–BGHL	GA-7 Cougar	Superprime Ltd.	
G–BGHM	Robin R1180T	Mitchell Dewey Associates	
G–BGHN	Agusta-Bell 47G-3B1	M. H. Wills	
G–BGHO	Agusta-Bell 47G-3B1	M. H. Wills	
G–BGHP	Beech 76 Duchess	Eagle Aircraft Services Ltd.	
G–BGHS	Cameron N-31 balloon	Balloon Stable Ltd.	
G–BGHT	Falconair F-12	T. K. Baillie	
G–BGHU	T-6G Harvard	Gladaircraft Ltd.	
G–BGHV	Cameron V-77 balloon	E. Davies	
G–BGHW	Thunder Ax8-90 balloon	Edinburgh University Balloon Group *James Tytler*	
G–BGHX	Chasle YC-12 Tourbillon	C. Clark	
G–BGHY	Taylor JT.1 Monoplane	J. Prowse	
G–BGHZ	FRED Srs. 2	A. G. Edwards	
G–BGIA	Cessna 152 II	A. G. Chrismas Ltd.	
G–BGIB	Cessna 152 II	K. D. Wickenden	
G–BGIC	Cessna 172N	T. F. Turner	
G–BGID	Westland-Bell 47G-3B1	Helicopter Supplies & Engineering Ltd.	
G–BGIE	EMB-121 Xingu	C.S.E. Aviation Ltd.	
G–BGIF	AS.350B Ecureuil	Cabair Ltd.	
G–BGIG	PA-38-112 Tomahawk	Western Flying Services Ltd.	
G–BGIH	Rand KR-2	G. & D. G. Park	
G–BGII	PA-32-300 Cherokee Six	Rosefair Electronics Ltd.	
G–BGIJ	Cameron 0–77 balloon	H. P. Carlton	
G–BGIK	Taylor JT.1 Monoplane	J. H. Medforth	
G–BGIL	AS.350B Ecureuil	McAlpine Helicopters Ltd.	
G–BGIM	AS.350B Ecureuil	Lord Glendyne	
G–BGIO	Benson B.8M	C. G. Johns	
G–BGIP	Colt 56A balloon	Lipton Export Ltd.	
G–BGIR	Boeing 707-321C	British Midland Airways Ltd.	
G–BGIS	Boeing 707-321C	Scimitar Airlines Ltd.	
G–BGIU	Cessna F.172H	Metro Equipment (Chesham) Ltd.	
G–BGIV	Bell 47G-5	Helicopter Farming Ltd.	
G–BGIW	Bell 47G-2	Autair Ltd.	
G–BGIX	H.295 Super Courier	Nordic Oil Services Ltd.	
G–BGIY	Cessna F.172N	Mercian Flange Co. Ltd.	
G–BGIZ	Cessna F.152	Telepoint Ltd.	
G–BGJA	Cessna FA.152	Light Planes (Lancashire) Ltd.	
G–BGJB	PA-44-180 Seminole	DFA (Aircraft) Ltd.	
G–BGJD	—	—	
G–BGJE	Boeing 737-236	British Airtours Ltd. *Sandpiper*	
G–BGJF	Boeing 737-236	British Airtours Ltd. *Skylark*	
G–BGJG	Boeing 737-236	British Airtours Ltd. *Kingfisher*	
G–BGJH	Boeing 737-236	British Airtours Ltd. *Osprey*	
G–BGJI	Boeing 737-236	British Airtours Ltd. *Swallow*	
G–BGJJ	Boeing 737-236	British Airtours Ltd. *Kestrel*	
G–BGJK	Boeing 737-236	British Airtours Ltd. *Firecrest*	
G–BGJL	Boeing 737-236	British Airtours Ltd. *Goldfinch*	
G–BGJM	Boeing 737-236	British Airtours Ltd. *Curlew*	
G–BGJN	—	—	
G–BGJO	—	—	
G–BGJP	—	—	
G–BGJR	—	—	
G–BGJS	—	—	
G–BGJT	—	—	
G–BGJU	Cameron V-56 Balloon	D. T. Watkins *Spoils*	
G–BGJV	H.S.748 Srs. 2B	British Aerospace	
G–BGJW	GA-7 Cougar	Glos-Air Ltd.	
G–BGJY	—	—	

Notes	Reg.	Type	Owner or Operator
	G–BGJZ	Airborne Industries AB.400 balloon	Balloon Stable Ltd. *Adva II*
	G–BGKA	P.56 Provost T.1 (XF690)	D. W. Mickleburgh
	G–BGKB	SOCATA Rallye 110ST	Air Touring Services Ltd.
	G–BGKC	SOCATA Rallye 110ST	Air Touring Services Ltd.
	G–BGKD	SOCATA Rallye 110ST	Air Touring Services Ltd.
	G–BGKE	BAC One-Eleven 539	British Airways *County of West Midlands*
	G–BGKF	BAC One-Eleven 539	British Airways *County of Stafford*
	G–BGKG	BAC One-Eleven 539	British Airways *County of Warwick*
	G–BGKH	Cessna A.188B AgTruck	A.G. Aviation (UK) Ltd.
	G–BGKI	Cessna A.188B AgTruck	A.G. Aviation (UK) Ltd.
	G–BGKJ	MBB Bo 105C	North Scottish Helicopters Ltd.
	G–BGKK	Westland-Bell 47G-3B1	GSM Helicopters Ltd.
	G–BGKM	SA.365C Dauphin	Management Aviation Ltd.
	G–BGKN	H.S.125 Srs. 600B	GKN Group Services Ltd.
	G–BGKO	GY-20 Minicab	R. B. Webber
	G–BGKP	MBB Bo 105C	North Scottish Helicopters Ltd.
	G–BGKR	PA-28-161 Warrior II	C.S.E. Aviation Ltd.
	G–BGKS	PA-28-161 Warrior II	C.S.E. Aviation Ltd.
	G–BGKT	Auster AOP.9	K. H. Wallis
	G–BGKU	PA-28R-201 Arrow III	Aircraft Mart
	G–BGKV	PA-28R-201 Arrow III	G. E. Salter Industrial Enterprises Ltd.
	G–BGKW	Evans VP-1	I. W. Black
	G–BGKX	PA-38-112 Tomahawk	Flamingo Aviation Ltd.
	G–BGKY	PA-38-112 Tomahawk	C.S.E. Aviation Ltd.
	G–BGKZ	J/5F Aiglet Trainer	R. C. H. Hibberd
	G–BGLA	PA-38-112 Tomahawk	Cambrian Flying Club
	G–BGLB	Bede BD-5B	W. Sawney
	G–BGLD	Beech 76 Duchess	Eagle Aircraft Services Ltd.
	G–BGLE	Saffrey S.330 Balloon	C. J. Dodd & ptnrs.
	G–BGLF	Evans VP-1	E. F. Fryer
	G–BGLG	Cessna 152	Skyviews & General Ltd.
	G–BGLH	Cessna 152	M. A. Lenihan
	G–BGLI	Cessna 152	M. A. Lenihan
	G–BGLK	Monnet Sonerai II	G. L. Kemp & J. Beck
	G–BGLN	Cessna FA.152	Flighttec Airline Support Ltd.
	G–BGLO	Cessna F.172N	K. W. Rourke
	G–BGLP	Cessna F.152	K. W. Rourke
	G–BGLR	Cessna F.152	Shirlstar Container Transport Ltd.
	G–BGLS	Baby Great Lakes	D. S. Morgan
	G–BGLV	—	—
	G–BGLW	PA-34-200 Seneca	Galaxy Aviation Ltd.
	G–BGLX	Cameron N-56 balloon	Sara A. G. Williams
	G–BGLY	—	—
	G–BGLZ	Stits SA-3A Playboy	Air Executive Ltd.
	G–BGMA	D.31 Turbulent	G. C. Masterson
	G–BGMB	Taylor JT.2 Titch	E. M. Bourne
	G–BGMC	D.H.C.-6 Twin Otter 310	Brymon Aviation Ltd.
	G–BGMD	D.H.C.-6 Twin Otter 310	Brymon Aviation Ltd.
	G–BGME	SIPA S.903	M. Emery
	G–BGMG	Bell 212	United Helicopters Ltd.
	G–BGMI	Bell 212	Bristow Helicopters (International) Ltd.
	G–BGMJ	GY-201 Minicab	H. P. Burrill
	G–BGMM	—	—
	G–BGMP	Cessna F.172G	Norvic Racing Engines Ltd.
	G–BGMR	GY-201 Minicab	T. J. D. Hodge & A. B. Holloway
	G–BGMS	Taylor JT.2 Titch	M. A. J. Spice
	G–BGMT	MS.894E Rallye 235GT	M. E. Taylor
	G–BGMU	Westland-Bell 47G-3B1	Midland Counties Helicopter (Services) Ltd.
	G–BGMV	SF.25B Falke	Worlds Gliding Club Ltd.
	G–BGMW	Edgley EA-7 Optica	Edgley Aircraft Ltd.
	G–BGMX	Enstrom F-280C	Barry Sheene Racing Ltd.
	G–BGNA	Short SD3-30	Loganair Ltd.
	G–BGND	Cessna F.172N	N. F. Duke
	G–BGNK	EMB-110P2 Bandeirante	Fairflight Ltd.
	G–BGNL	Hiway Super Scorpion powered hang glider	G. Breen
	G–BGNM	SA.365C Dauphin	Management Aviation Ltd.
	G–BGNN	AA-5A Cheetah	Curd & Green Ltd.

Reg.	Type	Owner or Operator	Notes
G–BGNO	AA-5A Cheetah	Cabair Ltd.	
G–BGNP	Saffery S.200 balloon	N. H. Ponsford	
G–BGNR	Cessna F.172N	A. Hamlin	
G–BGNS	Cessna F.172N	Solent Flight Centre	
G–BGNT	Cessna F.152	Solent Flight Centre	
G–BGNU	Beech E90 King Air	Norwich Union Fire Insurance Ltd.	
G–BGNV	GA-7 Cougar	H. Snelson (Engineers) Ltd.	
G–BGNW	Boeing 737-219ADV	Britannia Airways Ltd. *George Stephenson*	
G–BGNX	PA-28RT-201T Turbo Arrow IV	Dickson Bros. Ltd.	
G–BGNZ	Cessna FRA.150L	Renco (Aviation) Ltd.	
G–BGOA	Cessna FR.182Q	The Forestry Commission	
G–BGOB	—	—	
G–BGOC	Cessna F.152	Rogers Aviation Sales Ltd.	
G–BGOD	Colt 77A balloon	J. R. Gore *Harvey Wallbanger*	
G–BGOE	Beech 76 Duchess	Ackroyd Westwood & Associates Ltd.	
G–BGOF	Cessna F.152	Renco (Aviation) Ltd.	
G–BGOG	PA-28-161 Warrior II	Seven Seas Finance Ltd.	
G–BGOH	Cessna F.182Q	Zonex Ltd.	
G–BGOI	Cameron O-56 balloon	Balloon Stable Ltd. *Skymaster*	
G–BGOJ	Cessna F.150L	J. Edwards	
G–BGOL	PA-28R-201T Turbo Arrow III	M. G. Tyrrell & Co. Ltd.	
G–BGOM	PA-31-310 Navajo	Air Commuter Ltd.	
G–BGON	GA-7 Cougar	Cabair Ltd.	
G–BGOO	Colt 56 SS balloon	British Gas Corporation	
G–BGOP	Dassault Falcon 20F	Datsun (UK) Ltd.	
G–BGOR	AT-6D Harvard III	P. Mercer	
G–BGOS	—	—	
G–BGOU	AT-6C Harvard IIA	A. P. Snell & ptnrs.	
G–BGOV	AT-6C Harvard IIA	Aces High Ltd.	
G–BGOX	PA-31-350 Navajo Chieftain	Fairflight Ltd.	
G–BGOY	PA-31-350 Navajo Chieftain	Fairflight Ltd.	
G–BGOZ	Westland-Bell 47G-3B1	GSM Helicopters	
G–BGPA	Cessna 182L	Arch Motor & Manufacturing Co. Ltd.	
G–BGPB	AT-16 Harvard IV (385)	A. G. Walker & R. Lamplough	
G–BGPC	D.H.C.-6 Twin Otter 310	Loganair Ltd.	
G–BGPD	Piper L-4H Cub	E. P. Beck	
G–BGPE	Thunder Ax6-56 balloon	C. Wolstenholme *Sergeant Pepper*	
G–BGPF	Thunder Ax6-56Z balloon	Thunder Balloons Ltd.	
G–BGPG	AA-5B Tiger	Trehaven Trust Ltd.	
G–BGPH	AA-5B Tiger	B.I. Aviation Services Ltd.	
G–BGPI	Plumb BGP-1	B. G. Plumb	
G–BGPJ	PA-28-161 Warrior II	Truman Aviation Ltd.	
G–BGPK	AA-5B Tiger	Ann Green Manufacturing Co. Ltd.	
G–BGPL	PA-28-161 Warrior II	Burnthills Aviation Ltd.	
G–BGPM	Evans VP-2	T. G. Painter	
G–BGPN	PA-18-150 Super Cub	Moore Bros. (Builders) Ltd.	
G–BGPO	PA-25 Pawnee 235	CKS Air Ltd.	
G–BGPP	PA-25 Pawnee 235	Sprayfields (Scothern) Ltd.	
G–BGPS	Aero Commander 200D	K. Davison	
G–BGPT	Parker Teenie Two	P. N. Haigh	
G–BGPU	PA-28 Cherokee 140	Air Navigation & Trading Co. Ltd.	
G–BGPV	BN-2A Islander	BN (Bembridge) Ltd.	
G–BGPW	BN-2A Islander	BN (Bembridge) Ltd.	
G–BGPX	BN-2A Islander	BN (Bembridge) Ltd.	
G–BGPY	BN-2A Islander	BN (Bembridge) Ltd.	
G–BGPZ	MS.890A Rallye Commodore	M. J. Kirk	
G–BGRA	Taylor JT.2 Titch	J. R. C. Thompson	
G–BGRB	AA-5B Tiger	Fairprime Ltd.	
G–BGRC	PA-28 Cherokee 140	Snowdon Aviation Ltd.	
G–BGRE	Beech A200 Super King Air	Dowty Group Services Ltd.	
G–BGRF	Beech 95-58P Baron	Hitchins (Hatfield) Ltd.	
G–BGRG	Beech 76 Duchess	Lambson Holdings Ltd.	
G–BGRH	Robin DR.400 2+2	Headcorn Flying School Ltd.	
G–BGRI	Jodel DR.1051	R. M. McEwan	
G–BGRJ	Cessna T.310R	C.C.M. (Lancs) Ltd. & Gledhill Water Storage Ltd.	
G–BGRK	PA-38-112 Tomahawk	Goodwood Terrena Ltd.	
G–BGRL	PA-38-112 Tomahawk	Goodwood Terrena Ltd.	

Notes	Reg.	Type	Owner or Operator
	G–BGRM	PA-38-112 Tomahawk	Goodwood Terrena Ltd.
	G–BGRN	PA-38-112 Tomahawk	Goodwood Terrena Ltd.
	G–BGRO	Cessna F.172M	Citation Flying Services Ltd.
	G–BGRP	—	—
	G–BGRR	PA-38-112 Tomahawk	Cormack (Aircraft Services) Ltd.
	G–BGRS	Thunder Ax7-77Z balloon	P. Hassall
	G–BGRT	Steen Skybolt	R. C. Taverson
	G–BGRU	—	—
	G–BGRV	—	—
	G–BGRW	—	—
	G–BGRX	PA-38-112 Tomahawk	Flamingo Aviation Ltd.
	G–BGRY	SC.7 Skyvan Srs. 3	Short Bros. Ltd.
	G–BGRZ	BN-2A Islander	Capital Aviation Sales (UK) Ltd.
	G–BGSA	SOCATA Rallye 150GT	G. A. Schulz & ptnrs.
	G–BGSB	P.56 Provost T.1 (WV494)	J. Powell
	G–BGSC	Ayres Thrush Commander S2R-T34/500	Shoreham Flight Simulation Ltd.
	G–BGSD	Pitts S-2A Special	Rothmans International Ltd.
	G–BGSE	Pitts S-2A Special	Rothmans International Ltd.
	G–BGSF	Pitts S-2A Special	Rothmans International Ltd.
	G–BGSG	PA-44-180 Seminole	Spooner Aviation Ltd.
	G–BGSH	PA-38-112 Tomahawk	Western Flying Services Ltd.
	G–BGSI	PA-38-112 Tomahawk	Burnthills Plant Hire Ltd.
	G–BGSJ	Piper J-3C-65 Cub	H. A. Bridgman
	G–BGSK	AA-5A Cheetah	D. E. Nixon
	G–BGSL	AA-5A Cheetah	Cabair Ltd.
	G–BGSM	SOCATA Rallye 150GT	Tyre & Tune
	G–BGSN	Enstrom F-28C-2	Nicelynn Ltd.
	G–BGSO	PA-31-310 Navajo	Cranbury Estates Ltd.
	G–BGSS	PA-38-112 Tomahawk	RB Aviation Ltd.
	G–BGST	Thunder Ax7-65 balloon	L. H. T. Large & ptnrs. *Eclipse*
	G–BGSV	Cessna F.172N	Francis Mander Aviation
	G–BGSW	Beech F33 Debonair	Summers Transport Ltd.
	G–BGSX	Cessna F.152	Salair
	G–BGSY	GA-7 Cougar	M. Harland & Son
	G–BGSZ	GA-7 Cougar	Cabair Ltd.
	G–BGTA	Firebird Bunce B.500 balloon	S. J. Bunce
	G–BGTB	SOCATA TB.10 Tobago	Air Touring Services Ltd.
	G–BGTC	Auster AOP.9	P. J. Marsham
	G–BGTD	H.S.125 Srs. 700B	Rank Xerox (UK) Ltd.
	G–BGTE	R. Commander 114A	Fletcher Rentals Ltd.
	G–BGTF	PA-44-180 Seminole	Lincoln Street Motors Birmingham Ltd.
	G–BGTG	PA-23 Aztec 250	Pearson Aviation Ltd.
	G–BGTH	PA-23 Aztec 250F	Western Flying Services Ltd.
	G–BGTI	Piper J-3C-65 Cub	W. J. Clarke & A. C. Broad
	G–BGTJ	PA-28 Cherokee 180	Serendipity Aviation
	G–BGTK	Cessna FR.182RG	Kestrel Air Services Ltd.
	G–BGTL	GY-20 Minicab	A. K. Lang
	G–BGTM	Thunder Ax6-56Z balloon	Engineering Polymers Ltd.
	G–BGTN	—	—
	G–BGTO	—	—
	G–BGTP	Robin HR.100/210	A. E. James & P. Houghton
	G–BGTR	PA-28 Cherokee 140	Keenair Services Ltd.
	G–BGTS	PA-28 Cherokee 140	Keenair Services Ltd.
	G–BGTT	Cessna 310R	Air Atlantique
	G–BGTU	BAC One-Eleven 409	Turbo Union Ltd.
	G–BGTV	Boeing 737-2T5	Orion Airways Ltd.
	G–BGTW	Boeing 737-2T5	Orion Airways Ltd.
	G–BGTX	Jodel D.117	D. A. Davidson
	G–BGTY	Boeing 737-2Q8	Orion Airways Ltd.
	G–BGTZ	Westland-Bell 47G-3B1	GSM Helicopters Ltd.
	G–BGUA	PA-38-112 Tomahawk	Northern Executive Aviation Ltd.
	G–BGUB	PA-32-300 Cherokee Six	J. Beckers & ptnrs.
	NOTE: The G-BGUx sequence will not be issued unless specifically requested.		
	G–BGUY	Cameron V-56 balloon	G. V. Beckwith
	G–BGVA	Cessna 414A	A. Masters Ltd.
	G–BGVB	Robin DR.315	R. J. Fray
	G–BGVC	—	—
	G–BGVE	CP1310-C3 Super Emeraude	B. Arnall
	G–BGVF	Colt 77A balloon	Hot Air Balloon Co. Ltd.
	G–BGVG	Beech 76 Duchess	Eagle Aircraft Services Ltd.
	G–BGVH	Beech 76 Duchess	Laura Ashley Ltd.

Reg.	Type	Owner or Operator	Notes
G–BGVI	Cessna F.152	R. T. Haddow	
G–BGVJ	PA-28 Cherokee 180	H. Devonish	
G–BGVK	PA-28-161 Warrior II	W. E. Wordworth	
G–BGVL	PA-38-112 Tomahawk	Cambrian Flying Club	
G–BGVM	WLS Cassutt IIM	J. T. Mirley	
G–BGVN	PA-28RT-201 Arrow IV	Essex Aviation Ltd.	
G–BGVP	Thunder Ax6-56Z balloon	Hot Air Balloon Co. Ltd.	
G–BGVR	Thunder Ax6-56Z balloon	Hot Air Balloon Co. Ltd.	
G–BGVS	Cessna F.172M	P.D.A. Aviation Ltd.	
G–BGVT	Cessna R.182RG	T. Hayselden (Doncaster) Ltd.	
G–BGVU	PA-28 Cherokee 180	Solair Ltd.	
G–BGVV	AA-5A Cheetah	Peter Turnbull (York) Ltd.	
G–BGVW	AA-5A Cheetah	N. R. Harford	
G–BGVX	Cessna P.210N	Rogers Aviation Sales Ltd.	
G–BGVY	AA-5B Tiger	Mannix Aviation Ltd.	
G–BGVZ	PA-28-181 Archer II	McCullough Aviation Ltd.	
G–BGWA	GA-7 Cougar	Cabair Ltd.	
G–BGWC	Robin DR.400/180	E. F. Braddon	
G–BGWD	Robin HR.100/285	Hordell Engineering Ltd.	
G–BGWE	Scheibe SF.28B	J. R. S. Pirquet	
G–BGWF	PA-18-150 Super Cub	Farmair Ltd.	
G–BGWG	PA-18-150 Super Cub	Farmair Ltd.	
G–BGWH	PA-18-150 Super Cub	Farmair Ltd.	
G–BGWI	Cameron V-56 balloon	Army Balloon Club	
G–BGWJ	Sikorsky S-61N	Bristow Helicopters Ltd.	
G–BGWK	Sikorsky S-61N	Bristow Helicopters Ltd.	
G–BGWM	PA-28-181 Archer II	Western Flying Services Ltd.	
G–BGWN	PA-38-112 Tomahawk	Western Flying Services Ltd	
G–BGWO	Jodel D.112	J. R. Ramshaw	
G–BGWP	MBB Bo 105C	Management Aviation Ltd.	
G–BGWR	Cessna U.206A	Berrard Ltd.	
G–BGWS	Enstrom F-280C Shark	G. Firbank & N. M. Grimshaw	
G–BGWT	WS-58 Wessex 60 Srs. I	Bristow Helicopters Ltd.	
G–BGWU	PA-38-112 Tomahawk	Burnthills Aviation Ltd.	
G–BGWV	Aeronca 7AC Champion	RFC Flying Group	
G–BGWW	PA-23 Aztec 250E	Chargewell Ltd.	
G–BGWX	—	—	
G–BGWY	Thunder Ax6-56Z balloon	Thunder Balloons Ltd.	
G–BGWZ	Eclipse Super Eagle	J. S. Long	
G–BGXA	Piper J-3C-65 Cub	D. J. Dulborough	
G–BGXB	PA-38-112 Tomahawk	Cambrian Flying Club	
G–BGXC	SOCATA TB.10 Tobago	A. J. Halliday	
G–BGXD	SOCATA TB.10 Tobago	J. Traynor	
G–BGXE	Douglas DC-10-30	Laker Airways	
G–BGXF	Douglas DC-10-30	Laker Airways	
G–BGXG	Douglas DC-10-30	Laker Airways	
G–BGXH	Douglas DC-10-30	Laker Airways *Florida Belle*	
G–BGXI	Douglas DC-10-30	Laker Airways	
G–BGXJ	Partenavia P.68B	DK Aviation Ltd.	
G–BGXK	Cessna 310R	B. H. Turner & ptnrs.	
G–BGXL	Benson B.8MV	H. A. Bancroft-Wilson	
G–BGXN	PA-38-112 Tomahawk	Air Navigation & Trading Co. Ltd.	
G–BGXO	PA-38-112 Tomahawk	Travelwell Coach & Commercial Services Ltd.	
G–BGXP	Westland-Bell 47G-3B1	Specbridge Ltd.	
G–BGXR	Robin HR.200/100	J. H. Spanton	
G–BGXS	PA-28-236 Dakota	C.S.E. Aviation Ltd.	
G–BGXT	SOCATA TB.10 Tobago	County Aviation Ltd.	
G–BGXU	WMB-1 balloon	C. J. Dodd & ptnrs.	
G–BGXV	Piper J-3C-65 Cub	W. Scrope	
G–BGXW	Thunder Ax6-56 balloon	Thunder Balloons Ltd.	
G–BGXX	Jodel DR.1051M1	Gladaircraft Ltd.	
G–BGXY	Sikorsky S-76	Bristow Helicopters Ltd.	
G–BGXZ	Cessna FA.152	Renco (Aviation) Ltd.	
G–BGYA	Aerostar 601P	Mann Aviation Sales Ltd.	
G–BGYB	—	—	
G–BGYC	Aerostar 601P	Mann Aviation Sales Ltd.	
G–BGYE	—	—	
G–BGYF	Bell 206B JetRanger 2	Helicopter Hire Ltd.	
G–BGYG	PA-28-161 Warrior II	C.S.E. (Aircraft Services) Ltd.	
G–BGYH	PA-28-161 Warrior II	C.S.E. (Aircraft Services) Ltd.	
G–BGYJ	Boeing 737-204	Britannia Airways Ltd. *Sir Barnes Wallis*	

Notes	Reg.	Type	Owner or Operator
	G–BGYK	Boeing 737-204	Britannia Airways Ltd. *R. J. Mitchell*
	G–BGYL	Boeing 737-204	Britannia Airways Ltd. *Jean Batten*
	G–BGYM	PA-31-350 Navajo Chieftain	Airmore International Ltd.
	G–BGYN	PA-18-150 Super Cub	A. G. Walker
	G–BGYO	Thunder Ax6-56Z balloon	Wrangler-Bluebell Apparel Ltd.
	G–BGYP	GA-7 Cougar	Bambair & Nixon Aviation
	G–BGYR	H.S.125 Srs. 600B	British Aerospace
	G–BGYS	EMB-110P2 Bandeirante	Air UK
	G–BGYT	EMB-110P1 Bandeirante	Air UK
	G–BGYU	EMB-110P2 Bandeirante	Air UK
	G–BGYV	EMB-110P1 Bandeirante	Air UK
	G–BGYW	Hughes 269C	Minster Aviation Ltd.
	G–BGYZ	Hughes 269C	G. Hewson
	G–BGZA	ICA IS-28M2	F. A. Wright
	G–BGZB	Cessna F.177RG	P. C. Faber
	G–BGZC	C.A.S.A. I.131 Jungmann	J. E. Douglas
	G–BGZE	PA-38-112 Tomahawk	Merlin Service Stations Ltd.
	G–BGZF	PA-38-112 Tomahawk	C.S.E. Aviation Ltd.
	G–BGZG	PA-38-112 Tomahawk	Kilmartin Leasing & Finance
	G–BGZH	PA-38-112 Tomahawk	C.S.E. Aviation Ltd.
	G–BGZI	PA-38-112 Tomahawk	C.S.E. Aviation Ltd.
	G–BGZJ	PA-38-112 Tomahawk	RB Aviation Ltd.
	G–BGZK	Westland-Bell 47G-3B1	K. McDonald
	G–BGZL	Eiri PIK-20E	J. Hulme & ptnrs.
	G–BGZN	WMB.2 Windtracker balloon	S. R. Woolfries
	G–BGZO	M.S.880B Rallye Club	W. G. R. Wunderlich
	G–BGZP	D.H.C.-6 Twin Otter 310	Brymon Aviation Ltd.
	G–BGZR	Meagher Model balloon Mk. I	S. C. Meagher
	G–BGZS	Keirs Heated Air Tube	K. J. Faulkner
	G–BGZW	PA-38-112 Tomahawk	Truman Aviation Ltd.
	G–BGZX	PA-32 Cherokee Six 260	E. D. Brown
	G–BGZY	Jodel D.120	P. J. Sebastian
	G–BGZZ	Thunder Ax6-56 balloon	J. M. Robinson
	G–BHAA	Cessna 152	Herefordshire Aero Club Ltd.
	G–BHAB	Cessna 152	Herefordshire Aero Club Ltd.
	G–BHAC	Cessna A.152	Herefordshire Aero Club Ltd.
	G–BHAD	Cessna A.152	Shropshire Aero Club Ltd.
	G–BHAE	Cameron V-56 balloon	K. Norris
	G–BHAF	PA-38-112 Tomahawk	C.S.E. Aviation Ltd.
	G–BHAG	Scheibe SF.25E Super Falke	British Gliding Association
	G–BHAH	Sikorsky S-61N	Management Aviation Ltd.
	G–BHAI	Cessna F.152	Channel Islands Aero Holdings Ltd.
	G–BHAJ	Robin DR.400/180	Crocker Aviation Services
	G–BHAK	PA-28RT-201 Arrow IV	C.S.E. Aviation Ltd.
	G–BHAL	Rango Saffery S.200 SS	A. M. Lindsay *Anneky Panky*
	G–BHAM	Thunder Ax6-56 balloon	D. Sampson
	G–BHAO	Beech 76 Duchess	Eagle Aircraft Services Ltd.
	G–BHAR	Westland-Bell 47G-3B1	G. Greenall
	G–BHAT	Thunder Ax7-77 balloon	Thunder Balloons Ltd.
	G–BHAU	B.175 Britannia 253F	Redcoat Air Cargo Ltd.
	G–BHAV	Cessna F.152	A. G. Chrismas Ltd.
	G–BHAW	Cessna F.172N	Spooner Aviation Ltd.
	G–BHAX	Enstrom F-28C	Spooner Aviation Ltd.
	G–BHAY	PA-28RT-201 Arrow IV	M. A. Lenihan
	G–BHAZ	Cessna F.152	Edinburgh Flying Club
	G–BHBA	Campbell Cricket	S. M. Irwin
	G–BHBB	Colt 77C balloon	S. D. Bellew
	G–BHBC	GA-7 Cougar	Cabair Ltd. (G-MALA/G-BGCP)
	G–BHBD	Colt 77A balloon	Colt Balloons Ltd.
	G–BHBE	Westland Bell 47G-3B1	Heliwork Ltd.
	G–BHBF	Sikorsky S-76	Bristow Helicopters
	G–BHBG	PA-32R-300 Lance	Spooner Aviation Ltd.
	G–BHBH	Cessna 550 Citation II	RTZ Services Ltd.
	G–BHBI	Mooney M.20J	Express Aviation Services Ltd.
	G–BHBJ	Cameron D-96 airship	Blickframe Ltd.
	G–BHBK	Viscount V-5 balloon	B. Hargraves & B. King
	G–BHBL	L.1011-385 TriStar 200	British Airways
	G–BHBM	L.1011-385 TriStar 200	British Airways
	G–BHBN	L.1011-385 TriStar 200	British Airways
	G–BHBO	L.1011-385 TriStar 200	British Airways
	G–BHBP	L.1011-385 TriStar 200	British Airtours Ltd.
	G–BHBR	L.1011-385 TriStar 200	British Airtours Ltd.

G–BGTY (Top) Boeing 737-2Q8 of Orion Airways

G–BHCR (Centre) Pilatus PC-6/B2-H2 Porter

G–BHJY (Bottom) Embraer EMB-110P1 Bandeirante of Jersey European Airways

Reg.	Type	Owner or Operator	Notes
G–BHBS	PA-28RT-201T Turbo Arrow IV	Spooner Aviation Ltd.	
G–BHBT	MA.5 Charger	R. G. & C. J. Maidment	
G–BHBV	Westland Bell 47G-3B1 (Soloy)	Helicrops Ltd.	
G–BHBW	Westland Bell 47G-3B1	Heliwork Ltd.	
G–BHBX	Westland Bell 47G-3B1	Heliwork Ltd.	
G–BHBY	Westland Bell 47G-3B1	Heliwork Ltd.	
G–BHBZ	Partenavia P.68B	C. A. Church Ltd.	
G–BHCA	Fokker D.VIII Replica	Leisure Sport Ltd.	
G–BHCB	AA-5A Cheetah	Cabair Ltd.	
G–BHCC	Cessna 172M	Air Executive Ltd.	
G–BHCD	—	—	
G–BHCE	Jodel D.112	G. F. M. Garner	
G–BHCF	WMB.2 Windtracker balloon	C. J. Dodd & ptnrs.	
G–BHCI	Jodel D.140	Kent Gliding Club Ltd.	
G–BHCJ	H.S.748 Srs. 2	Dan-Air Services Ltd.	
G–BHCK	AA-5A Cheetah	Cabair Ltd.	
G–BHCL	Boeing 737-200ADV	Orion Airways Ltd.	
G–BHCM	Cessna F.172H	The English Connection Ltd.	
G–BHCO	SOCATA TB.10 Tobago	Butler Heating Associates Ltd.	
G–BHCP	Cessna F.152	Neptune Securities Ltd.	
G–BHCR	Pilatus PC-6/B2-H2 Porter	Peterborough Parachute Centre Ltd.	
G–BHCS	Cameron N-77 balloon	Cameron Balloons Ltd.	
G–BHCT	PA-23 Aztec 250	Colt Transport Ltd.	
G–BHCW	PA-22 Tri-Pacer 150	Tredair	
G–BHCX	Cessna F.152	Westair Flying Services Ltd.	
G–BHCZ	PA-38-112 Tomahawk	Capital Aviation Sales (UK) Ltd.	
G–BHDA	Shultz balloon	G. F. Fitzjohn	
G–BHDB	Maule M5-235 Lunar Rocket	Cleanacres Ltd.	
G–BHDC	Slingsby T.65A Vega	Slingsby Engineering Ltd.	
G–BHDD	V.668 Varsity T.1 (WL626)	Loughborough & Leicester Aircraft Museum	
G–BHDE	SOCATA TB.10 Tobago	G. Brookes	
G–BHDF	Beech 95-58P Baron	Eagle Aircraft Services Ltd.	
G–BHDH	Douglas DC-10-30	British Caledonian Airways *Sir Walter Scott*	
G–BHDI	Douglas DC-10-30	British Caledonian Airways	
G–BHDJ	Douglas DC-10-30	British Caledonian Airways	
G–BHDK	Boeing B-29A-BN (461748)	Imperial War Museum	
G–BHDM	Cessna F.152 II	Tayside Aviation Ltd.	
G–BHDO	Cessna F.182Q II	Air Service Training Ltd.	
G–BHDP	Cessna F.182Q II	Air Service Training Ltd.	
G–BHDR	Cessna F.152 II	Air Service Training Ltd.	
G–BHDS	Cessna F.152 II	Air Service Training Ltd.	
G–BHDT	SOCATA TB.10 Tobago	A. W. Prior	
G–BHDU	Cessna F.152 II	P. N. Voysey	
G–BHDV	Cameron V-77 balloon	D. E. P. Price	
G–BHDW	Cessna F.152	Spooner Aviation Ltd.	
G–BHDX	Cessna F.172N	R. J. D. Rimmer	
G–BHDZ	Cessna F.172N	Northair Aviation Ltd.	
G–BHEC	Cessna F.152	Vectstar Ltd.	
G–BHED	Cessna FA.152	Northair Aviation Ltd.	
G–BHEE	Partenavia P.68B	DK Aviation Services Ltd.	
G–BHEF	Partenavia P.68B	DK Aviation Services Ltd.	
G–BHEG	Jodel D.150	E. J. Horsfall	
G–BHEH	Cessna 310G	P. D. Higgs	
G–BHEI	Cessna T.182RG	Northair Aviation Ltd.	
G–BHEJ	—	—	
G–BHEK	CP.1315C-3 Super Emeraude	D. B. Winstanley	
G–BHEL	Jodel D.117	L. L. Vickers	
G–BHEM	Benson B.8M	E. Kenny	
G–BHEN	Cessna FA.152	Leicestershire Aero Club Ltd.	
G–BHEO	Cessna F.182RG	Cosworth Engineering Ltd.	
G–BHEP	Cessna 172 RG Cutlass	Memec Systems Ltd.	
G–BHER	SOCATA TB.10 Tobago	W. R. M. Dury	
G–BHET	SOCATA TB.10 Tobago	Potissue Ltd.	
G–BHEU	Thunder Ax7-65 balloon	M. H. R. Govett *Polo Moche*	
G–BHEV	PA-28R Cherokee Arrow 200	AF Aviation Ltd.	
G–BHEW	Sopwith Triplane Replica (N5430)	The Hon. Patrick Lindsay	
G–BHEX	Colt 56A balloon	A. S. Dear & ptnrs. *Super Wasp*	
G–BHEY	Pterodactyl O.R.	High School of Hang Gliding Ltd.	

Notes	Reg.	Type	Owner or Operator
	G–BHEZ	Jodel D.150	E. J. Horsfall
	G–BHFA	Pterodactyl O.R.	High School of Hang Gliding Ltd.
	G–BHFB	Pterodactyl O.R.	High School of Hang Gliding Ltd.
	G–BHFC	Cessna F.152	Trehaven Trust Ltd.
	G–BHFD	D.H.C.-6 Twin Otter 310	Loganair Ltd.
	G–BHFE	PA-44-180 Seminole	N. Clayton
	G–BHFF	Jodel D.112	J. A. Scott
	G–BHFG	SNCAN SV-4C	R. J. Godfrey
	G–BHFH	PA-34-200T-II Seneca	Heltor Ltd.
	G–BHFI	Cessna F.152	The BAe (Warton) Flying Group
	G–BHFJ	PA-28RT-201T Turbo Arrow IV	C.S.E. Aviation Ltd.
	G–BHFK	PA-28-151 Warrior	L. Henderson
	G–BHFL	PA-28 Cherokee 180	R. & C. Lord
	G–BHFM	Murphy S.200 balloon	M. Murphy
	G–BHFN	Eiri PIK-20E	B. A. Eastwell
	G–BHFP	Eiri PIK-20E	H. C. Mackinnon & ptnrs.
	G–BHFR	Eiri PIK-20E	G. Mackie
	G–BHFS	Robin DR.400/180	Flair (Soft Drinks) Ltd.
	G–BHFU	Saffery S.330 balloon	N. H. Ponsford *Jennie Toyhill*
	G–BHFV	Rango-Saffery S.200 balloon	N. H. Ponsford *Cloud Sculptor*
	G–BHFW	Rango NA-1000 gas balloon	N. H. Ponsford *Alecto*
	G–BHFY	Beech 95-B58 Baron	Kebbell Holdings Ltd.
	G–BHFZ	Saffery S.200 balloon	D. Morris
	G–BHGA	PA-31-310 Navajo	Maimot Ltd.
	G–BHGB	Colt 77A balloon	Colt Balloons Ltd.
	G–BHGC	PA-18-150 Super Cub	Herefordshire Gliding Club Ltd.
	G–BHGD	Cessna 421C	Shuimpex (Services) Ltd.
	G–BHGE	Boeing 720-051B	Monarch Airlines Ltd.
	G–BHGF	Cameron V-56 balloon	I. T. & H. Seddon *Biggles*
	G–BHGG	Cessna F.172N	Tosair Ltd.
	G–BHGH	—	—
	G–BHGI	—	—
	G–BHGJ	Jodel D.120	B. Brooks
	G–BHGK	Sikorsky S-76	North Scottish Helicopters Ltd.
	G–BHGM	Beech 76 Duchess	Eagle Aircraft Services Ltd.
	G–BHGN	Evans VP-1	A. R. Cameron
	G–BHGO	PA-32 Cherokee Six 260	L. J. Steward
	G–BHGP	SOCATA TB.10 Tobago	G-Plane Ltd.
	G–BHGR	Robin DR.315	Headcorn Flying School Ltd.
	G–BHGS	PA-31-350 Navajo Chieftain	Jersey European Airways
	G–BHGT	Beech B90 King Air	Airmore Aviation Ltd.
	G–BHGU	WMB.2 Windtracker balloon	I. D. Bamber & ptnrs.
	G–BHGV	Keirs captive balloon	K. J. Faulkner
	G–BHGW	Colt 12A balloon	Colt Balloons Ltd.
	G–BHGX	Colt 56B balloon	Colt Balloons Ltd. *Vomibag*
	G–BHGY	PA-28R Cherokee Arrow 200	Suca Marketing Ltd.
	G–BHGZ	Bucker Bu.131 Jungmann	Sywell Aviation Services Ltd.
	G–BHHA	EMB-110P1 Bandeirante	Loganair Ltd.
	G–BHHB	Cameron V-77 balloon	I. G. N. Franklin
	G–BHHC	—	—
	G–BHHD	—	—
	G–BHHE	Jodel DR.1051/M1	H. J. W. Ellison
	G–BHHF	—	—
	G–BHHG	Cessna F.152	Rogers Aviation Sales Ltd.
	G–BHHH	Thunder Ax7-65 balloon	Thunder Balloons Ltd.
	G–BHHI	Cessna F.152	Rogers Aviation Sales Ltd.
	G–BHHJ	Cessna F.152	Rogers Aviation Sales Ltd.
	G–BHHK	Cameron N-77 balloon	S. Bridge & ptnrs.
	G–BHHN	Cameron V-77 balloon	Itchen Valley Balloon Group
	G–BHHO	PA-28 Cherokee 180	Peter Clifford Aviation Ltd.
	G–BHHP	Cameron 0-105 balloon	A. J. Machinery
	G–BHHR	Robin DR.400/180R	R. Jones
	G–BHHX	Jodel D.112	S. C. Delwarte
	G–BHHY	G-164 Ag Cat D	Miller Aerial Spraying Ltd.
	G–BHHZ	Rotorway Scorpion 133	P. A. Gunn & D. Willingham
	G–BHIA	Cessna F.152	Rogers Aviation Sales Ltd.
	G–BHIB	Cessna F.182Q	Northair Aviation Ltd.
	G–BHIC	Cessna F.182Q	Northair Aviation Ltd.
	G–BHID	SOCATA TB.10 Tobago	W. B. Pinckney & Sons Farming Co. Ltd.
	G–BHIF	Colt 160A balloon	Colt Balloons Ltd.
	G–BHIG	Colt 31A balloon	Colt Balloons Ltd.

Reg.	Type	Owner or Operator	Notes
G–BHIH	Cessna F.172N	Watkiss Group Aviation Ltd.	
G–BHII	Cameron V-77 balloon	R. M. Davies	
G–BHIJ	Eiri PIK-20E-1	R. W. Hall & ptnrs.	
G–BHIK	Adam RA-14 Loisirs	C. C. Lovell	
G–BHIL	PA-28-161 Warrior II	C.S.E Aviation Ltd.	
G–BHIM	Jodel D.112	G. P. Badham	
G–BHIN	Cessna F.152	Denham Flying Training School Ltd.	
G–BHIO	H.S.125 Srs. 700B	San Miguel Corporation	
G–BHIP	Thunder Ax3-1 balloon	Thunder Balloons Ltd.	
G–BHIR	PA-28R Cherokee Arrow 200	J. P. Aviation	
G–BHIS	Thunder Ax7-65 balloon	Thunder Balloons Ltd. *Yo-Yo*	
G–BHIT	SOCATA TB.9 Tampico	Air Touring Services Ltd.	
G–BHIV	AS.350B Ecureuil	Marley Tile Co. Ltd.	
G–BHIW	—	—	
G–BHIX	Cameron V-56 balloon	P. S. Richardson	
G–BHIY	Cessna F.150K	W. H. Cole	
G–BHIZ	PA-31 Navajo	L. J. Steward	
G–BHJA	Cessna A.152	Capital Aviation Sales Ltd.	
G–BHJC	Cessna A.152	Capital Aviation Sales Ltd.	
G–BHJE	Beech 95-58P Baron	Eagle Aircraft Services Ltd.	
G–BHJF	SOCATA TB.10 Tobago	G. S. Goodsir & A. M. Fricker	
G–BHJG	Cessna 172RG Cutlass	Northair Aviation Ltd.	
G–BHJH	Cessna 421C	Northair Aviation Ltd.	
G–BHJI	Mooney 201	T. R. Bamber & B. Refson	
G–BHJK	Maule M5-235C Lunar Rocket	A. J. Machinery Ltd.	
G–BHJL	—	—	
G–BHJN	Fournier RF-4D	G. G. Milton	
G–BHJO	PA-28-161 Warrior II	Cormack (Aircraft Services) Ltd.	
G–BHJP	Partenavia P-68C	Shirlstar Container Transport Ltd.	
G–BHJR	Saffery S.200 balloon	P. N. S. Bussey	
G–BHJS	Partenavia P-68B	Fairoaks Aviation Services Ltd.	
G–BHJT	Cessna 414A	Northair Aviation Ltd.	
G–BHJU	Robin DR.400/2+2	Headcorn Flying School Ltd.	
G–BHJV	Cameron Rectangular SS 60 balloon	R. J. Reynolds Tobacco Co. Ltd.	
G–BHJW	Cessna F.152	Leicestershire Aero Club Ltd.	
G–BHJX	Partenavia P-68C	Partenavia Sales Ltd.	
G–BHJY	EMB-110P1 Bandeirante	Jersey European Airways	
G–BHJZ	EMB-110P1 Bandeirante	Jersey European Airways	
G–BHKA	Evans VP-1	M. L. Perry	
G–BHKB	Westland-Bell 47G-3B1 (Soloy)	Heliwork Ltd.	
G–BHKC	Westland-Bell 47G-3B1 (Soloy)	Heliwork Finance Ltd.	
G–BHKD	Westland-Bell 47G-3B1	Heliwork Ltd.	
G–BHKE	Benson B.8MV	V. C. Whitehead	
G–BHKG	Cessna F.172N	Westair Flying Services Ltd.	
G–BHKH	Cameron 0-65 balloon	D. G. Body	
G–BHKI	Cessna 402C	Northair Aviation Ltd.	
G–BHKJ	Cessna 421C	Northair Aviation Ltd.	
G–BHKL	Colt Flying Bottle 1A balloon	Hot-Air Balloon Co. Ltd.	
G–BHKM	Colt 12A balloon	Hot-Air Balloon Co. Ltd.	
G–BHKN	Colt 12A balloon	Hot-Air Balloon Co. Ltd.	
G–BHKO	Colt 12A balloon	Hot-Air Balloon Co. Ltd.	
G–BHKP	Colt 12A balloon	Hot-Air Balloon Co. Ltd.	
G–BHKR	Colt 12A balloon	Hot-Air Balloon Co. Ltd.	
G–BHKS	Beech E90 King Air	Eagle Aircraft Services Ltd.	
G–BHKT	Jodel D.112	G. V. Harfield & J. R. Notter	
G–BHKU	AA-5A Cheetah	Cabair Ltd.	
G–BHKV	AA-5A Cheetah	Cabair Ltd.	
G–BHKW	Westland-Bell 47G-3B1	Specbridge Ltd.	
G–BHKX	Beech 76 Duchess	Eagle Aircraft Services Ltd.	
G–BHKY	Cessna 310R II	Air Service Training Ltd.	
G–BHKZ	Cessna 172N	C. C. Donald	
G–BHLA	Cessna 421C	Northair Aviation Ltd.	
G–BHLD	—	—	
G–BHLE	Robin DR.400/180	V-G Instruments Ltd.	
G–BHLF	H.S.125 Srs. 700B	The Marconi Co. Ltd.	
G–BHLG	Sinnett X3-4 Windrifter	D. Sinnett	
G–BHLH	Robin DR.400/180	Trinecare Ltd.	
G–BHLI	R. Commander 690B	Commander Jetprop Ltd.	
G–BHLJ	Saffery-Rigg S.200 balloon	I. A. Rigg	

Notes	Reg.	Type	Owner or Operator
	G–BHLK	GA-7 Cougar	Cabair Ltd.
	G–BHLL	Cessna 421C	Northair Aviation Ltd.
	G–BHLM	Cessna 421C	Northair Aviation Ltd.
	G–BHLN	Cessna 441	Career Care Group Ltd.
	G–BHLO	—	—
	G–BHLP	—	—
	G–BHLT	D.H.82A Tiger Moth	R. L. Godwin
	G–BHLU	Fournier RF-3	G. G. Milton
	G–BHLV	CP.301A Emeraude	G. G. Milton
	G–BHLW	Cessna 120	J. E. Cummings
	G–BHLX	AA-5B Tiger	Executive Wings Ltd.
	G–BHLY	Sikorsky S-76A	Bristow Helicopters Ltd.
	G–BHLZ	GY-30 Supercab	M. Hutton
	G–BHMA	SIPA 903	Fairwood Flying Club
	G–BHMB	Rango NA.6 balloon	A. M. Lindsay *Daedelus*
	G–BHMC	M.S.880B Rallye Club	Bodmin Aviation
	G–BHMD	Rand KR-2	W. D. Francis
	G–BHME	WMB.2 Windtracker balloon	I. R. Bell & ptnrs.
	G–BHMF	Cessna FA.152	Denham Flying Training School Ltd.
	G–BHMG	Cessna FA.152	Denham Flying Training School Ltd.
	G–BHMH	Cessna FA.152	Denham Flying Training School Ltd.
	G–BHMI	Cessna F.172N	Denham Flying Training School Ltd.
	G–BHMJ	Avenger T.200-2112 balloon	R. Light
	G–BHMK	Avenger T.200-2112 balloon	P. Kinder
	G–BHML	Avenger T.200-2112 balloon	L. Caulfield
	G–BHMM	Avenger T.200-2112 balloon	M. Murphy
	G–BHMN	SOCATA TB.10 Tobago	S. E. Shah & ptnrs.
	G–BHMO	PA-20M Cerpa Special (Pacer)	N. J. Mathias & R. J. Maxey
	G–BHMR	Stinson 108-3	C. C. Lovell
	G–BHMS	PA-34-200T II Seneca	BHF Petroleum Ltd.
	G–BHMT	Evans VP-1	P. E. J. Sturgeon
	G–BHMU	Colt 21A balloon	Hot Air Balloon Co. Ltd.
	G–BHMV	Bell 206A JetRanger	Bristow Helicopters Ltd.
	G–BHMW	F.27 Friendship Mk. 200	Air UK
	G–BHMX	F.27 Friendship Mk. 200	Air UK
	G–BHMY	F.27 Friendship Mk. 200	Air UK
	G–BHMZ	F.27 Friendship Mk. 200	Air UK
	G–BHNA	Cessna F.152	W. E. B. Wordsword
	G–BHNC	Cameron O-65 balloon	D. & C. Bareford
	G–BHND	Cameron N-65 balloon	Hunter & Sons (Mells) Ltd.
	G–BHNE	Boeing 727-2J4	Dan-Air Services Ltd.
	G–BHNF	Boeing 727-2J4	Dan-Air Services Ltd.
	G–BHNG	PA-23 Aztec 250	Southern Wings Ltd.
	G–BHNH	Cessna 404 Titan	Northair Aviation Ltd.
	G–BHNJ	—	—
	G–BHNK	Jodel D.120A	A. Nash
	G–BHNL	Jodel D.112	R. North & J. Ware
	G–BHNM	PA-44-180 Seminole	C.S.E. Aviation Ltd.
	G–BHNN	PA-32R-301 Saratoga	Diagold Ltd.
	G–BHNO	PA-28-181 Archer II	C.S.E. Aviation Ltd.
	G–BHNP	Eiri PIK-20E-1	Astley's Ltd.
	G–BHNR	Cameron N-77 balloon	Bath University Hot-Air Balloon Club
	G–BHNS	PA-28R-201 Arrow III	Cluebond Ltd.
	G–BHNT	Cessna F.172N	Kestrel Air Services Ltd.
	G–BHNU	Cessna F.172N	Westair Flying Services Ltd.
	G–BHNV	Westland-Bell 47G-3B1	Specbridge Ltd.
	G–BHNW	—	—
	G–BHNX	Jodel D.117	H. Bosworth
	G–BHNY	—	—
	G–BHNZ	AA-5B Tiger	Cabair Ltd.
	G–BHOA	Robin DR.400/160	Headcorn Flying School Ltd.
	G–BHOB	Cessna 404	Euroair Transport Ltd.
	G–BHOC	R. Commander 112A	Tuscany Ltd.
	G–BHOD	—	—
	G–BHOE	—	—
	G–BHOF	Sikorsky S-61N	Bristow Helicopters Ltd.
	G–BHOG	Sikorsky S-61N	Bristow Helicopters Ltd.
	G–BHOH	Sikorsky S-61N	Bristow Helicopters Ltd.
	G–BHOI	Westland-Bell 47G-3B1	Helicopter Hire Ltd.
	G–BHOJ	Colt 12A balloon	Hot Air Balloon Co. Ltd.
	G–BHOK	Mooney M.20J	Express Aviation Services Ltd.
	G–BHOL	Jodel DR.1050	B. D. Deubelbeiss
	G–BHOM	PA-18 Super Cub 95	W. J. C. Scrope

G–BHLY (Top) Sikorsky S-76A/*A. S. Wright*

G–BIJB (Centre) Piper PA-18-150 Super Cub

G–BJHK (Bottom) EAA Acro Sport

Reg.	Type	Owner or Operator	Notes
G–BHON	Rango NA.6 balloon	Rango Kite Co. *Rosamund Hine*	
G–BHOO	Thunder Ax7 balloon	D. Livesey & J. M. Purves	
G–BHOP	Thunder Ax3 balloon	Thunder Balloons Ltd.	
G–BHOR	PA-28-161 Warrior II	C.S.E. Aviation Ltd.	
G–BHOT	—	—	
G–BHOU	Cameron V-65 balloon	D. I. Gray-Fisk	
G–BHOV	Partenavia P.68C	Lowland Aero Services Ltd.	
G–BHOW	Beech 95-58P Baron	Anglo-African Machinery Ltd.	
G–BHOX	Boeing 707-123B	Monarch Airlines Ltd. (G–TJAB)	
G–BHOY	Boeing 707-123B	Monarch Airlines Ltd. (G–TJAC)	
G–BHOZ	SOCATA TB.9 Tampico	Air Touring Services Ltd.	
G–BHPA	Sikorsky S-61N	Management Aviation Ltd.	
G–BHPB	Sikorsky S-61N	Management Aviation Ltd.	
G–BHPC	Mooney M.20E	Executive Wings Ltd.	
G–BHPJ	Eagle Microlite	G. Breen	
G–BHPK	Piper J-3C-65 Cub	A. R. Johnston	
G–BHPL	C.A.S.A. 1.131 Jungmann	M. G. Jefferies	
G–BHPM	PA-18 Super Cub 95	P. Morgans	
G–BHPN	Colt 12A balloon	Colt Balloons Ltd.	
G–BHPO	Colt 12A balloon	Colt Balloons Ltd.	
G–BHPP	—	—	
G–BHPR	—	—	
G–BHPS	Jodel D.120A	C. J. & S. E. Francis	
G–BHPT	Piper J-3C-65 Cub	Sywell Aviation Services Ltd.	
G–BHPU	Sikorsky S-61N	British Caledonian Helicopters Ltd.	
G–BHPV	Cessna U.206G	Balfour Beatty Construction Ltd.	
G–BHPW	Aerostar 601P	C.S.E. Aviation Ltd.	
G–BHPX	Cessna 152	A. G. Chrismas Ltd.	
G–BHPY	Cessna 152	Titlecraft Ltd.	
G–BHPZ	Cessna 172N	Titlecraft Ltd.	
G–BHRA	R. Commander 114A	Rockwell Aviation Ltd.	
G–BHRB	Cessna F.152	Light Planes (Lancashire) Ltd.	
G–BHRC	PA-28-161 Warrior II	C.S.E. Aviation Ltd.	
G–BHRD	D.H.C.1 Chipmunk 22 (WP977)	B. C. Heywood & ptnrs.	
G–BHRE	Persephone S.200 balloon	Cupro-Sapphire Ltd.	
G–BHRF	Airborne Industries AB400 gas balloon	Balloon Stable Ltd.	
G–BHRG	—	—	
G–BHRH	Cessna FA.150K	Citation Flying Services Ltd.	
G–BHRI	Can-Can S.200 balloon	Cupro-Sapphire Ltd.	
G–BHRK	Colt Saucepan balloon	Colt Balloons Ltd.	
G–BHRL	Thunder Ax6-56Z balloon	Thunder Balloons Ltd.	
G–BHRM	Cessna F.152	Rogers Aviation Sales Ltd.	
G–BHRN	Cessna F.152	Rogers Aviation Sales Ltd.	
G–BHRO	R. Commander 112A	John Raymond Transport Ltd.	
G–BHRP	PA-44-180 Seminole	Rosemount Aviation	
G–BHRR	CP.301A Emeraude	W. H. Cole	
G–BHRS	ICA IS-28M2	Slingsby Engineering Ltd.	
G–BHRU	Petunia S.1000 balloon	Cupro-Sapphire Ltd.	
G–BHRV	Mooney M.20J	Express Aviation Services Ltd.	
G–BHRW	Jodel DR.221	J. M. Bisco	
G–BHRX	Cessna 340A	Martin Butler Associates Ltd.	
G–BHRY	Colt 56A balloon	Hot Air Balloon Co. Ltd.	
G–BHRZ	Colt 77B balloon	Colt Balloons Ltd.	
G–BHSA	Cessna 152	Skyviews & General Ltd.	
G–BHSB	Cessna 172N	W. R. Craddock & Son Ltd.	
G–BHSC	—	—	
G–BHSD	Scheibe SF.25E Super Falke	Lasham Gliding Soc. Ltd.	
G–BHSE	R. Commander 114	Magram Holdings Ltd.	
G–BHSF	AA-5A Cheetah	Cabair Ltd. (G–BHAS)	
G–BHSG	AB-206A JetRanger	Heliwork Ltd.	
G–BHSI	Jodel D.9	J. H. Betton	
G–BHSJ	CP.301A Emeraude	P. G. Phelps	
G–BHSL	C.A.S.A. 1.131 Jungmann	Cotswold Flying Group	
G–BHSM	AB-206B JetRanger	Heliwork Ltd.	
G–BHSN	Cameron N-56 balloon	Ballooning Endeavours Ltd.	
G–BHSO	PA-23 Aztec 250	Truman Aviation Ltd.	
G–BHSP	Thunder Ax7-77Z balloon	Chicago Instruments Ltd.	
G–BHSR	—	—	
G–BHSS	Pitts S-1C Special	T. R. & E. A. Wiltshire	
G–BHST	Hughes 369D	Abbey Hill Vehicle Services	
G–BHSU	H.S.125 Srs. 700B	Shell Aircraft Ltd.	
G–BHSV	H.S.125 Srs. 700B	Shell Aircraft Ltd.	

Notes	Reg.	Type	Owner or Operator
	G–BHSW	H.S.125 Srs. 700B	Shell Aircraft Ltd.
	G–BHSX	—	—
	G–BHSY	Jodel DR.1050	Mercury Aircraft & Engineering
	G–BHSZ	Cessna 152	Millet Shipping Ltd.
	G–BHTA	PA-28-236 Dakota	N. T. & Mrs. M. Smith
	G–BHTB	Monnet Sonerai II	R. H. Thring
	G–BHTC	Jodel DR.1050/M1	T. A. Carpenter
	G–BHTD	Cessna T.188C AgHusky	Dallah-ADS Ltd.
	G–BHTE	PA-23 Aztec 250	Southern Wings Ltd.
	G–BHTF	Enstrom F-28C	Spooner Aviation Ltd.
	G–BHTG	Thunder Ax6-56 balloon	Thunder Balloons Ltd.
	G–BHTH	T-6G Texan (2807)	Keenair Services Ltd.
	G–BHTI	SA.102.5 Cavalier	R. Cochrane
	G–BHTK	D.H.C.-6 Twin Otter 310	Loganair Ltd.
	G–BHTM	Cameron Can SS balloon	BP Oil Ltd.
	G–BHTN	Rango-Saffery S.330 balloon	Rango Kite Co. *Mango Santa M.*
	G–BHTO	—	—
	G–BHTP	PA-31T-500 Cheyenne I	C.S.E. Aviation Ltd.
	G–BHTR	Bell 206B JetRanger 3	C.S.E. Aviation Ltd.
	G–BHTS	—	—
	G–BHTT	Cessna 500 Citation	Lucas Industries Ltd.
	G–BHTU	—	—
	G–BHTV	Cessna 310R	Air Atlantique Ltd.
	G–BHTW	Cessna FR.172J	Rogers Aviation Sales Ltd.
	G–BHTZ	AA-5A Cheetah	W. H. Wilkins Ltd.
	G–BHUA	Douglas C-47	Aces High Ltd.
	G–BHUB	Douglas C-47	Aces High Ltd.
	G–BHUC	Douglas C-47	Aces High Ltd.
	G–BHUE	Jodel DR.1050	G. A. Mason
	G–BHUF	PA-18-150 Super Cub	T. A. McMullin
	G–BHUG	Cessna 172N	B. Giles
	G–BHUH	Cremer PC.14 balloon	P. A. Cremer
	G–BHUI	Cessna 152	Lattey Aviation Ltd.
	G–BHUJ	Cessna 172N	Three Counties Aero Club Ltd.
	G–BHUK	Rutan Quickie	D. B. Robinson & ptnrs.
	G–BHUL	Beech E90 King Air	Airmore Aviation Ltd. (G–BBKM)
	G–BHUM	D.H.82A Tiger Moth	S. G. Towers
	G–BHUN	PZL-104 Wilga 35	W. Radwanski
	G–BHUO	Evans VP-2	R. A. Povall
	G–BHUP	Cessna F.152	Westair Flying Services Ltd.
	G–BHUR	Thunder Ax3 balloon	Thunder Balloons Ltd.
	G–BHUS	Beech F90 King Air	Eagle Aircraft Services Ltd.
	G–BHUT	Beech F90 King Air	Eagle Aircraft Services Ltd.
	G–BHUU	PA-25 Pawnee 235	Miller Aerial Spraying Ltd.
	G–BHUV	PA-25 Pawnee 235	Miller Aerial Spraying Ltd.
	G–BHUW	Boeing N2S-5 Kaydet	Keenair Services Ltd.
	G–BHUX	—	—
	G–BHUZ	—	—
	G–BHVA	—	—
	G–BHVB	PA-28-161 Warrior II	Express Aviation Services Ltd.
	G–BHVC	Cessna 172RG Cutlass	Rogers Aviation Sales Ltd.
	G–BHVE	Saffery S.330 balloon	P. M. Randles
	G–BHVF	Jodel D.150A	C. A. Parker
	G–BHVG	Boeing 737-2T5	Orion Airways Ltd.
	G–BHVH	Boeing 737-2T5	Orion Airways Ltd.
	G–BHVI	Boeing 737-2T5	Orion Airways Ltd.
	G–BHVM	Cessna 152	Three Counties Aero Club Ltd.
	G–BHVN	Cessna 152	Three Counties Aero Club Ltd.
	G–BHVO	PA-34-200 Seneca	G.R.T. Catering Ltd.
	G–BHVP	Cessna 182Q	F. & S. E. Horridge
	G–BHVR	Cessna 172N	Air Tows
	G–BHVS	Enstrom F-28A	Spooner Aviation Ltd.
	G–BHVT	Boeing 727-212	Dan-Air Services Ltd.
	G–BHVU	Cessna 414A	Milton Bradley Ltd.
	G–BHVV	Piper J-3C-65 Cub	Mrs. J. Kirk
	G–BHVY	AA-5B Tiger	Amex Automobile Exports Ltd.
	G–BHVZ	Cessna 180	R. Moore
	G–BHWA	Cessna F.152	Wickenby Aviation Ltd.
	G–BHWB	Cessna F.152	Wickenby Aviation Ltd.
	G–BHWC	—	—
	G–BHWD	Hughes 369D	J. P. D. Heyward
	G–BHWE	Boeing 737-204ADV	Britannia Airways Ltd. *Sir Sidney Camm*

Reg.	Type	Owner or Operator	Notes
G–BHWF	Boeing 737-204ADV	Britannia Airways Ltd. *Lord Brabazon of Tara*	
G–BHWG	Mahatma S.200SR balloon	H. W. Gandy *Spectrum*	
G–BHWH	Weedhopper JC-24A	G. A. Clephane	
G–BHWI	AA-5B Tiger	Cabair Ltd.	
G–BHWJ	AA-5A Cheetah	Cabair Ltd.	
G–BHWK	M.S.880B Rallye Club	Tredair	
G–BHWL	—	—	
G–BHWM	M.S.880B Rallye Club	Tredair	
G–BHWN	WMB.3 Windtracker 200 balloon	C. J. Dodd & G. J. Luckett	
G–BHWO	WMB.4 Windtracker II balloon	C. J. Dodd	
G–BHWP	AA-5B Tiger	Cabair Ltd.	
G–BHWR	AA-5A Cheetah	Gulfstream American (Denham) Ltd.	
G–BHWS	Cessna F.152	Rogers Aviation Sales Ltd.	
G–BHWW	Cessna U.206G	Citation Flying Services Ltd.	
G–BHWX	Cessna 404	DSI Engineering Ltd.	
G–BHWY	PA-28R-200 Cherokee Arrow	F. Johnson (Westwood Garages) Ltd.	
G–BHWZ	PA-28-181 Archer II	Zitair Flying Club Ltd.	
G–BHXC	R. Commander 112TCA	Commander Jetprop Ltd.	
G–BHXD	Jodel D.120	P. R. Powell	
G–BHXE	Thunder Ax3 balloon	Thunder Balloons Ltd.	
G–BHXF	—	—	
G–BHXG	D.H.C.-6 Twin Otter 310	Loganair Ltd.	
G–BHXH	BN-2B Islander	Pilatus BN Ltd.	
G–BHXI	BN-2B Islander	Pilatus BN Ltd.	
G–BHXJ	Nord 1203/2 Norecrin	R. E. Coates	
G–BHXK	PA-28 Cherokee 140	I. R. F. Hammond	
G–BHXL	Evans VP-2	J. J. Fitzgerald	
G–BHXM	Electra Flyer Eagle	C. G. Wrzesiew	
G–BHXN	Van's RV.3	P. R. Hing	
G–BHXO	Colt 12A balloon	Colt Balloons Ltd.	
G–BHXP	—	—	
G–BHXR	Thunder Ax7-65 balloon	Thunder Balloons Ltd.	
G–BHXS	Jodel D.120	K. Fox & ptnrs.	
G–BHXT	Thunder Ax6-56Z balloon	Thunder Balloons Ltd.	
G–BHXU	AB-206B JetRanger	Alan Mann Helicopters Ltd.	
G–BHXV	AB-206B JetRanger	Alan Mann Helicopters Ltd.	
G–BHXX	PA-23 Aztec 250	Hockstar Ltd.	
G–BHXY	Piper J-3C-65 Cub	R. K. Haldenby & T. S. Warren	
G–BHXZ	Sikorsky S-76A	British Airways Helicopters Ltd.	
G–BHYA	Cessna R.182RG II	MLP Aviation Ltd.	
G–BHYB	Sikorsky S-76A	British Airways Helicopters Ltd.	
G–BHYC	Cessna 172RG Cutlass	Westair Flying Services Ltd.	
G–BHYD	Cessna R.172K	Sylmar Aviation & Services Ltd.	
G–BHYE	PA-34-200T II Seneca	C.S.E. Aviation Ltd.	
G–BHYF	PA-34-200T II Seneca	C.S.E. Aviation Ltd.	
G–BHYG	PA-34-200T II Seneca	C.S.E. Aviation Ltd.	
G–BHYH	AS.350B Ecureuil	McAlpine Helicopters Ltd.	
G–BHYI	Stampe SV-4A	D. E. Starkey & A. T. Fraser	
G–BHYN	Evans VP-2	A. B. Cameron	
G–BHYO	Cameron N-77 balloon	C. Sisson	
G–BHYP	Cessna F.172M	Executive Wings Ltd.	
G–BHYR	Cessna F.172M	Executive Wings Ltd.	
G–BHYS	PA-28-181 Archer II	C.S.E. Aviation Ltd.	
G–BHYT	EMB-110P2 Bandeirante	C.S.E. Aviation Ltd.	
G–BHYU	Beech A200 Super King Air	Kenton Utilities & Developments Ltd.	
G–BHYV	Evans VP-1	L. Chiappi	
G–BHYW	AB-206A JetRanger	James Templeton & Co.	
G–BHYX	Cessna 152	Tradecliff Ltd.	
G–BHYY	PA-28-161 Warrior II	Shirlstar Container Transport Ltd.	
G–BHYZ	Cameron V-77 balloon	The Hon. A. G. Ritchie-Calder	
G–BHZA	Piper J-3C-65 Cub	A. J. Hyatt & J. R. Wraight	
G–BHZB	—	—	
G–BHZC	R. Commander 690C	Rialto Builders Ltd.	
G–BHZD	—	—	
G–BHZE	PA-28-181 Archer II	Cormack (Aircraft Services) Ltd.	
G–BHZF	Evans VP-2	D. Silsbury	
G–BHZG	Monnet Sonerai II	R. A. Gardiner & B. Chapman	
G–BHZH	Cessna F.152	Rogers Aviation Sales Ltd.	
G–BHZI	Thunder Ax3 balloon	Thunder Balloons Ltd.	
G–BHZJ	Hughes Stratosphere 150 balloon	P. J. Hughes	

Notes	Reg.	Type	Owner or Operator
	G–BHZK	AA-5B Tiger	L. Henderson
	G–BHZL	AA-5A Cheetah	L. Henderson
	G–BHZM	Jodel DR.1050	P. B. Kingsford
	G–BHZN	AA-5B Tiger	Cabair Ltd.
	G–BHZO	AA-5A Cheetah	Cabair Ltd.
	G–BHZU	Piper J-3C-65 Cub	S. R. Clarke
	G–BHZV	Jodel D.120A	W. C. Forster
	G–BHZW	—	—
	G–BHZX	Thunder Ax7-65A balloon	Thermask (Plastic Processing) Ltd.
	G–BHZY	Monnet Sonerai II	C. A. Keech
	G–BHZZ	Air Britain Manchester Eagle 8 balloon	Air Britain Manchester Branch
	G–BIAA	SOCATA TB.9 Tampico	Air Touring Services Ltd.
	G–BIAB	SOCATA TB.9 Tampico	Air Touring Services Ltd.
	G–BIAC	M.S.894E Rallye Minerva	Air Touring Services Ltd.
	G–BIAD	BAe 146-100	British Aerospace
	G–BIAE	BAe 146-100	British Aerospace
	G–BIAF	BAe 146-100	British Aerospace
	G–BIAG	BAe 146-100	British Aerospace
	G–BIAH	Jodel D.112	N. M. Bloom
	G–BIAI	WMB.2 Windtracker balloon	I. Chadwick
	G–BIAJ	BAe 146-100	British Aerospace
	G–BIAK	SOCATA TB.10 Tobago	Dukeries Aviation Ltd.
	G–BIAL	Rango NA.8 balloon	A. M. Lindsay
	G–BIAM	SOCATA TB.10 Tobago	Jabrest Ltd.
	G–BIAO	Evans VP-2	J. Stephenson
	G–BIAP	PA-16 Clipper	I. M. Callier & P. J. Bish
	G–BIAR	Rigg Skyliner II balloon	I. A. Rigg
	G–BIAS	Douglas DC-8-55F	—
	G–BIAT	—	—
	G–BIAU	—	—
	G–BIAV	Sikorsky S-76A	British Airways Helicopters Ltd.
	G–BIAW	Sikorsky S-76A	British Airways Helicopters Ltd.
	G–BIAX	Taylor JT.2 Titch	G. F. Rowley
	G–BIAY	AA-5 Traveler	Lorien Textiles (UK) LTD.
	G–BIBA	SOCATA TB.9 Tampico	Air Touring Services Ltd.
	G–BIBB	Mooney M.20C	S. F. & G. G. Dray
	G–BIBC	Cessna 310R	Air Service Training Ltd.
	G–BIBD	Rotec Rally 2B	A. Clarke
	G–BIBE	EMB-110P1 Bandeirante	Loganair Ltd.
	G–BIBF	A12 Sport balloon	T. J. Smith
	G–BIBG	Sikorsky S-76A	British Caledonian Helicopters Ltd.
	G–BIBH	—	—
	G–BIBJ	Enstrom F-280C Shark	Spooner Aviation Ltd.
	G–BIBK	Taylor JT.2 Titch	T. Corbett
	G–BIBL	Taylor JT.2 Titch	J. Sharp
	G–BIBM	AA-5A Cheetah	Ace Star Ltd.
	G–BIBN	Cessna FA.150K	P. H. Lewis
	G–BIBO	Cameron V-65 balloon	Southern Balloon Group
	G–BIBP	AA-5A Cheetah	Cabair Ltd.
	G–BIBR	—	—
	G–BIBS	Cameron P-20 balloon	Cameron Balloons Ltd.
	G–BIBT	AA-5B Tiger	Tradecliff Ltd.
	G–BIBU	Morris Ax7-77 balloon	K. Morris
	G–BIBV	WMB.3 Windtracker 200 balloon	P. B. Street
	G–BIBW	Cessna F.172N	Northair Aviation Ltd.
	G–BIBX	WMB.2 Windtracker 200 balloon	C. J. Dodd
	G–BIBY	Beech F33A Bonanza	Carl Peterson Ltd.
	G–BIBZ	Thunder Ax3 balloon	Thunder Balloons Ltd.
	G–BICA	—	—
	G–BICB	Rotec Rally 2B	J. D. Lye & A. P. Jones
	G–BICC	Vulture Tx3 balloon	P. Clitheroe
	G–BICD	Auster 5	J. A. S. Baldry & ptnrs.
	G–BICE	AT-6C Harvard IIA	C. M. L. Edwards
	G–BICF	GA-7 Cougar	Cabair Ltd.
	G–BICG	Cessna F.152	R. M. Clarke
	G–BICH	WMB.2 Windtracker balloon	D. F. Moth
	G–BICI	—	—
	G–BICJ	Monnet Sonerai II	J. R. S. Heaton
	G–BICK	H.S.748 Srs. 2A	British Aerospace
	G–BICL	—	—
	G–BICM	Colt 56A balloon	T. A. R. & S. Turner
	G–BICN	F.8L Falco	R. J. Barber

Reg.	*Type*	*Owner or Operator*	*Notes*
G–BICO	Neale Mitefly Model balloon	T. J. Neale	
G–BICP	Robin DR.360	I. H. Thompson	
G–BICR	Jodel D.120A	S. W. C. Hall & ptnrs.	
G–BICS	—	—	
G–BICT	Evans VP-1	A. C. Coombe & D. L. Tribe	
G–BICU	Cameron V-56 balloon	I. S. Clark	
G–BICV	Boeing 737-2L9	Dan-Air Services Ltd.	
G–BICW	PA-28-161 Warrior II	Blestes Aviation	
G–BICX	—	—	
G–BICY	PA-23 Apache 160	Allen Technical Services Ltd.	
G–BICZ	Hill Hummer	J. Simpson	
G–BIDA	M.S.880B Rallye Club	Oakley Motor Units Ltd.	
G–BIDB	BAe 167 Strikemaster	British Aerospace	
G–BIDC	Evans VP-1	I. D. Daniels	
G–BIDD	Evans VP-1	I. D. Daniels	
G–BIDE	CP.301A Emeraude	D. Elliott	
G–BIDF	Cessna F.172P	Rogers Aviation Sales Ltd.	
G–BIDG	Jodel D.150A	D. R. Gray	
G–BIDH	Cessna 152	G. Lewis	
G–BIDI	PA-28R-201 Arrow III	Spooner Aviation Ltd.	
G–BIDJ	PA-18-150 Super Cub	Light Aircraft Engineering Ltd.	
G–BIDK	PA-18-150 Super Cub	Light Aircraft Engineering Ltd.	
G–BIDL	—	—	
G–BIDM	Cessna F.172H	D. H. MacDonald	
G–BIDN	P.57 Sea Prince T.1	T. Warren	
G–BIDO	—	—	
G–BIDP	—	—	
G–BIDR	—	—	
G–BIDS	—	—	
G–BIDT	Cameron A375 balloon	Ballooning Endeavours Ltd.	
G–BIDU	—	—	
G–BIDV	Colt Ax2 balloon	International Distillers & Vintners (House Trade) Ltd.	
G–BIDW	Sopwith 1½ strutter replica	V. H. Bellamy	
G–BIDX	Jodel D.112	H. N. Nuttall & R. P. Walley	
G–BIDY	WMB.2 Windtracker balloon	D. M. Campion	
G–BIDZ	Colt 21A balloon	Hot Air Balloon Co. Ltd.	
G–BIEB	Agusta-Bell 47G-3B1	Autair Helicopters Ltd.	
G–BIEC	AB-206A JetRanger	Autair Helicopters Ltd.	
G–BIED	Beech F90 King Air	Eagle Aircraft Services Ltd.	
G–BIEE	Beech C90 King Air	Eagle Aircraft Services Ltd.	
G–BIEF	Cameron V-77 balloon	D. S. Bush	
G–BIEG	PA-32-301 Saratoga	Spooner Aviation Ltd.	
G–BIEH	Sikorsky S-76A	Management Aviation Ltd.	
G–BIEI	D.H.C.-6 Twin Otter 310	Bristow Helicopters Ltd.	
G–BIEJ	Sikorsky S-76A	Bristow Helicopters Ltd.	
G–BIEK	WMB.4 Windtracker balloon	P. B. Street	
G–BIEL	WMB.4 Windtracker balloon	A. T. Walden	
G–BIEM	D.H.C.-6 Twin Otter 310	Loganair Ltd.	
G–BIEN	—	—	
G–BIEO	—	—	
G–BIEP	PA-28-181 Archer II	L. V. Henderson	
G–BIER	Rutan Long-Eze	V. Mossor	
G–BIES	—	—	
G–BIET	Cameron O-77 balloon	G. M. Westley	
G–BIEU	AA-5A Cheetah	Cabair Ltd.	
G–BIEV	AA-5A Cheetah	Cabair Ltd.	
G–BIEW	Cessna U.206G	G. D. Atkinson	
G–BIEX	Andreasson BA-4B	H. P. Burrill	
G–BIEY	PA-28-151 Warrior	A. F. Aviation Ltd.	
G–BIEZ	Beech F90 King Air	Eagle Aircraft Services Ltd.	
G–BIFA	—	—	
G–BIFB	PA-28 Cherokee 150	C. J. Reed	
G–BIFC	Colt 14A balloon	Colt Balloons Ltd.	
G–BIFD	—	—	
G–BIFE	Cessna A.185F	Northair Aviation Ltd.	
G–BIFF	AA-5A Cheetah	Cabair Ltd.	
G–BIFG	Short SD3-30	Short Bros. Ltd.	
G–BIFH	Short SD3-30	Short Bros. Ltd.	
G–BIFI	Short SD3-30	Short Bros. Ltd.	
G–BIFJ	Short SD3-30	Short Bros. Ltd.	
G–BIFK	Short SD3-30	Short Bros. Ltd.	
G–BIFL	SC.7 Skyvan Srs. 3	Short Bros. Ltd.	

Notes	Reg.	Type	Owner or Operator
	G–BIFM	—	—
	G–BIFN	Benson B.8M	K. Willows
	G–BIFO	Evans VP-1	P. Raggett
	G–BIFP	Colt 56C balloon	J. Philp
	G–BIFR	—	—
	G–BIFS	Beech C90 King Air	Airmore Aviation Ltd.
	G–BIFT	Cessna F.150L	Practavia Ltd.
	G–BIFU	Short Skyhawk balloon	D. K. Short
	G–BIFV	Jodel D.150	J. H. Kirkham
	G–BIFW	Scruggs BL.2 Wunda balloon	D. Morris
	G–BIFX	Westland WG.13 Lynx	Westland Helicopters Ltd.
	G–BIFY	Cessna F.150L	Practavia Ltd.
	G–BIFZ	—	—
	G–BIGA	Short SD3-30	Short Bros. Ltd.
	G–BIGB	Bell 212	B.E.A.S. Ltd.
	G–BIGC	Cameron O-42 balloon	C. M. Moroney
	G–BIGD	Cameron N-77 balloon	Cameron Balloons Ltd.
	G–BIGE	Champion Cloudseeker balloon	A. Foster
	G–BIGF	Thunder Ax7-77 balloon	Thunder Balloons Ltd.
	G–BIGG	Saffery S.200 balloon	A. F. Longhelt
	G–BIGH	—	—
	G–BIGI	Mooney M.20J	Kingswinford Engineering Co. Ltd.
	G–BIGJ	Cessna F.172M	Colton Aviation Ltd.
	G–BIGK	Taylorcraft BC-12D	J. Rowell
	G–BIGL	Cameron 0-65 balloon	A. H. L. Alpin
	G–BIGM	Avenger T.200-2112 balloon	M. Murphy
	G–BIGN	Attic Srs. 1 balloon	G. Nettleship
	G–BIGO	AB-206B JetRanger 3	Alan Mann Helicopters Ltd.
	G–BIGP	Benson B.8.	R. H. S. Cooper
	G–BIGR	Avenger T.200-2112 balloon	R. Light
	G–BIGS	AB-206B JetRanger 3	Alan Mann Helicopters Ltd.
	G–BIGT	Colt 77A balloon	D. M. Winder
	G–BIGU	Benson B.8M	J. R. Martin
	G–BIGV	AA-5B Tiger	Cabair Ltd.
	G–BIGW	Westland WG.13 Lynx	Westland Helicopters Ltd.
	G–BIGX	Benson B.8M	J. R. Martin
	G–BIGY	Cameron V-65 balloon	Dante Balloon Group
	G–BIGZ	Scheibe SF.25B Falke	K. Ballington
	G–BIHA	—	—
	G–BIHB	Scruggs BL.2 Wunda balloon	D. Morris
	G–BIHC	Scruggs BL.2 Wunda balloon	P. D. Kiddell
	G–BIHD	Robin DR.400/160	Headcorn Flying School Ltd.
	G–BIHE	Cessna FA.152	Northair Aviation Ltd.
	G–BIHF	SE-5A Replica	K. J. Garrett
	G–BIHG	PA-28 Cherokee 140	T. Parmenter
	G–BIHH	Sikorsky S-61N	British Caledonian Helicopters Ltd.
	G–BIHI	Cessna 172M	Cardiff Aviation Ltd.
	G–BIHJ	Westland WG.13 Lynx Mk. 88	Westland Helicopters Ltd.
	G–BIHK	Westland WG.13 Lynx Mk. 88	Westland Helicopters Ltd.
	G–BIHL	Westland WG.13 Lynx Mk. 88	Westland Helicopters Ltd.
	G–BIHM	Westland WG.13 Lynx Mk. 88	Westland Helicopters Ltd.
	G–BIHN	Skyship 500 airship	Airship Industries Ltd.
	G–BIHO	—	—
	G–BIHP	Gas (free) balloon	J. J. Harris
	G–BIHR	WMB.2 Windtracker balloon	A. F. Longhelt
	G–BIHS	T.6G Harvard	Aces High Ltd.
	G–BIHT	—	—
	G–BIHU	Saffery S.200 balloon	A. F. Longhelt
	G–BIHV	WMB.2 Windtracker balloon	T. H. W. Bradley
	G–BIHW	Aeronca A65TAC	W. F. Crozier & ptnrs.
	G–BIHX	Benson B.8M	C. C. Irvine
	G–BIHY	Isaacs Fury	D. E. Olivant
	G–BIHZ	—	—
	G–BIIA	Fournier RF-3	M. K. Field
	G–BIIB	Cessna F.172M	Citation Flying Services Ltd.
	G–BIIC	Scruggs BL.2 Wunda Balloon	S. J. Hodder & D. Cockerill
	G–BIID	PA-18 Super Cub 95	Nolson Aviation Ltd.
	G–BIIE	Cessna F.172P	A & G Aviation Ltd.
	G–BIIF	Fournier RF-4D	M. J. Revill (G–BVET)
	G–BIIG	Thunder Ax-6-56Z balloon	Thunder Balloons Ltd.
	G–BIIH	Scruggs BL.2T Turbo balloon	B. M. Scott
	G–BIII	—	—
	G-BIIJ	Cessna F.152	A & G Aviation Ltd.

Reg.	Type	Owner or Operator	Notes
G–BIIK	MS883 Rallye 115	G. A. Bentley	
G–BIIL	Thunder Ax6-56 balloon	Thunder Balloons Ltd.	
G–BIIM	Scruggs BL.2A Wunda Balloon	K. D. Head	
G–BIIN	BN-2B Islander	Pilatus BN Ltd.	
G–BIIO	BN-2B Islander	Pilatus BN Ltd.	
G–BIIP	BN-2B Islander	Pilatus BN Ltd.	
G–BIIR	BN-2B Islander	Pilatus BN Ltd.	
G–BIIS	BN-2B Islander	Pilatus BN Ltd.	
G–BIIT	PA-28-161 Warrior II	C.S.E. Aviation Ltd.	
G–BIIU	PA-28-181 Archer II	C.S.E. Aviation Ltd.	
G–BIIV	PA-28-181 Archer II	R. Watkins	
G–BIIW	Rango NA.10 balloon	Rango Kite Co.	
G–BIIX	Rango NA.12 balloon	Rango Kite Co.	
G–BIIY	PA-31-350 Navajo Chieftain	Aravca Ltd.	
G–BIIZ	—	—	
G–BIJA	Scruggs BL.2A Wunda Balloon	P. Edwards	
G–BIJB	PA-18-150 Super Cub	Essex Gliding Club	
G–BIJC	AB-206B JetRanger 3	Hunting Aviation Ltd.	
G–BIJD	Bo 208C Junior	D. J. Dulborough	
G–BIJE	—	—	
G–BIJF	Bell 212	Bristow Helicopters Ltd.	
G–BIJS	Luton L.A.4A Minor	I. J. Smith	
G–BIJT	—	—	
G–BIJU	CP.301 Emeraude	C. J. Norman & M. Howard (G–BHTX)	
G–BIJV	Cessna F.152	Citation Flying Services Ltd.	
G–BIJW	Cessna F.152	Citation Flying Services Ltd.	
G–BIJX	Cessna F.152	Citation Flying Services Ltd.	
G–BIJY	—	—	
G–BIJZ	Skyventurer Mk. I balloon	R. Sweeting	
G–BIKE	PA-28R Cherokee Arrow 200	R. V. Webb Ltd.	
G–BILA	—	—	
G–BILB	—	—	
G–BILC	Thunder Ax3 balloon	Thunder Balloons Ltd.	
G–BILD	Piper J-3C-65 Cub	W. R. Pickett (G–KERK)	
G–BILE	—	—	
G–BILF	Practavia Sprite 125	G. Harfield	
G–BILG	—	—	
G–BILH	—	—	
G–BILI	—	—	
G–BILJ	Cessna FA.152	A & G Aviation Ltd.	
G–BILK	—	—	
G–BILL	PA-25 Pawnee 235	Bowker Air Services Ltd.	

NOTE: Registrations G–BIJG to G–BIJR, G–BIKA to G–BIKD and G–BIKF to G–BIKZ are reserved for British Airways.

Out-of-sequence registrations

Reg.	Type	Owner or Operator	Notes
G–BJAL	C.A.S.A. 1.131E Jungmann	Buccaneer Aviation Ltd.	
G–BJAM	Westland Bell 47G-3B1	Trent Helicopters Ltd.	
G–BJAN	SA.102.5 Cavalier	J. Powlesland	
G–BJAY	Piper J-3C-65 Cub	J. A. Yates	
G–BJCB	H.S.125 Srs. 600B	J. C. Bamford Excavators Ltd.	
G–BJCM	FRED Srs. 2	J. C. Miller	
G–BJDM	SA.102.5 Cavalier	J. D. McCracken	
G–BJFH	Boeing 737-2S3	Air Europe Ltd. *Sandy*	
G–BJHK	EAA Acro Sport	J. H. Kimber	
G–BJKW	Willis Aera II	J. K. S. Wills	
G–BJLS	Cessna 340A	J. L. Shaw	
G–BJMR	Cessna 310R	Citation Flying Services Ltd.	
G–BJOI	Isaacs Special	J. O. Isaacs	
G–BJRW	Cessna U.206G	Rogers Aviation Sales Ltd.	
G–BJVS	CP.1315C-3 Super Emeraude	Victor Sierra Aero Club	
G–BKAT	Pitts S-1C Special	I. M. G. Senior & J. G. Harper	
G–BKCM	Bell 206B JetRanger III	Patgrove Ltd.	
G–BKDE	Kendrick I	J. K. Rushton	
G–BKDF	Kendrick II	J. K. Rushton	
G–BKGF	Saxon II	Saxon Aircraft Co. Ltd.	
G–BKJW	PA-23 Aztec 250	Iceni Aviation Ltd.	
G–BKPC	Cessna A.185F	Air Service Training Ltd.	
G–BKTI	Beech A200 Super King Air	Tube Investments Ltd.	
G–BLAC	Cessna FA.152	Westair Flying Services Ltd.	

Notes	Reg.	Type	Owner or Operator
	G–BLAL	Cessna F.150K	Bevan Lynch Aviation Ltd.
	G–BLCE	Cessna 402C	Cecil Aviation Ltd.
	G-BLEW	Cessna F.182Q	Bennett & Wright Ltd.
	G–BLGS	SOCATA Rallye 180GT	Lasham Gliding Soc. Ltd.
	G–BLGW	F.27 Friendship Mk. 200	Air UK
	G–BLHL	CP.301A Emeraude	W. N. O'Brien & J. W. Cameron
	G–BLHN	Robin HR.100/285	H. M. Bonquiere
	G–BLJM	Beech 95-B55 Baron	Malcolm Enames Ltd.
	G–BLPP	Cameron V-77 balloon	L. P. Purfield *Merlin*
	G–BLRJ	Jodel DR.1051	M. P. Hallam
	G–BLSL	Cessna 310R	Rassler Aero Services
	G–BLST	Cessna 421C	Cecil Aviation Ltd.
	G–BLUE	Colting Ax-77 balloon	Lighter-Than-Air Ltd. *Bluebird*
	G–BMAA	Douglas DC-9-15	British Midland Airways Ltd. *Dovedale* (G–BFIH)
	G–BMAB	Douglas DC-9-15	British Midland Airways Ltd. *Merseyside*
	G–BMAC	Douglas DC-9-15	British Midland Airways Ltd.
	G–BMAV	AS.350B Ecureuil	Masselaz Helicopters Ltd.
	G–BMAX	FRED Srs. 2	P. Cawkwell
	G–BMCA	Beech A200 Super King Air	Marchwiel Plant & Engineering Co. Ltd.
	G–BMCL	Cessna 550 Citation II	Micro Consultants Ltd.
	G–BMCP	Maule M5-235C Lunar Rocket	M. C. Peirce
	G–BMEC	Boeing 737-2S3	Air Europe Ltd. *Eve*
	G–BMEL	PA-23 Aztec 250	Eagle Surface Coatings Ltd.
	G–BMFD	PA-23 Aztec 250	Express Aviation Services Ltd. (G–BGYY)
	G–BMHC	Cessna U.206F	Mechanical Handling Consultants (Aviation) Ltd.
	G–BMHG	Boeing 737-2S3	Air Europe Ltd. *Adam*
	G–BMIC	R. Turbo Commander 690B	Micro Consultants Ltd.
	G–BMIP	Jodel D.112	M. T. Kinch
	G–BMJH	Hughes 369D	Hughes of Beaconsfield Ltd.
	G–BMLM	Beech 95-58 Baron	J. Mowlem & Co. Ltd. (G–BBJF)
	G–BMMP	Boeing 737-2S3	Air Europe Ltd.
	G–BMON	Boeing 737-2K9	Monarch Airlines Ltd.
	G–BMOR	Boeing 737-2S3	Air Europe Ltd. *Joy*
	G–BMSB	V.S. Spitfire IX	M. S. Bayliss
	G–BMSF	PA-38-112 Tomahawk	Apache Aircraft Services Ltd.
	G–BMSM	Boeing 737-2S3	Air Europe Ltd. *Roma*
	G–BMSP	Boeing 737-2S3	Air Europe Ltd.
	G–BMVV	Rutan Vari-Viggen	G. B. Morris
	G–BMYU	Jodel D.120	G. H. Wilde
	G–BNHP	Saffery S.330 balloon	N. H. Ponsford *Alpha II*
	G–BNJF	PA-32RT-300 Turbo Lance II	N. J. Froment & Co. Ltd.
	G–BNNH	PA-31-350 Navajo Chieftain	Central Manufacturing & Trading Group
	G–BNOC	EMB-110P1 Bandeirante	Fairflight Ltd.
	G–BNPD	PA-23 Aztec 250	Northern Pig Development Co. Ltd.
	G–BOAA	Concorde 102	British Airways (G–N94AA)
	G–BOAB	Concorde 102	British Airways (G–N94AB)
	G–BOAC	Concorde 102	British Airways (G–N81AC)
	G–BOAD	Concorde 102	British Airways (G–N94AD)
	G–BOAE	Concorde 102	British Airways (G–N94AE)
	G–BOAF	Concorde 102	British Airways (G–N94AF/G–BFKX)
	G–BOAT	Cessna 310R	Birchwood Boat International Ltd.
	G–BOBI	Cessna 152	Capital Aviation Sales Ltd. (G–BHJD)
	G–BOBY	Monnet Sonerai II	R. G. Hallam
	G–BOLT	R. Commander 114	Hooper & Jones Ltd.
	G–BOMB	Cassutt Racer	R. W. L. Breckell
	G–BOND	Sikorsky S-76	North Scottish Helicopters Ltd.
	G–BOOB	Cameron N-65 balloon	E. M. Ten Houten
	G–BOOK	Pitts S-1S	B. K. Lecomber
	G–BOOM	Hunter T.7	B. R. Kay
	G–BOSL	Boeing 737-2U4ADV	Britannia Airways Ltd *Sir Frank Whittle*
	G–BOSS	PA-34-200T Seneca	Elmsleigh Autos (Contract Hire) Ltd.
	G–BOST	PA-23 Aztec 250	Boston Deep Sea Fisheries Ltd.
	G–BPAJ	D.H.82A Tiger Moth	P. A. Jackson (G–AOIX)

Reg.	Type	Owner or Operator	Notes
G–BPAM	Jodel D.150A	M. Hughes & P. D. Yapp	
G–BPAR	PA-31-350 Navajo Chieftain	Burnthills Plant Hire Ltd.	
G–BPAV	FRED Srs. 2	P. A. Valentine	
G–BPAZ	Pazmany PL.2	T. Davies & P. H. Chamberlain	
G–BPPN	BN-2T Islander	Pilatus BN Ltd. (G–BCMY)	
G–BPBP	Brugger Colibri Mk. II	B. Perkins	
G–BPDW	Eezybuild Envoy Mk. 2	P. D. Wheatland	
G–BPEG	Currie Wot	D. M. Harrington	
G–BPFA	Swallow GK-2	G. Knight & D. G. Pridham	
G–BPMB	Maule M5-235C Lunar Rocket	P. M. Breton	
G–BPMN	Super Coot Model A	P. Napp	
G–BPPN	Cessna F.182Q	Hunt Norris Ltd.	
G–BPUF	Thunder Ax6-56Z balloon	Buf-Puf Balloon Group	
G–BPYN	Piper J-3C-65 Cub	D. W. Stubbs & ptnrs.	
G–BPZD	Nord NC.854S	J. C. Voyzey	
G–BRAD	Beech 95-B55 Baron	Finrad Ltd.	
G–BRAF	V.S. Spitfire 18	D. W. Arnold	
G–BRAG	Taylor JT.2 Titch	A. R. Greenfield	
G–BRAL	G.159 Gulfstream I	Ford Motor Co. Ltd.	
G–BRAT	Jodel D.120A	A. J. Diack	
G–BRED	Canadair CL-44D-4	Redcoat Air Cargo *James*	
G–BREF	Cessna 421C	Refair Ltd.	
G–BREL	Cameron 0-77 balloon	BICC Research & Engineering Ltd.	
G–BREN	Cameron Jeans Special Shape balloon	Leaveglen Ltd.	
G–BREW	PA-31-350 Navajo Chieftain	Whitbread & Co. Ltd.	
G–BRFC	P.57 Sea Prince T.I	Rural Flying Corps	
G–BRFG	Cessna 340A	Planet Gloves Ltd.	
G–BRGH	FRED Srs. 2	F. G. Hallam	
G–BRGV	PA-31-350 Navajo Chieftain	Centreline Air Services Ltd.	
G–BRGW	GY-201 Minicab	R. G. White	
G–BRHD	PA-23 Aztec 250F	Omega Consultants Ltd.	
G–BRIK	Tipsy Nipper 3	C. W. R. Piper	
G–BRIT	Cessna 421C	Britannia Airways Ltd.	
G–BRMA	*WS-51 Dragonfly Mk. 5 (WG719)	British Rotorcraft Museum	
G–BRMB	*B.192 Belvedere Mk. I (XG452)	British Rotorcraft Museum	
G–BRMC	Stampe SV-4A	A. Cullen	
G–BROM	ICA IS-28M2	British Aircraft Corporation	
G–BRON	Beech A200 Super King Air	Owledge Ltd. (G–BFEA)	
G–BRSL	Cameron N-56 balloon	Balloon Stable Ltd. *Boris*	
G–BRUX	PA-44-180 Seminole	Hambrair Ltd.	
G–BRWG	Maule M5-235C Lunar Rocket	R. W. Gaskell	
G–BRWS	PA-31-350 Navajo Chieftain	R.W.S. & Co. (Engineers) Ltd.	
G–BSAA	H.S.125 Srs. 3B	Aravco Ltd.	
G–BSAL	G.1159 Gulfstream 2	Shell Aviation Ltd.	
G–BSAM	M.S.880B Rallye Club	M. W. Salmon	
G–BSBH	Short SD3-30	Short Bros. Ltd.	
G–BSDL	SOCATA TB.10 Tobago	Systems Designers Ltd.	
G–BSEL	Slingsby T-61G	Slingsby Engineering Ltd.	
G–BSFC	PA-38-112 Tomahawk	Sherwood Flying Club Ltd.	
G–BSFL	PA-23 Aztec 250	Bevan Lynch Aviation Ltd.	
G–BSFT	PA-31-300 Navajo	Simulated Flight Training Ltd. (G–AXYC)	
G–BSFZ	PA-25 Pawnee 235	Skegness Air Taxi Services Ltd. (G–ASFZ)	
G–BSIS	Pitts S-1S Special	R. A. Mills	
G–BSKY	Douglas DC-8-55F		
G–BSMC	PA-32-301T Turbo Saratoga SP	Deemglow Ltd.	
G–BSSL	Beech B80 Queen Air	Parker & Heard Ltd. (G–BFEP)	
G–BSST	*Concorde 002	Fleet Air Arm Museum	
G–BSTN	PA-31-350 Navajo Chieftain	Boston Deep Sea Fisheries Ltd.	
G–BSUS	Taylor JT.1 Monoplane	R. Parker	
G–BSVP	PA-23 Aztec 250F	Sun Valley Products Ltd.	
G–BSVT	EMB-110P2 Bandeirante	Fairflight Ltd. (G–BWTV)	
G–BTAL	Cessna F.152	R. & R. Aviation	
G–BTBM	*Grumman TBM-3W2 Avenger	Strathallan Aircraft Collection	
G–BTDK	Cessna 421B	Texalon International Ltd.	
G–BTEA	Cameron A-105 balloon	Southern Balloon Group	
G–BTFC	Cessna F.152 II	Tayside Aviation Ltd.	
G–BTFH	Cessna 414A	TBF (Transport) Ltd.	
G–BTGS	Smyth Sidewinder	T. G. Soloman	

Notes	Reg.	Type	Owner or Operator
	G–BTHL	PA-31-350 Navajo Chieftain	Burnthills Plant Hire Ltd.
	G–BTHS	PA-23 Aztec 250F	Mooney Aviation Ltd.
	G–BTJM	Taylor JT.2 Titch	T. J. Miller
	G–BTLE	PA-31-350 Navajo Chieftain	Merlix Air Ltd.
	G–BTOM	PA-38-112 Tomahawk	Channel Aviation Ltd.
	G–BTSC	Evans VP-2	D. W. Burrell
	G–BTSF	Evans VP-2	D. W. Burrell
	G–BTSH	Evans VP-2	B. P. Irish
	G–BTUG	SOCATA Rallye 180T	Lasham Gliding Soc. Ltd.
	G–BTWA	Bell 206B JetRanger	C. Hughesdon
	G–BTWT	D.H.C.-6 Twin Otter 310	Loganair Ltd.
	G–BUCC	C.A.S.A. 1.131E Jungmann	Buccaneer Aviation Ltd.
	G–BUCK	C.A.S.A. 1.131E Jungmann	Buccaneer Aviation Ltd.
	G–BUFF	Jodel D.112	D. J. Buffham
	G–BUMP	PA-28-181 Archer II	Autofarm Ltd.
	G–BURD	Cessna F.172N	Volstatic Aviation Ltd.
	G–BUSY	Thunder Ax6-56A balloon	B. R. & Mrs. M. Boyle *Busy Bodies*
	G–BUZZ	AB-206B JetRanger 2	Western Air (Scotland) Ltd.
	G–BVET	Fournier RF-4D	Kirk Aviation Ltd.
	G–BVMM	Robin HR.200/100	Stately & Co. Ltd.
	G–BVPI	Evans VP-1	N. L. E. & R. A. Dupee
	G–BVPM	Evans VP-2	P. Marigold
	G–BWEC	Cassutt-Colson Variant	W. E. Colson
	G–BWFJ	Evans VP-1	W. F. Jones
	G–BWJB	Thunder Ax8-105 balloon	Justerini & Brooks Ltd. *Whiskey J. & B.*
	G–BWKK	Auster AOP.9	G. F. Kilsby
	G–BWKS	Lake LA.4-200 Buccaneer	Newell Aircraft & Tool Co. Ltd. (G–BDDI)
	G–BWMB	Jodel D.119	D. P. Holmyard
	G–BWRB	D.H.C.-6 Twin Otter 310	Brymon Aviation Ltd.
	G–BXBX	PA-31-350 Navajo Chieftain	Alexandra Aviation Ltd.
	G–BXNW	SNCAN SV-4C	S. W. Atkins
	G–BXYB	PA-31-300 Navajo	Culinair Ltd. (G–AXYB)
	G–BXYZ	R. Turbo Commander 690C	Glos-Air (Sales) Ltd.
	G–BZAC	Sikorsky S-76	British Airways Helicopters Ltd.
	G–BZBH	Thunder Ax6-65 balloon	R. S. Whittaker & P. E. Sadler
	G–BZBY	Colt 56 Buzby balloon	Post Office Communications
	G–BZKK	Cameron V-56 balloon	P. J. Green & C. Beasley *Gemini II*
	G–CAFE	Cessna 401	Maxwells Restaurants Ltd. (G–AWXM)
	G–CALL	PA-23 Axtec 250F	Aircraft Mart
	G–CBBI	H.S.125 Srs. 700B	Eagle Leasing Ltd.
	G–CBIA	BAC One-Eleven 416	Air UK (G–AWXJ) *Island Ensign*
	G–CBIL	Cessna 182K	Conway Brown International Ltd.
	G–CCAR	Cameron N-77 balloon	Colt Car Co. Ltd. *Colt*
	G–CCCC	Cessna 172H	P. D. Higgs
	G–CCOZ	Monnet Sonerai II	P. R. Cozens
	G–CDAH	Taylor Coot A	D. A. Hood
	G–CDBI	PA-23 Aztec 250	Spacegrand Ltd.
	G–CDBL	FRED Srs. 2	D. M. Limbert
	G–CDGA	Taylor JT.1 Monoplane	D. G. Anderson
	G–CDGL	Saffery S.330 balloon	C. J. Dodd & G. J. Luckett *Penny*
	G–CELT	EMB-110P2 Bandeirante	Fairflight Ltd.
	G–CENT	Cessna T.210L	CGH Managements Ltd.
	G–CFLY	Cessna 172F	M. J. Stephens
	G–CGFC	PA-38-112 Tomahawk	Cormack (Aircraft Services) Ltd.
	G–CGHM	PA-28 Cherokee 140	CGH Managements Ltd.
	G–CHEV	EMB-110P2 Bandeirante	Fairflight Ltd.
	G–CHOC	Bell 206B JetRanger 2	P. Cadbury
	G–CHOP	Westland Bell 47G-3B1	Alec Wortley Ltd.
	G–CITY	PA-31-350 Navajo Chieftain	City Air Links Ltd.
	G–CJAN	PA-28-181 Archer II	Cormack (Aircraft Services) Ltd.
	G–CJHH	Cessna 550 Citation II	Armstrong Aviation Ltd.
	G–CJIM	Taylor JT.1 Monoplane	J. Crawford
	G–CLAN	PA-31-350 Navajo Chieftain	British Caledonian Airways
	G–CLEA	PA-28-161 Warrior II	Cleaver Marine Ltd.
	G–CLEF	Beech 95-58P Baron	Spanair Ltd.
	G–CLUT	Clutton Special	E. Clutton & A. Tabenor
	G–CLUX	Cessna F.172N	N. J. Hebditch
	G–CNSI	Beech 200 Super King Air	Conoco (UK) Ltd. (G-OSKA)
	G–COAL	Bell 206B JetRanger 3	NSM Aviation Ltd.
	G–COCO	Cessna F.172M	Citation Flying Services Ltd.

Reg.	Type	Owner or Operator	Notes
G–COOL	Cameron O-31 balloon	Swire Bros. *Sprite*	
G–COOP	Cameron N-31 balloon	Balloon Stable Ltd. *Co-op*	
G–COPS	Piper J-3C-65 Cub	P. B. Hanson	
G–CPFC	Cessna F.152	Central Air Services	
G–CPTS	AB-206B JetRanger 2	A. R. B. Aspinall	
G–CRAN	Robin R.1180T	Headcorn Flying School Ltd.	
G–CRDA	Cessna 421C	Cowick Hall Aviation Ltd.	
G–CRIL	R. Commander 112B	Castle Reeves Investments Ltd.	
G–CRIS	Taylor JT.1 Monoplane	C. J. Bragg	
G–CSBM	Cessna F.150M	Coventry (Civil) Aviation Ltd.	
G–CSFC	Cessna 150L	Civil Service Flying Club	
G–CSNA	Cessna 421C	Armstrong Aviation Ltd.	
G–CSSC	Cessna F.152	Cleansing Service Group Aviation Ltd.	
G–CSZA	V.807 Viscount	Southern International Air Transport Ltd.	
G–CSZB	V.804 Viscount	Southern International Air Transport Ltd. (G–AOXU)	
G–CTLN	EMB-110P1 Bandeirante	Centreline Air Services Ltd.	
G–CTRN	Enstrom F-28C	W. E. Taylor & Son Ltd.	
G–CUBB	PA-18-150 Super Cub	W. B. Hosie	
G–CUBI	PA-18-150 Super Cub	G. McLean & ptnrs.	
G–CWOT	Currie Wot	D. A. Lord	
G–CXMF	G.1159 Gulfstream 2	Fay Air (Jersey) Ltd.	
G–DAAH	PA-28R-201T Turbo Arrow IV	A. A. Hunter	
G–DACA	P.57 Sea Prince T.1	Atlantic & Caribbean Aviation Ltd.	
G–DAJW	K & S Jungster 1	A. J. Walters	
G–DAKS	Dakota 3 (TS423)	Aces High Ltd.	
G–DAND	SOCATA TB.10 Tobago	Dandy Caravans	
G–DART	Rollason Beta B2	M. G. Ollis	
G–DATS	Cessna 310R	Denham Air Taxi Ltd.	
G–DAVE	Jodel D.112	D. A. Porter	
G–DAVI	Cessna TU.206G	G. A. Dommett	
G–DAVY	Evans VP-2	D. Morris	
G–DAWN	Cessna T.210M	The Earl Jermyn	
G–DCAN	PA-38-112 Tomahawk	Airways Aero Associations Ltd.	
G–DCKK	Cessna F.172N	Citation Flying Services Ltd.	
G–DCOL	PA-30 Twin Comanche 160	Airmore Aviation Ltd. (G–BGSU)	
G–DEBI	Beech E36 Bonanza	R. J. Higgins	
G–DELI	Thunder Ax7-77 balloon	C. Delius	
G–DEVA	PA-23 Aztec 250	Eyelure Ltd.	
G–DFLY	PA-38-112 Tomahawk	Airways Aero Associations Ltd.	
G–DFTS	Cessna FA.152	Denham Flying Training School Ltd.	
G–DFUB	Boeing 737-2K9	Monarch Airlines Ltd.	
G–DHLD	Beech B60 Duke	T. S. Grimshaw Ltd.	
G–DICK	Thunder Ax6-56Z balloon	Thunder Balloons Ltd.	
G–DIPS	Taylor JT.1 Monoplane	B. J. Halls	
G–DIVE	BN-2A-26 Islander	Secretary of State (G–BEXA)	
G–DJBE	Cessna 550 Citation II	DJB Engineering Ltd.	
G–DJIM	DHCA-1	J. Crawford	
G–DJMS	PA-28-181 Archer II	D. J. McSorley	
G–DMAN	H.S.125 Srs. 600B	McAlpine Aviation Ltd.	
G–DOGS	Cessna R.182RG	Newbranch Ltd.	
G–DOVE	Cessna 182Q	Dove Air Ltd.	
G–DRAY	Taylor JT.1 Monoplane	L. J. Dray	
G–DTOO	PA-38-112 Tomahawk	Airways Aero Associations Ltd.	
G–DUET	Wood Duet	C. Wood	
G–DUKE	Beech B60 Duke	New Equipment Ltd.	
G–DUNN	Zenair CH.200	A. Dunn	
G–DUVL	Cessna F.172N	Duval Studios Ltd.	
G–DWMI	Bell 206L-1 LongRanger	Glenwood Helicopters Ltd.	
G–DYOU	PA-38-112 Tomahawk	Airways Aero Associations Ltd.	
G–EAGL	Cessna 421C	Voxson Audio Ltd.	
G–EBJI	Hawker Cygnet Replica	A. V. Francis	
G–EBWJ	SOCATA TB.10 Tobago	Crocker Air Services	
G–ECCO	GA-7 Cougar	J. E. Clark	
G–ECGC	Cessna F.172N-II	Leicestershire Aero Club Ltd.	
G–ECMA	PA-31-325 Navajo	Elliott Bros. (London) Ltd.	
G–ECOX	Pietenpol Aircamper GN.1	H. C. Cox	
G–EDDY	PA-28RT-201 Arrow IV	Supaglide Ltd.	
G–EDEN	SOCATA TB.10 Tobago	Tobago Air Services	
G–EDIF	Evans VP-2	R. Simpson	

Notes	Reg.	Type	Owner or Operator
	G–EEEE	Slingsby T.31 Motor Glider	R. F. Selby
	G–EENY	GA-7 Cougar	Martin Butler Associates Ltd.
	G–EEUP	SNCAN SV-4C	Gladaircraft Co. Ltd.
	G–EEZE	Rutan Vari-Eze	A. J. Nurse
	G–EFPT	H.S.125 Srs. 700B	Save Energy Services Ltd. (G–BFVN)
	G–EGGS	Robin DR.400/180	R. Foot
	G–EHAP	Sportavia-Pützer RF.7	M. J. Revill
	G–EIIR	Cameron N-77 balloon	Major C. J. T. Davey *Silver Jubilee*
	G–ELLY	AT-6D Harvard III	E. D. Sallingboe
	G–EMKM	Jodel D.120A	C. R. Davey
	G–EMMA	Cessna F.182Q	Rogers Aviation Ltd.
	G–EMMS	PA-38-112 Tomahawk	Surrey & Kent Flying Club Ltd.
	G–EMMY	Rutan Vari-Eze	M. J. Tooze
	G–ENIE	Nipper T.66 Srs. 3	A. J. Waller
	G–ENII	Cessna F.172M	M. S. Knight & Varlex Ltd.
	G–ENSI	Beech F33A Bonanza	F. B. Gibbons & Son
	G–EOFF	Taylor JT.2 Titch	G. Wylde
	G–EORG	PA-38-112 Tomahawk	Cormack (Aircraft Services) Ltd.
	G–EPDI	Cameron N-77 balloon	E.P.D. Containers & Supply Co. Ltd. & Pegasus Aviation Ltd. *Pegasus*
	G–ERDB	Hawker Cygnet Replica	R. D. Bertram
	G–ERIC	R. Commander 112TC	DJF (Jersey) Ltd.
	G–ERMS	Thunder AS33 Airship	Thunder Balloons Ltd.
	G–ESAL	AB-206B JetRanger 3	ESAL Commodities Ltd. (G.BHXW)
	G–ETUP	Cessna F.150L	Citation Flying Services Ltd.
	G–EURO	Cessna 310R	Carex Exhaust Centres Ltd.
	G–EVAN	Taylor JT.2 Titch	E. Evans
	G–EWBJ	SOCATA TB.10 Tobago	Crocker Air Services
	G–EXEC	PA-34-200 Seneca	Staverton Flying School Ltd.
	G–EXEX	Cessna 404	Owledge Ltd.
	G–EXIT	SOCATA Rallye 235GT	G-Exit Ltd.
	G–EZLT	Rutan Vari-Eze	L. J. Turner
	G–EZOS	Rutan Vari-Eze	O. Smith
	G–FADS	PA-23 Aztec 250	A. C. Stanley Ltd.
	G–FALL	Cessna 182L	K. J. Toyer
	G–FANL	Cessna FR.172K	Francis & Lewis Ltd.
	G–FANS	BN-2A-3 Islander	Dowty Rotol Ltd.
	G–FANZ	PA-23 Aztec 250	Fanz International Ltd.
	G–FARM	SOCATA Rallye 235GT	M. J. Jardine-Paterson
	G–FAST	Cessna F.337G	Shirlstar Container Transport Ltd.
	G–FAYE	Cessna F.150M	Citation Flying Services Ltd.
	G–FBWH	PA-28R Cherokee Arrow 180	B. W. Hopkins
	G–FCAS	PA-23 Aztec 250	Alnwick Storage Haulage Ltd.
	G–FERG	AS.350B Ecureuil	Ferguson Aviation Ltd. (G–BGCW)
	G–FERY	Cessna 550 Citation II	European Ferries Ltd. (G–DJBI)
	G–FFEN	Cessna F.150M	E. P. Collier
	G–FIHL	Cessna 421C	Fitair Ltd.
	G–FILM	SE.313B Alouette II	Alan Mann Helicopters Ltd. (G–BANR)
	G–FINE	Aerostar 601P	Oldham Aviation Ltd.
	G–FIRE	V. S. Spitfire	S. Flack
	G–FIVE	H.S.125 Srs. 1	British Air Ferries Ltd. (G–ASEC)
	G–FIZZ	PA-28-161 Warrior II	Ballandra Airways Ltd.
	G–FLPI	R. Commander 112A	Tuscany Ltd.
	G–FMFC	EMB-110P2 Bandeirante	Fairflight Ltd.
	G–FOCK	Focke-Wulf Fw.190-A	P. R. Underhill
	G–FOIL	PA-31-310 Navajo	Air Foyle Ltd.
	G–FORD	SNCAN SV-4C	P. Meeson
	G–FOTO	PA-23 Aztec 250	Davis Gibson Advertising Ltd. (G–BJDH/G–BDXV)
	G–FOUR	H.S.125 Srs. 3B	Group 4 Aviation Ltd. (G–AVRE)
	G–FOXY	Cessna F.172M	Hungerford Garages Ltd.
	G–FOYL	PA-23 Aztec 250	Air Foyle Ltd. (G–AVNK)
	G–FRAG	PA-32-300 Cherokee Six	R. Goodwin & Co. Ltd.
	G–FRED	FRED Srs. 2	R. Cox
	G–FSPL	PA-32R-300 Lance	R. E. Husband & A. D. Widdows
	G–FTTA	PA-31-350 Navajo Chieftain	Tyne-Tees Airways
	G–FUJI	Fuji FA.200-180	S. T. N. Rollesby
	G–FURY	Sea Fury T.XI	S. Flack
	G–FUZZ	PA-19 Super Cub	C. W. Cline
	G–FVEE	Monnet Sonerai I	D. R. Sparke
	G–FXIV	V. S. Spitfire FR.XIV	M. Wickenden & ptnrs.

Reg.	Type	Owner or Operator	Notes
G–GACA	P.57 Sea Prince T.1	Atlantic & Caribbean Aviation Ltd.	
G–GALE	PA-34-200 Seneca	Gale Construction Co. Ltd.	
G–GAYE	Cessna 421C	K. A. C. Thorogood	
G–GBSC	Beech E90 King Air	Kenton Utilities & Developments Ltd.	
G–GCKI	Mooney M.20K	Express Aviation Services Ltd.	
G–GEAR	Cessna FR.182Q	Beechwood Garage Ltd.	
G–GEEP	Robin R.1180T	Organic Concentrates Ltd.	
G–GEES	Cameron N-77 balloon	Mark Jarvis Ltd. *Mark Jarvis*	
G–GEOF	Pereira Osprey 2	G. Crossley	
G–GFAL	Douglas DC-10-10	Laker Airways Ltd *Northern Belle*	
G–GFLY	Cessna F.150K	Citation Flying Services Ltd.	
G–GGAE	H.S.125 Srs. 3B/RA	Associated Engineering Ltd.	
G–GGGG	Thunder Ax7-77A balloon	Test Valley Balloon Group	
G–GHNC	AA-5A Cheetah	G. Humphrey & N. Chamberlain	
G–GILL	Cessna 402C	Gill Aviation Ltd.	
G–GINA	AS.350B Ecureuil	Endeavour Aviation Ltd.	
G–GIRL	Cessna 421C	Lewis Shops Group	
G–GKNB	Beech 200 Super King Air	GKN Group Services Ltd.	
G–GLEN	Bell 212	Valley of Gleneagles Helicopters Ltd.	
G–GMSI	SOCATA TB.9 Tampico	Dukeries Aviation Ltd.	
G–GODS	Brugger MB.2 Colibri	F. Sharples	
G–GOGO	Hughes 369D	Sloane Aviation Ltd.	
G–GOLD	Thunder Ax6-56A balloon	John Terry & Sons Ltd.	
G–GOMM	PA-32R-300 Lance	Aerospares Ltd.	
G–GOOS	Cessna F.182Q	Rogers Aviation Sales Ltd.	
G–GOSH	Cessna 404	Euroair Transport Ltd.	
G–GRAY	Cessna 172N	Buddale Ltd.	
G–GSKY	Douglas DC-10-10	Laker Airways Ltd. *California Belle*	
G–GSMP	Eiri PIK-20E-1	GSMP Ltd.	
G–GTPL	Mooney M.20K	Valencienne Ltd. (G-BHOS)	
G–HADI	G.1159 Gulfstream 2	Arab Express Ltd.	
G–HALL	PA-22 Tri-Pacer 160	F. P. Hall (G–ARAH)	
G–HANS	Robin DR.400 2+2	Stahl Engineering Co. Ltd.	
G–HAPR	*B.171 Sycamore HC.14 (XG547)	British Rotorcraft Museum	
G–HARV	PA-23 Aztec 250	T. S. Grimshaw Ltd.	
G–HAWK	H.S. 1182 Hawk	British Aerospace	
G–HEAT	Bell 206B JetRanger 3	Warmco Spaceheating (Manchester) Ltd.	
G–HELI	*Saro Skeeter Mk.12 (XM556)	British Rotorcraft Museum	
G–HELY	Agusta 109A	Barratt Developments Ltd.	
G–HGGS	EMB-110P1 Bandeirante	Euroair Transport Ltd.	
G–HHOI	H.S.125 Srs. 700B	Trust House Forte Airport Services Ltd. (G–BHTJ)	
G–HIGH	Cessna FT.337GP	British Cybercom Ltd.	
G–HILL	Cessna U.206F	Citation Flying Services Ltd.	
G–HILR	Hiller UH-12E	G. & S. G. Neal (Helicopters) Ltd.	
G–HLUB	Beech A200 Super King Air	United Biscuits Ltd.	
G–HMCG	BN-2A-26 Islander	Renco (Aviation) Ltd. (G–BESH)	
G–HOLS	Warner Special	J. O. C. Warner	
G–HOLT	Taylor JT.1 Monoplane	K. D. Holt	
G–HOME	Colt 77A balloon	Anglia Balloon School *Tardis*	
G. HOOK	Hughes 369D	Auto Alloys (Helicopters) Ltd.	
G–HOOV	Cameron N-56 balloon	Balloon Stable Ltd. *Hoover*	
G–HOPE	Beech F33A Bonanza	Hope Jones (Electrical Spares) Ltd.	
G–HORN	Cameron V-77 balloon	G. J. E. Horn	
G–HOSE	Cessna 152 II	Toon Ghose Aviation Ltd.	
G–HOST	Cameron N-77 balloon	S. Williams & ptnrs. *Blythe Spirit*	
G–HOUL	FRED Srs. 2	D. M. M. Richardson	
G–HOUS	Double Sky Chariot balloon	Anglia Balloons Ltd.	
G–HRLM	Brugger MB.2 Colibri	R. A. Harris	
G–HUFF	Cessna 182P	Lighthouse Investments Ltd.	
G–HUGH	PA-32RT-300T Turbo Lance II	Mann Aviation Sales Ltd. (G–IFLY)	
G–HULL	Cessna F.150M	Citation Flying Services Ltd.	
G–HUNT	Hunter F.51	S. Flack	
G–HUSH	Hughes 269C	Auto Alloys (Helicopters) Ltd.	
G–HYDE	AB-206B JetRanger 3	Hyde International Ltd.	
G–IAIN	Cessna P.210N	Lorne Stewart Ltd.	
G–IANS	R. Turbo Commander 690B	Parahold Ltd.	
G–IANT	Cessna 404	Lorren Textiles Ltd.	

Notes	Reg.	Type	Owner or Operator
	G–IBFW	PA-28R-201 Arrow III	B. Walker & Co. (Dursley) Ltd. & J. & C. Ward (Holdings) Ltd.
	G–IBIS	H.S.125 Srs. 3BRA	Ibis Ltd. (G–AXPU)
	G–ICES	Thunder Ax6-56 balloon	Thunder Balloons Ltd.
	G–ICRU	Bell 206A JetRanger	Warmco (Manchester) Ltd.
	G–IDDY	D.H.C.-I Super Chipmunk	N. A. Brendish
	G–IIIA	Swearingen Merlin IIIB	Willowbrook International Ltd.
	G–IIIB	Swearingen Merlin IIIB	Willowbrook International Ltd.
	G–IKIS	Cessna 210M	Lorne Stewart Ltd.
	G–ILLI	Cessna FR. 182RG	Rogers Aviation Sales Ltd.
	G–ILLY	PA-28-181 Archer II	A. G. & K. M. Spiers
	G–INNY	SE-5A Replica	R. M. Ordish
	G–IOSI	Jodel DR.1051	R. G. E. Simpson & A. M. Alexander
	G–IPPM	SA.102-5 Cavalier	I. D. Perry & P. S. Murfitt
	G–IPRA	Beech A200 Super King Air	J. H. Ritblat (G–BGRD)
	G–IPSY	Rutan Vari-Eze	R. A. Fairclough
	G–IRLS	Cessna FR.172J	Citation Flying Services Ltd.
	G–IVAN	Rutan Vari-Eze	I. Shaw
	G–IWPL	Cessna F172M	Ian Willis Publicity Ltd.
	G–JADE	Beech 58 Baron	Liaison & Consultant Services Ltd.
	G–JAKE	D. H. C.-I Chipmunk 22	Ewart & Co. (Studio) Ltd. (G–BBMY)
	G–JAKO	Cessna TU.206G	New Hathesley Garage
	G–JAKY	PA-31-325 Navajo	Appin Air Ltd.
	G–JAMI	Bell 206L LongRanger	Heart of England Helicopters Ltd.
	G–JANE	Cessna 340A	Standard Aviation Ltd.
	G–JANS	Cessna FR.172J	Jans of London
	G–JANY	AS.350B Ecureuil	P. E. Cadbury (G–BHCN)
	G–JASP	PA-23 Aztec 250	J. A. Storey & ptnrs.
	G–JAWS	Enstrom F-280C Shark	Earl Developments Ltd.
	G–JBUS	FRED Srs. 2	R. V. Joyce
	G–JCWW	F.28 Fellowship 4000	Air Alsace/Air France
	G–JDST	PA-31-350 Navajo Chieftain	Jack Tighe Ltd.
	G–JEAN	Cessna 500 Citation	Castlewood Air Services Ltd.
	G–JEFF	PA-38-112 Tomahawk	Channel Aviation Ltd.
	G–JENS	SOCATA Rallye 100ST	B. H. Burnet (G–BDEG)
	G–JENY	Baby Great Lakes	J. M. C. Pothecary
	G–JETA	Cessna 550 Citation II	IDS Aircraft Ltd.
	G–JFWI	Cessna F.172N	Citation Flying Services Ltd.
	G–JGCL	Cessna 414A	Westair Flying Services Ltd.
	G–JGFF	AB-206B JetRanger 3	Mailam Ltd.
	G–JILL	R. Commander 112TCA	Kennair (General Aviation) Ltd.
	G–JJSG	Learjet 35A	Smurfit Ltd.
	G–JMCA	PA-31-350 Navajo Chieftain	Fluorocarbon Ltd.
	G–JMCC	Beech 95-58 Baron	Mowlem Construction (Plant Hire) Ltd.
	G–JMFW	Taylor JT.2 Titch	G. J. M. F. Winder
	G–JOAN	AA-5B Tiger	Oldment Ltd. (G–BFML)
	G–JOHN	PA-28R-201T Turbo Arrow III	Fairoaks Flight Centre
	G–JONS	PA-31-350 Navajo Chieftain	Berdan Enterprises Ltd.
	G–JOSE	Cessna U.206G	Safari Skylink Enterprises Ltd.
	G–JRCM	Hawker Fury Mk. I Replica	J. R. C. Morgan
	G–JRCT	Cessna 550 Citation II	IDS Aircraft Ltd.
	G–JRMM	R. Commander 690B	Colt Car Co. Ltd.
	G–JSSD	SA. Jetstream 3001	British Aerospace (G–AXJZ)
	G–JUDI	AT-6 Harvard III (FX301)	A. Haig-Thomas
	G–JUDY	AA-5A Cheetah	Pioneer Welders and Fabrications
	G–JUMP	AB-206B JetRanger 3	Hickstead Ltd.
	G–JUNE	PA-28-161 Warrior II	Allen Technical Services Ltd.
	G–JURG	R. Commander 114	Maronco Ltd.
	G–JWIV	Jodel DR.1051	J. W West
	G–KACT	Cessna 421B	K. A. C. Thorogood
	G–KAIR	PA-28-161 Warrior II	R. Keys & Co.
	G–KATH	Cessna P.210N	Northair Aviation Ltd.
	G–KCAV	Cessna 414A II	Northair Aviation Ltd.
	G–KCIG	Sportavia RF5B	Executive Air Sport Ltd.
	G–KDIX	Jodel D.9 Bebe	K. Barlow
	G–KEEN	Stolp Starduster Too	Keenair Ltd.
	G–KENS	PA-32-300 Cherokee Six	Ken Heanes Ltd.
	G–KENT	Cessna 414A	British Car Auctions (Aviation) Ltd.
	G–KERC	Nord NC.854S	Kirk Aviation
	G–KERK	Piper J-3C-65 Cub	Kirk Aviation
	G–KERR	Cessna FR.172K-XP	Rogers Aviation Sales Ltd.

Reg.	Type	Owner or Operator	Notes
G–KEYS	PA-23 Aztec 250	Robert Keys & Co. Ltd.	
G–KILO	Boeing 747-236F	British Airways Cargo *British Trader*	
G–KING	PA-38-112 Tomahawk	Gordon King Aviation Ltd.	
G–KIRK	Piper J-3C65 Cub	M. Kirk	
G–KLAY	Enstrom F-280C Shark	Klay Electronics Ltd. (G–BGZD)	
G–KUKU	Pfalzkuku (BS676)	A. D. Lawrence	
G–KWAX	Cessna 182E Skylane	E. W. Duck	
G–KWIK	Partenavia P.68B	Ryburn Polythene Ltd.	
G–KYAK	Yakolev C-11	R. Lamplough	
G–LACY	R. Turbo Commander 690B	G. F. Lacey & ptnrs.	
G–LADA	PA-32 Cherokee Six 300D	Euroway Car Centre Ltd. (G–AYWK)	
G–LAKI	Jodel DR.1051	V. Panteli	
G–LANE	Cessna F.172N	Mark Laing Aviation Ltd.	
G–LASH	Monnet Sonerai II	A. Lawson	
G–LASS	Rutan Vari-Eze	P. J. Callert & ptnrs.	
G–LATC	EMB-110P1 Bandeirante	Euroair Transport Ltd.	
G–LAZE	Jodel DR.1050	Culmor Windows Ltd.	
G–LEAM	PA-28-236 Dakota	Clutchstar Ltd. (G–BHLS)	
G–LEAR	Learjet 35A	C.S.E. Aviation Ltd. (G–ZEST)	
G–LENS	Thunder Ax7-77A balloon	Keith Johnston Photographic Ltd.	
G–LEON	PA-31-350 Navajo Chieftain	Fairflight Ltd.	
G–LEXI	Cameron N-77 balloon	Lex Mead Bristol Ltd. *Lex*	
G–LFIX	V.S. Spitfire LF.IX	Island Trading Ltd.	
G–LFCA	Cessna F.152	Aerolease	
G–LIDE	PA-31-350 Navajo Chieftain	Rockville Investments Ltd.	
G–LIFE	Thunder Ax6-56Z balloon	Schroder Life Assurance Ltd.	
G–LIFT	Bell 47G-3	National Car Parks Ltd.	
G–LIMA	R. Commander 114	W. M. P. Miller	
G–LINK	Sikorsky S-61N	British Caledonian Airways	
G–LION	PA-18-135 Super Cub	Citation Flying Services Ltd.	
G–LITE	R. Commander 112A	Executive Wings Ltd.	
G–LOAG	Cameron N-77 balloon	Matthew Gloag & Son Ltd.	
G–LOCK	AB-206B JetRanger 3	Lovaux Ltd.	
G–LONG	Bell 206L LongRanger	Air Hanson Ltd.	
G–LOOK	Cessna F.172M	Laarbruch Flying Club	
G–LOOP	Pitts S-1C Special	P. Meeson	
G–LOTI	Bleriot XI (replica)	M. L. Beach	
G–LOWE	Monnet Sonerai II	J. A. Lowe	
G–LRII	Bell 206L LongRanger	Castle Motors (Trebrown) Ltd.	
G–LSMI	Cessna F.152	Rogers Aviation Sales Ltd.	
G–LUCK	Cessna F.150M	Citation Flying Services Ltd.	
G–LUNA	PA-32RT-300T Turbo Lance II	Everest Aviation Ltd.	
G–LYDE	Eiri PIK-20E	G. S. Wilson	
G–LYNN	PA-32RT-300 Lance II	B. S. Sales Ltd. (G–BGNY)	
G–LYNX	Westland WG.13 Lynx	Westland Helicopters Ltd.	
G–MACH	SIAI-Marchetti SF.260	R. A. Sareen	
G–MACK	PA-28R Cherokee Arrow 200	John Mack Ltd.	
G–MADI	Cessna 310R II	Talley Aviation Ltd.	
G–MAGI	AS.350B Ecureuil	Micro Consultants Ltd. (G–BHLR)	
G–MAGS	Cessna 340A	Rogers Aviation Sales Ltd.	
G–MAIL	D.H.C.-6 Twin Otter 310	Fairflight Ltd.	
G–MALC	AA-5 Traveler	Air Coventry Ltd. (G–BCPM)	
G–MANX	FRED Srs. 2	P. Williamson	
G–MARK	Cessna F.337H	Denis Ferranti Hoverknights Ltd.	
G–MARY	Cassutt Special I	J. Chadwick	
G–MAST	PA-28R Cherokee Arrow 180	F. Bennister (G–AZGM)	
G–MAUL	Maule M5-235C Lunar Rocket	Capital Aviation Sales (UK) Ltd.	
G–MAXY	Cessna 210L	Ripmax Models Ltd.	
G–MCDS	Cessna 210N	Merseyside Car Delivery (G–BHNB)	
G–MDAS	PA-31-310 Navajo	Milford Haven Dry Dock Co. Ltd. (G–BCJZ)	
G–MDRB	PA-31-350 Navajo Chieftain	Iceni Aviation Ltd.	
G–MERI	PA-28-181 Archer II	Spooner Aviation Ltd.	
G–META	Bell 212	The Metropolitan Police	
G–MFCI	Cessna F.150L	Citation Flying Services Ltd.	
G–MFEU	H.S.125 Srs. 600B	Clartacrest Ltd.	
G–MHGI	Cessna 414A	Northair Aviation Ltd. (G–BHKK)	
G–MICK	Cessna F.172N	J. W. Barwell	
G–MIKE	Hornet Gyroplane	M. H. J. Goldring	
G–MILK	SOCATA TB.10 Tobago	Accent Aviation Ltd.	

Notes	Reg.	Type	Owner or Operator
	G–MINI	Currie Wot	D. Collinson
	G–MISS	Taylor JT.2 Titch	A. Brenan
	G–MKAN	AB-206B JetRanger 3	Mackan Group (UK) Ltd. (G–DOUG)
	G–MKEE	EAA Acro Sport	G. M. McKee
	G–MLAS	Cessna 182E	Mark Luton Aviation Services
	G–MMOI	H.S.125 Srs. 700	Trust House Forte Airport Services Ltd. (G–BHTJ)
	G–MOBL	EMB-110P2 Bandeirante	Fairflight Ltd.
	G–MOGG	Cessna F.172N	R. M. W. & B. N. C. Mogg (G–BHDY)
	G–MOLY	PA-23 Apache 160	Steer Aviation Ltd. (G–APFV)
	G–MONA	M.S.880B Rallye Club	J. D. M Wilson (G–AWJK)
	G–MONO	Taylor JT.1 Monoplane	A. Doughty
	G–MORR	AS.350B Ecureuil	Colt Car Co. Ltd. (G–BHIU)
	G–MOTH	D.H.82A Tiger Moth	M. C. Russell & ptnrs.
	G–MOVE	Aerostar 601P	Knight Programming Support Ltd.
	G–MOZY	D.H.98 (replica)	J. Beck & G. L. Kemp
	G–MPWA	Wassmer WA-54 Atlantic	MPW Aviation Ltd.
	G–MPWI	Robin HR.100/200	MPW Aviation Ltd.
	G–MSDS	Cessna 404	Lorien Textiles Ltd.
	G–MUST	Commonwealth CA-18 Mustang 22	Fairoaks Aviation Services Ltd.
	G–MXIV	V.S. Spitfire FR.xiv	M. & K. Wickenden
	G–NASH	AA-5A Cheetah	T. R. Bamber
	G–NATS	R. Commander 690B	Tembo Records Ltd.
	G–NATT	R. Commander 114A	Tembo Music Ltd.
	G–NBSI	Cameron N-77 balloon	Nottingham Building Soc.
	G–NDNI	NDN-1 Firecracker	Norman Marsh Aircraft Ltd.
	G–NESS	R. Commander 685	Air Factors
	G–NEUS	Brugger MB.2 Colibri	G. S. Smeaton
	G–NEVA	PA-31-310 Navajo	Solitaire Flight Management Ltd. (G–BGVO/G–AZAC)
	G–NEWR	PA-31-350 Navajo Chieftain	Burnthills Plant Hire Ltd.
	G–NEWS	Bell 206B JetRanger 3	Peter Press Ltd.
	G–NHVH	Maule M5-235C Lunar Rocket	Commercial Go-Karts Ltd.
	G–NICK	PA-19 Super Cub 95	N. J. Cole
	G–NILE	Colt 77A balloon	J. M. Donnell
	G–NITE	PA-31-350 Navajo Chieftain	WT Shipping Group Ltd.
	G–NIUS	Cessna F.172N	Horizon Lighting Products Ltd.
	G–NJAG	Cessna 207	G. H. Nolan Ltd.
	G–NOEL	AB-206B JetRanger 2	N. Edmunds & P. R. McEnhill (G–BCWN)
	G–NOME	Baby Great Lakes	J. B. Scott & ptnrs.
	G–NORD	Nord NC.854	R. G. E. Simpson & A. M. Alexander
	G–NORX	Cessna 421C	Norcross Transport Ltd.
	G–NOVA	Cessna T.337H	Northair Aviation Ltd.
	G–NWPB	Thunder Ax7-77Z balloon	Thunder Balloons Ltd.
	G–OABI	Cessna 421C	ABI Caravans Ltd.
	G–OAHB	T-33 Mk. 3 Silver Star	—
	G–OAIR	EMB-110P1 Bandeirante	Air UK
	G–OAKS	Cessna 421C	Barratt Developments Ltd.
	G–OAMH	Agusta 109A	Willowbrook International Ltd.
	G–OAST	Cessna T.182RG	Airwork Services Ltd.
	G–OATS	PA-38-112 Tomahawk	C.S.E. Aviation Ltd.
	G–OBAE	H.S.125 Srs. 700B	British Aerospace
	G–OBAT	Cessna F.152	Renco Aviation Ltd.
	G–OBCA	Cessna 421C	British Car Auctions Ltd.
	G–OBEY	PA-23 Aztec 250	Keller Bryant & Co. Ltd. (G–BAAJ)
	G–OBIA	EMB-110P1 Bandeirante	Air UK
	G–OBLE	C.A.S.A. 1.131 Jungmann	A.J.D. Securities Ltd.
	G–OBMW	AA-5 Traveler	Automania Airways (G–BDFV)
	G–OCAL	Partenavia P.68B	Dowson Car Sales Ltd. (G–BCMY)
	G–OCAT	Eiri PIK-20E	D. S. Innes
	G–OCPC	Cessna FA.152	Denham Flying Training School Ltd.
	G–ODAY	Cameron N-56 balloon	C. O. Day (Estate Agents)
	G–ODEL	Falconar F-II-3	J. R. D. Bygraves
	G–OEZE	Rutan Vari-Eze	S. Stride & ptnrs.
	G–OFAR	Cessna 402C	Wm. Leach (Builders) Ltd.
	G–OFLY	Cessna 210L	Northair Aviation Ltd.
	G–OFRL	Cessna 414A	Flight Refuelling Ltd.
	G–OGDN	Beech A200 Super King Air	A. Ogden & Sons Ltd.
	G–OHTL	Sikorsky S-76	Air Hanson Ltd.

Reg.	Type	Owner or Operator	Notes
G–OILS	Cessna T.210L	Spectro Analysis Co. (G–BCZP)	
G–OIML	AB-206B JetRanger 3	IML Group of Companies	
G–OJCB	AB-206B JetRanger 2	J. C. Bamford Excavators Ltd.	
G–OJCW	PA-32RT-300 Lance II	J. C. Walsh Contractors & Plant Hire Ltd.	
G–OJEA	D.H.C.-6 Twin Otter 310	Jersey European Airways	
G–OJEE	Bede BD-4	G. Hodges	
G–OJMA	Cessna 421B	Pant House Ltd.	
G–OJOE	Partenavia P.68B	Drive House Ltd.	
G–OJON	Taylor JT.2 Titch	J. H. Fell	
G–OKAY	Pitts S-1E Special	Skyfever (Aviation Enterprises)	
G–OLDY	Luton LA-5 Major	M. P. & A. P. Sargent	
G–OLEE	Cessna F.152	Rogers Aviation Sales Ltd.	
G–OLLI	Cameron O-31 balloon	Robertson Foods Ltd. *Golly III*	
G–OLLY	PA-31-350 Navajo Chieftain	Robertson Foods Ltd. (G–BCES)	
G–OLVR	FRED Srs. 2	A. R. Oliver	
G–OMET	Beech C90 King Air	Comet Radiovision Services Ltd. (G–COTE/G–BBKN)	
G–ONPN	H.S.125 Srs. 1B	Shirlstar Container Transport Ltd. (G–BAXG)	
G–OODE	SNCAN SV-4A	V. S. E. Norman (G–AZNN)	
G–OODY	PA-28R Cherokee Arrow 200	J. Traynor	
G–OOFY	Rollason Beta	G. Staples	
G–OOSE	Rutan Vari-Eze	J. A. Towers	
G–ORAV	Cessna 337D	Rogers Aviation Sales Ltd. (G–AXGJ)	
G–ORAY	Cessna F.182Q-11	Ray Hoult (Land Drainage) Ltd. (G–BHDN)	
G–ORMC	Beech A200 Super King Air	RMC Group Services Ltd. (G–BEST)	
G–OSHH	Cessna 404	Northair Aviation Ltd.	
G–OSKY	Cessna 172M	Bevan Lynch Aviation Ltd.	
G–OSPT	PA-31-350 Navajo Chieftain	Escas Ltd.	
G–OSRF	Cessna 421C	Simpson Ready Foods Ltd.	
G–OTOM	Cessna FR.172J	J. T. L. Jones	
G–OTRG	Cessna TR.182RG	G. F. Holdings (Contractors) Ltd.	
G–OVFR	Cessna F.172N	M. Harrison & Co. (Leeds) Ltd.	
G–OVMC	Cessna F.152 II	R. W. Godwin	
G–OWAC	Cessna F.152	Northair Aviation Ltd. (G–BHEB)	
G–OWAK	Cessna F.152	Northair Aviation Ltd. (G–BHEA)	
G–OWEN	K & S Jungster	R. C. Owen	
G–PACE	Robin R.1180	Millicron Instruments Ltd.	
G–PACY	Rutan Vari-Eze	E. Pace	
G–PADY	R. Commander 114	Wyndley Nurseries Ltd.	
G–PAGE	Cessna F.150L	Citation Flying Services Ltd.	
G–PALS	Enstrom F-280C Shark	Spooner Aviation Ltd.	
G–PARA	Cessna 207	R. Goodwin	
G–PARI	Cessna 172RG Cutlass	Airwork Service Training Ltd.	
G–PARS	Evans VP-2	A. Parsfield	
G–PATT	Cessna 404 Titan	Casair Aviation Sales Ltd. (G–BHGL)	
G–PAUL	Partenavia P.68B	C. A. Church Ltd.	
G–PDON	WMB.2 Windtracker balloon	P. Donnellan	
G–PEET	Cessna 401A	Forest Aviation Ltd.	
G–PENN	AA-5B Tiger	Curd & Green Ltd.	
G–PENY	Sopwith LC-1T Triplane	J. S. Penny	
G–PEPD	PA-31-350 Navajo Chieftain	Lainmiche Ltd.	
G–PETE	Cessna 340 II	Northair Aviation Ltd.	
G–PFAA	EAA Model P biplane	P. E. Barker	
G–PFAB	FRED Srs. 2	P. E. Barker	
G–PFAC	FRED Srs. 2	R. Bennett	
G–PFAD	Wittman W.8 Tailwind	M. R. Stamp	
G–PFAE	Taylor JT.1 Monoplane	G. Johnson	
G–PFAF	FRED Srs. 2	K. Fern	
G–PFAG	Evans VP-1	N. S. Giles-Townsend	
G–PFAH	Evans VP-1	J. A. Scott	
G–PFAI	Clutton EC.2 Easy Too	E. Clutton & A. Tabenor	
G–PFAK	Kendal Mayfly	R. Y. Kendal	
G–PFAL	FRED Srs. 2	H. Pugh	
G–PFAM	FRED Srs. 2	W. C. Rigby	
G–PFAN	Avro 558 (replica)	N. P. Harrison	
G–PFAO	Evans VP-1	P. W. Price	
G–PFAP	Currie Wot	P. G. Abbey	

Notes	Reg.	Type	Owner or Operator
	G–PFAR	Isaacs Fury II	C. J. Repik
	G–PFAS	GY-20 Minicab	J. Sprouston & F. W. Speed
	G–PFAT	Monnet Sonerai II	H. B. Carter
	G–PFAU	Evans VP-2	D. E. Peace
	G–PFAV	D.31 Turbulent	B. A. Luckins
	G–PFAW	Evans VP-1	R. F. Shingler
	G–PFAX	FRED Srs. 2	A. J. Dunston
	G–PFAY	EAA Biplane	A. K. Lang & A. L. Young
	G–PFAZ	Evans VP-1	B. Kylo
	G–PHIL	Hornet Gyroplane	A. J. Philpotts
	G–PIED	PA-23 Aztec 250	Air London (Executive Travel) Ltd.
	G–PIES	Thunder Ax7-77Z balloon	Porkfarms Ltd.
	G–PINK	Cameron N-77 balloon	Leaveglen Ltd.
	G–PIPE	Cameron N-56 balloon	Carreras Rothmans Ltd.
	G–PISA	Thunder Ax7-77 balloon	Thunder Balloons Ltd.
	G–PLAN	Cessna F.150L	Wycombe Air Centre Ltd.
	G–PLAY	Robin R.2100	Miss A. C. Peacock
	G–PLIV	Pazmany PL.4	B. P. North
	G–PLUS	PA-34-200T-11 Seneca	Skycabs Ltd.
	G–PMCN	Monnet Sonerai II	P. J. McNamee
	G–PNUT	Cameron Special Shape balloon	Balloon Stable Ltd.
	G–POLO	PA-31-350 Navajo Chieftain	Rowntree Mackintosh Ltd.
	G–POLY	Cameron N-77 balloon	A. J. Bingley Ltd. *Polywallets*
	G–POOH	Piper J-3C-65 Cub	P. Robinson
	G–POPE	Eiri PIK-20E-1	J. T. Pope
	G–POST	EMB-110PI Bandeirante	Fairflight Ltd.
	G–POWA	PA-24 Comanche 400	G. F. Miller
	G–PPLI	Pazmany PL.1	G. Anderson
	G–PRAG	Brugger MB.2 Colibri	P. Russell
	G–PRES	Cessna 441 Conquest	Northair Aviation Ltd. (G.BHFX)
	G–PRIX	Cessna 414A	Group Lotus Car Co. Ltd.
	G–PUBS	Colt 56 balloon	Colt Balloons Ltd.
	G–PUFF	Thunder Ax7-77 balloon	Intervarsity Balloon Club *Puffin*
	G–PVAF	PA-44-180 Seminole	Pincus Vidler Arthur Fitzgerald Ltd.
	G–PVAM	Port Victoria 7 Grain Kitten	A. J. Manning
	G–PYRO	Cameron N-65 balloon	Masts Engineering Ltd.
	G–RACA	P.57 Sea Prince T.1	Atlantic & Caribbean Aviation Ltd.
	G–RACE	Aerostar 601P	Aerotime Ltd.
	G–RAFC	Robin R.2112	Headcorn Flying School Ltd.
	G–RAFE	Thunder Ax7-77 balloon	Thunder Balloons Ltd. *Billboard II*
	G–RAIN	Maule M5-235C Lunar Rocket	Velcourt (East) Ltd.
	G–RAMS	PA-32R-301 Saratoga	Peacock & Archer Ltd.
	G–RAND	Rand KR-2	R. L. Wharmby
	G–RANT	Enstrom F-28C	Sir Philip Grant-Suttie (G–BGKL)
	G–RARE	Thunder Ax4 balloon	Thunder Balloons Ltd.
	G–RASC	Evans VP-2	R. A. Codling
	G–RAYS	Zenair CH Srs.	R. E. Delves
	G–RBBE	Cessna 421C	Ruston Bucyrus Ltd.
	G–RBIN	Robin DR.400/2+2	Headcorn Flying School Ltd.
	G–RBLA	D.H.C.-6 Twin Otter 310	Loganair Ltd.
	G–RCCL	Beech C90 King Air	Reckitt & Colman Products Ltd.
	G–REAT	GA-7 Cougar	E. C. Heathcote & D. E. Nixon
	G–REEK	AA-5A Cheetah	JT Aviation Ltd.
	G–REES	Jodel D.140C	J. D. Rees
	G–REIS	PA-28R-201T Turbo Arrow III	H. Reis (Hard Chrome) Ltd.
	G–RETA	C.A.S.A. 1.131 Jungmann	C. Maron
	G–REXS	PA-28-181 Archer II	R. W. Willoughby
	G–RHCN	Cessna FR.182RG	R. H. C. Neville
	G–RHHT	PA-32RT-300 Lance II	H.T. Air Freight Ltd.
	G–RIDE	Stephens Akro	R. Mitchell
	G–RIGS	Aerostar 601P	Rigs Design Services Ltd.
	G–RILL	Cessna 421C	Maxwell Restaurants Ltd. (G–BGZM)
	G–RILY	Monnet Sonerai II	K. D. Riley
	G–RIND	Cessna 335	ATA Grinding Processes
	G–RKSF	Pitts S-2A Special	Anvil Aviation Ltd.
	G–RMAM	Musselwhite MAM.1	M. A. Musselwhite
	G–RMKM	R. Commander 112TC	Maynard Heels Ltd.
	G–RNAV	PA-31-350 Navajo Chieftain	Daleview Investments Ltd.
	G–ROAN	Boeing B.75N-1 Stearman	R. & A. Windley
	G–ROBN	Robin R.1180	F. J. Franklin
	G–RODI	Isaacs Fury	J. R. C. Morgan
	G–ROLL	Pitts S-2A Special	G. Lynn

G–HUNT (Top) Hawker Hunter F.51/*S. G. Richards*

G–ORAV (Centre) Cessna 337D, previously G–AXGJ/*C. P. Wright*

G–VIKE (Bottom) Bellanca 1730A Viking

Notes	Reg.	Type	Owner or Operator
	G–RONW	FRED Srs. 2	P. J. D. Granow
	G–RORO	Cessna 337B	Walter Edmunson (Haulage) Ltd. (G–AVIX)
	G–ROSE	Evans VP-1	W. K. Rose
	G–ROSS	Practavia Pilot Sprite	F. M. T. Ross
	G–ROUS	PA-34-200 Seneca	Linskill Air Charter Ltd.
	G–ROWL	AA-5B Tiger	Process Machine Ltd.
	G–ROWS	PA-28-161 Warrior II	S. C. Rowsell
	G–ROYL	Taylor JT.1 Monoplane	R. L. Wharmby
	G–ROYS	D.H.C.1 Chipmunk T.10	R. W. & S. Pullan
	G–RPAH	Rutan Vari-Eze	B. Hanson
	G–RRRR	Privateer Motor Glider	R. F. Selby
	G–RTHL	Leivers Special	R. Leivers
	G–RUIA	Cessna F.172M	Delamere & Norley Finance Ltd.
	G–RUMN	AA-1A Trainer	Colton Aviation Ltd.
	G–RUSS	Cessna 172N	Leisure Lease
	G–SAAB	R. Commander 112TC	Continental Cars (Stansted) Ltd. (G–BEFS)
	G–SAAS	Ayres S2R-T34 Thrush Commander	K. W. Norvis
	G–SABA	PA-28R-201T Turbo Arrow III	Newcastle Avionics Ltd. (G–BFEN)
	G–SAFE	Cameron N-77 balloon	Derbyshire Building Soc.
	G–SAIL	Boeing 707-323C	Tradewinds Ltd.
	G–SALA	PA-32-300 Cherokee Six	M. A. Lenihan
	G–SALL	Cessna F.150L	Citation Flying Services Ltd.
	G–SAMI	SAM-1	Trago Mills Ltd.
	G–SAMS	M.S.880B Rallye Club	M. W. Salmon
	G–SARO	Saro Skeeter Mk.12	F. F. Chamberlain
	G–SATC	Cessna F.150L	SAF Ltd.
	G–SATO	PA-23 Aztec 250	Colt Car Co. Ltd. (G–BCXP)
	G–SAVE	PA-31-350 Navajo Chieftain	Jetsave Ltd.
	G–SBRV	BRV Special	B. R. Vickers
	G–SCOT	PA-31-350 Navajo Chieftain	British Caledonian Airways
	G–SCUB	PA-18-135 Super Cub (542447)	N. D. Needham Farms
	G–SEAR	Pazmany PL.4	A. J. Sear
	G–SEED	Piper J-3C-65 Cub	J. H. Seed
	G–SEJW	PA-28-161 Warrior II	Grosvenor Homes Ltd.
	G–SHAW	PA-30 Twin Comanche 160	Micro Metalsmiths Ltd.
	G–SHEL	Cameron O-56 balloon	The Shell Company of Hong Kong Ltd.
	G–SHIP	PA-23 Aztec 250	Shetland Line Air Services Ltd.
	G–SHOK	Cessna 421C II	Armstrong Aviation Ltd.
	G–SHOW	M.S.733 Alcyon	L. M. Walton
	G–SIGN	PA-39 Twin Comanche C/R	K. W. Hawes (Electrical) Ltd.
	G–SILK	Aerostar 601P	R. Yorke
	G–SILV	Cessna 340A	Superprime Ltd.
	G–SIXA	Douglas DC-6B	Air Atlantique (G–ARXZ)
	G–SIXB	Douglas DC-6B	Air Atlantique
	G–SKYA	FH.227B Friendship	Skyways Cargo Airline
	G–SKYB	FH.227B Friendship	Skyways Cargo Airline
	G–SKYE	Cessna TU.206G	Rogers Aviation Sales Ltd.
	G–SKYH	Cessna 172N	CGH Managements Ltd.
	G–SKYM	Cessna F.337E	A. P. Mothew & B. J. Ixer (G–AYHW)
	G–SLIK	Taylor JT.2 Titch	J. Jennings
	G–SMIT	Messerschmitt Bf.109G	Fairoaks Aviation Services Ltd.
	G–SNOW	Cameron V-77 balloon	M. J. Snow
	G–SOAR	Eiri PIK-20E	C. P Witter Ltd.
	G–SOLO	Pitts S-2S Special	Rothmans International Ltd.
	G–SOLY	Westland Bell 47G-3B1 (Soloy)	Heliwork Ltd.
	G–SONA	SOCATA TB.10 Tobago	Sonadyne Ltd. (G–BIBI)
	G–SPIN	Pitts S-2A Special	R. N. Goode
	G–SPIT	V.S. Spitfire XIV	D. W. Arnold (G–BGHB)
	G–SPOP	Thunder Ax7-77 balloon	Thunder Balloons Ltd.
	G–SPTS	Beech C90 King Air	Seabourne Express Ltd. (G–BHAP)
	G–SPUD	F.27 Friendship Mk.100	Air UK
	G–SSBS	Colting Ax77 balloon	Lighter Than Air Ltd.
	G–STAN	F.27 Friendship Mk.200	Air UK
	G–STAR	Aerostar 601P	J. F. Ayre (Leasing) Ltd.
	G–STAT	Cessna U.206F	Corbett Farms Ltd.
	G–STIO	ST.10 Diplomate	Rogers Autos Ltd.
	G–STOL	M.S.894A Rallye Minerva	Seven Valley Aviation Group
	G–STUD	D.H.C.-6 Twin Otter 310	Fairflight Ltd.

Reg.	Type	Owner or Operator	Notes
G–SUES	AT-6D Harvard III	Mark Campbell Air Freight Charters Ltd.	
G–SUPA	PA-18-150 Super Cub	Yorkshire Gliding Club (Pty) Ltd.	
G–SUZY	Taylor JT.1 Monoplane	S. A. Kanick	
G–SWAN	R. Turbo Commander 690B	Fairflight Ltd.	
G–SWOT	Currie Super Wot	A. C. Walker	
G–TACA	P.57 Sea Prince T.1	Atlantic & Caribbean Aviation Ltd.	
G–TALY	AB-206B JetRanger 3	Alan Mann Helicopters Ltd.	
G–TAMY	Cessna 421B	Abbergail Ltd.	
G–TATT	GY-20 Minicab	L. Tattershall	
G–TAXI	PA-23 Aztec 250	Northern Executive Aviation Ltd.	
G–TAXY	PA-31 Navajo	Solitair Ltd.	
G–TBCA	Bell 206L LongRanger	British Car Auctions Ltd. (G–BFAL)	
G–TCAT	G.164D Ag-Cat	Miller Aerial Spraying Ltd.	
G–TDAA	Cessna U.206G	Rogers Aviation Sales Ltd.	
G–TEAC	AT-6C Harvard IIA (MC280)	Trans Europe Ltd.	
G–TEFC	PA-28 Cherokee 140	Thames Estuary Flying Club	
G–TFCI	Cessna FA.152	Tayside Aviation Ltd.	
G–THAM	Cessna F.182R	German Tourist Facilities Ltd.	
G–THOR	Thunder Ax8-105 balloon	N. C. Faithfull *Turncoat*	
G–THSL	PA-28R-201 Arrow II	Thistle Metallics Ltd.	
G–TIME	Aerostar 601P	Marlborough Fine Art (London) Ltd.	
G–TINA	SOCATA TB.10 Tobago	A. Lister	
G–TJCB	H.S.125 Srs. 700B	J. C. Bamford (Excavators) Ltd.	
G–TLOL	Cessna 421C	Littlewoods Organisation Ltd.	
G–TOMS	PA-38-112 Tomahawk	Channel Aviation Ltd.	
G–TONI	Cessna 421C	T.H. Knitwear Ltd.	
G–TOUR	Robin R.2000	F. J. Franklin	
G–TREV	Saffery S.330 balloon	T. W. Gurd	
G–TRIX	V.S. Spitfire T.IX	S. Atkins	
G–TROT	PA-31-350 Navajo Chieftain	Trottair Airways Ltd.	
G–TSIX	AT-6C Harvard IIA	D. Taylor	
G–TTAM	Taylor JT.2 Titch	A. J. Manning	
G–TTWO	Colt 56 balloon	D. M. Winder	
G–TWEL	PA-28-181 Archer II	T.W. Electrical Ltd.	
G–TWIN	PA-44-180 Seminole	J. D. Apthorp	
G–TYPE	PA-28RT-201T Turbo Arrow IV	F. F. Guinness & B. Dempsey (G–BIEA)	
G–TYRE	Cessna F.172M	Watts Aviation Ltd.	
G–UBKP	Beech 95-58P Baron	United Biscuits (UK) Ltd.	
G–UCPA	Eiri PIK-20E	B. J. Willson	
G–UESS	Cessna 500 Citation	Osiwel Ltd.	
G–UIDE	Jodel D.120	S. T. Gilbert	
G–VALE	AT-6C Harvard IIA	Kayvale Finance Ltd. (G–RBAC)	
G–USTY	FRED Srs. 2	S. Styles	
G–VAMP	Thunder Ax6-56 balloon	Thunder Balloons Ltd.	
G–VAUN	Cessna 340	Vaughan Associates Ltd.	
G–VEGA	Slingsby T.65A Vega	Vickers Slingsby Ltd.	
G–VEZE	Rutan Vari-Eze	P. J. Henderson	
G–VIKE	Bellanca 1730A Viking	MLP Aviation Ltd.	
G–VITE	Robin R.1180	Dismore Aviation Ltd.	
G–VIVA	Thunder Ax6-56 balloon	Thunder Balloons Ltd.	
G–VIZZ	Sportavia RS.180 Sportsman	Executive Air Sport Ltd.	
G–VMDE	Cessna P.210N	V. S. Evans & Horne & Sutton Ltd.	
G–VPTO	Evans VP-2	J. Cater	
G–VRES	Beech A200 Super King Air	Vernair Transport Services	
G–VTII	D.H.115 Vampire T.11 (WZ507)	J. Turnbull & S. Topin	
G–VTOL	H.S. Harrier T.52	British Aerospace	
G–VWGB	Cessna 404 Titan	Volkswagon (GB) Ltd.	
G–VWSE	Cessna 404 Titan	Northair Aviation Ltd.	
G–WAAC	Cameron N-56 balloon	B. A. World Arts & Adventure Club	
G–WAGY	Cessna F.172N	Rogers Aviation Sales Ltd.	
G–WASP	Brantly B.2B	Wasp Helicopter Hire Ltd. (G–ASKE)	
G–WETI	Cameron N-31 balloon	Balloon Stable Ltd. *Weti*	
G–WGHB	T-33 Mk.3 Silver Star	—	
G–WHIT	Westland Bell 47G-3B1	C. G. Whittaker	
G–WHIZ	Pitts S-1 Special	K. M. McLeod	
G–WHIZ	*V.732 Viscount	S. Wales Museum, Rhoose (fuselage only) (G–ANRS)	
G–WICH	FRED Srs. 2	R. H. Hearn	

Notes	Reg.	Type	Owner or Operator
	G–WICK	Partenavia P.68B	European Ferries Ltd. (G–BGFZ)
	G–WILL	AB-206B JetRanger 3	Anglian Double Glazing (Kent) Ltd.
	G–WIND	Boeing 707-323C	Tradewinds Ltd.
	G–WING	Cessna 404 Titan	Euroair Transport Ltd.
	G–WIZZ	AB-206B JetRanger 2	Wm. Monks (Builders Merchants) Ltd.
	G–WJMN	R. Commander 114	Shoreham Flight Simulation Ltd.
	G–WOOD	Beech 95-B55 Baron	Woods Management Services Ltd.
	G–WOLF	PA-28 Cherokee 140	Citation Flying Services Ltd.
	G–WOSP	Bell 206B JetRanger 3	Wasp Helicopter Hire Ltd.
	G–WPUI	Cessna P.172D	Oldham & Crowther (Promotions) Ltd. (G–AXPI)
	G–WRAY	PA-32RT-300T Turbo Lance II	Essex Insulation Ltd.
	G–WSSC	PA-31-350 Navajo Chieftain	Spacegrand Ltd.
	G–WSSL	PA-31-350 Navajo Chieftain	G. W. Sparrow & Sons Ltd.
	G–WTVA	Cessna 404 Titan	Executive Express Ltd.
	G–WTVB	Cessna 404 Titan	Northair Aviation Ltd.
	G–WTVC	Cessna 404 Titan	B.I.A.-Air West (for sale)
	G–WTVE	Cessna 404 Titan	Executive Express Ltd.
	G–WTVF	Cessna 402B	Westward Television Ltd.
	G–WULF	Focke-Wulf Fw.190 (04)	SBV Aeroservices Ltd.
	G–WWII	V.S. Spitfire 18	D. W. Arnold
	G–WWJC	F.28 Fellowship 4000	Air Alsace/Air France
	G–XING	EMB-121A Xingu	C.S.E. Aviation Ltd.
	G–XMAS	PA-32RT-300 Lance II	A. G. Chrismas
	G–YIII	Cessna F.150L	Citation Flying Services Ltd.
	G–YKIV	Cessna F.150L	Citation Flying Services Ltd.
	G–YORK	Cessna F.172M	Sherburn Aero Club Ltd.
	G–YPSY	Andreasson BA-4B	J. Thomas
	G–YTWO	Cessna F.172M	Sherburn Aero Centre Ltd.
	G–YULL	PA-28 Cherokee 180E	Lansdowne Chemical Co. (G–BEAJ)
	G–ZEAL	Learjet 35A	C.S.E. Aviation Ltd.
	G–ZEPP	Cameron D-96 balloon	Leaveglen Ltd.
	G–ZERO	AA-5B Tiger	Service Photography & Display Ltd.
	G–ZIPP	Cessna 310Q	Citation Flying Services Ltd. (G–BAYU)
	G–ZOOM	Learjet 35A	Falmer Aircraft Ltd.
	G–ZUMP	Cameron N-77 balloon	M. J. Allen *Gazump*
	G–ZZZZ	Point Maker Mk.1 balloon	M. J. Wakelin

MILITARY TO CIVIL CROSS-REFERENCE

Serial carried	Civil identity	Serial carried	Civil identity
04 (Luftwaffe)	G–WULF	Z2033	G–ASTL
14 (USSR)	G–AYAK	Z7197	G–AKZN
282 (Swiss A.F.)	G–BBMI	Z7258	G–AHGD
385 (RCAF)	G–BGPB	AB910	G–AISU
422	G–AVJO	AP507	G–ACWP
2345	G–ATVP	AR213 (QG-A)	G–AIST
2807 (VE-111 USN)	G–BHTH	AR501 (NN-A)	G–AWII
3066	G–AETA	BB814	G–AFWI
5964	G–BFVH	BS676 (K-U)	G–KUKU
8499M	G–ASWJ	DE208	G–AGYU
18393 (C.A.F.)	G–BCYK	DE419	G–APMM
329417 (USAAF)	G–BDHK	DE992	G–AXXV
329601 (USAAF)	G–AXHR	DF130	G–BACK
461748	G–BHDK	DF155	G–ANFV
480308	G–BDCD	DF198	G–BBRB
485784	G–BEDF	DG590	G–ADMW
542447	G–SCUB	EM903	G–APBI
A16–199 (RAAF)	G–BEOX	EV851	G–AJPI
B7270	G–BFCZ	FE992	G–BDAM
C1701	G–AWYY	FT229	G–AZKI
D8096	G–AEPH	FT323	G–AZSC
E449	G–EBKN	FT391	G–AZBN
E3404	G–ADEV	FX301 (FD-EQ)	G–JUDI
F904	G–EBIA	HB751	G–BCBL
F938	G–EBIC	LB312	G–AHXE
F939	G–EBIB	LS326	G–AJVH
F1425	G–BEFR	LZ766	G–ALCK
F8010	G–BDWJ	MC280	G–TEAC
F8614	G–AWAU	MH434 (AC-S)	G–ASJV
G-29-1 (Class B)	G–APRJ	MP425	G–AITB
G-48-1 (Class B)	G–ALSX	MT360	G–AKWT
H2311	G–ABAA	MW100	G–AGNV
J9941	G–ABMR	NF875	G–AGTM
K123	G–EACN	NJ695	G–AJXV
K1786	G–AFTA	NJ703	G–AKPI
K2050	G–ASCM	NM140	G–APGL
K3215	G–AHSA	NM181	G–AZGZ
K4235	G–AHMJ	NP181	G–AOAR
L2301	G–AIZG	NP184	G–ANYP
L8032	G–AMRK	NP303	G–ANZJ
N220	G–BDFF	NX611	G–ASXX
N1854	G–AIBE	PG617	G–AYVY
N3788	G–AKPF	PG651	G–AYUX
N4877 (VX-F)	G–AMDA	PZ865	G–AMAU
N5054	G–AWSA	RG333	G–AIEK
N5180	G–EBKY	RG333	G–AKEZ
N5182	G–APUP	RH377	G–ALAH
N5430	G–BHEW	RH378	G–AJOE
N6848	G–BALX	RL962	G–AHED
N6985	G–AHMN	RM221	G–ANXR
N9191	G–ALND	RM619 (AP-D)	G–ALGT
N9238	G–ANEL	RR299 (HT-E)	G–ASKH
P3308 (UP-A)	G–AWLW	RS712	G–ASKB
P6382	G–AJDR	TA634	G–AWJV
P7350	G–AWIJ	TA719	G–ASKC
R1914	G–AHUJ	TS423	G–DAKS
R4907	G–ANCS	VM360	G–APHV
R7524	G–AIWA	VR249	G–APIY
S1287	G–BEYB	VS356	G–AOLU
T5424	G–AJOA	VS610	G–AOKL
T5493	G–ANEF	VX302	G–BCOV
T5854	G–ANKK	VZ728	G–AGOS
T5879	G–AXBW	WA576	G–ALSS
T6553	G–APIG	WD413	G–BFIR
T6818	G–ANKT	WG307	G–BCYJ
T7187	G–AOBX	WG316	G–BCAH
T7281	G–ARTL	WG348	G–BBMV
T7404	G–ANMV	WG719	G–BRMA
T9707	G–AKKR	WH589	G–AGHB
T9738	G–AKAT	WJ897	G–BDFT
V3388	G–AHTW	WJ945	G–BEDV
V9281 (RU-M)	G–BCWL	WL626	G–BHDD
V9441 (AR-A)	G–AZWT	WM261	G–ARCX

FX301 AT-6 Harvard III, civil identity G–JUDI

Serial carried	*Civil identity*	*Serial carried*	*Civil identity*
WP800	G–BCXN	XF796	G–BFKK
WP808	G–BDEU	XF877	G–AWVF
WP857	G–BDRJ	XG452	G–BRMB
WP977	G–BHRD	XG547	G–HAPR
WV493	G–BDYG	XJ389	G–AJJP
WV494	G–BGSB	XK417	G–AVXY
WV783	G–ALSP	XK655	G–AMXA
WZ507	G–VTII	XL717	G–AOXG
WZ672	G–BDER	XM529	G–BDNS
WZ711	G–AVHT	XM553	G–AWSV
WZ868	G–BCIW	XM556	G–HELI
WZ873	G–BBNE	XR240	G–BDFH
XB733	G–ATBF	XR241	G–AXRR
XF690	G–BGKA	XR269	G–BDXY
XF785	G–ALBN	6J+PR (Luftwaffe)	G–AWHB

OVERSEAS AIRLINER REGISTRATIONS

(Aircraft included in this section are those most likely to be seen at UK airports on scheduled or charter services.)

A40 (Oman)

Reg.	*Type*	*Owner or Operator*	*Notes*
A40–TV	L.1011-385 TriStar 100	Gulf Air	
A40–TW	L-1011-385 TriStar 100	Gulf Air	
A40–TX	L-1011-385 TriStar 100	Gulf Air	
A40–TY	L-1011-385 TriStar 100	Gulf Air	
A40–TZ	L-1011-385 TriStar 100	Gulf Air	
A40–	L.1011-385 TriStar 200	Gulf Air	
A40–	L.1011-385 TriStar 200	Gulf Air	

AP (Pakistan)

AP–ATQ	Boeing 720-051B	Pakistan International Airlines	
AP–AWU	Boeing 707-373C	Pakistan International Airlines	
AP–AWY	Boeing 707-340C	Pakistan International Airlines	
AP–AXA	Boeing 707-340C	Pakistan International Airlines	
AP–AXC	Douglas DC-10-30	Pakistan International Airlines	
AP–AXD	Douglas DC-10-30	Pakistan International Airlines	
AP–AXE	Douglas DC-10-30	Pakistan International Airlines	
AP–AXG	Boeing 707-340C	Pakistan International Airlines	
AP–AXK	Boeing 720-047B	Pakistan International Airlines	
AP–AXL	Boeing 720-047B	Pakistan International Airlines	
AP–AXM	Boeing 720-047B	Pakistan International Airlines	
AP–AYM	Douglas DC-10-30	Pakistan International Airlines	
AP–AYV	Boeing 747-282B	Pakistan International Airlines	
AP–AYW	Boeing 747-282B	Pakistan International Airlines	
AP–AZP	Boeing 720-030B	Pakistan International Airlines	
AP–AZW	Boeing 707-351B	Pakistan International Airlines	
AP–BAA	Boeing 707-351B	Pakistan International Airlines	
AP–BAF	Boeing 720-047B	Pakistan International Airlines	
AP–BAK	Boeing 747-240B	Pakistan International Airlines	
AP–BAT	Boeing 747-240B	Pakistan International Airlines	

B (China)

B–2402	Boeing 707-3J6B	CAAC	
B–2404	Boeing 707-3J6B	CAAC	
B–2406	Boeing 707-3J6B	CAAC	
B–2408	Boeing 707-3J6B	CAAC	
B–2410	Boeing 707-3J6C	CAAC	
B–2412	Boeing 707-3J6C	CAAC	
B–2414	Boeing 707-3J6C	CAAC	
B–2416	Boeing 707-3J6C	CAAC	
B–2418	Boeing 707-3J6C	CAAC	
B–2420	Boeing 707-3J6C	CAAC	

C–F and C–G (Canada)

Notes	Reg.	Type	Owner or Operator
	C–FCPF	Douglas DC-8-43 (601)	CP Air *Empress of San Diego*
	C–FCPG	Douglas DC-8-43 (602)	CP Air *Empress of Buenos Aires*
	C–FCPH	Douglas DC-8-43 (603)	CP Air *Empress of Lima*
	C–FCPI	Douglas DC-8-43 (604)	CP Air *Empress of Amsterdam*
	C–FCPJ	Douglas DC-8-43 (605)	CP Air *Empress of Mexico City*
	C–FCPL	Douglas DC-8-63 (805)	CP Air *Empress of Manitoba*
	C–FCPM	Douglas DC-8-53 (607)	CP Air *Empress of Lisbon*
	C–FCPO	Douglas DC-8-63 (801)	CP Air *Empress of Quebec*
	C–FCPP	Douglas DC-8-63 (802)	CP Air *Empress of Alberta*
	C–FCPQ	Douglas DC-8-63 (803)	CP Air *Empress of Ontario*
	C–FCPS	Douglas DC-8-63 (804)	CP Air *Empress of Sydney*
	C–FCRA	Boeing 747-217B (741)	CP Air *Empress of Japan*
	C–FCRB	Boeing 747-217B (742)	CP Air *Empress of Canada*
	C–FCRD	Boeing 747-217B (743)	CP Air *Empress of Australia*
	C–FCRE	Boeing 747-217B (744)	CP Air *Empress of Italy*
	C–FDJC	Boeing 747-1D1	Wardair Canada *Phil Garratt*
	C–FFUN	Boeing 747-1D1	Wardair Canada *Romeo Vachon*
	C–FNWF	L-100-30 Hercules	North West Territorial Airways
	C–FNWY	L-100-30 Hercules	North West Territorial Airways
	C–FPWK	L-100-20 Hercules (386)	Pacific Western
	C–FPWN	L-100-20 Hercules (383)	Pacific Western
	C–FPWR	L-100-20 Hercules (385)	Pacific Western
	C–FTIK	Douglas DC-8-63F (867)	Air Canada
	C–FTIL	Douglas DC-8-63 (868)	Air Canada
	C–FTIM	Douglas DC-8-63 (869)	Air Canada
	C–FTIN	Douglas DC-8-63 (870)	Air Canada
	C–FTIO	Douglas DC-8-63 (871)	Air Canada
	C–FTIP	Douglas DC-8-63 (872)	Air Canada
	C–FTIQ	Douglas DC-8-63 (873)	Air Canada
	C–FTIR	Douglas DC-8-63 (874)	Air Canada
	C–FTIS	Douglas DC-8-63 (875)	Air Canada
	C–FTIU	Douglas DC-8-63 (876)	Air Canada
	C–FTIV	Douglas DC-8-63 (877)	Air Canada
	C–FTIX	Douglas DC-8-63 (879)	Air Canada
	C–FTJL	Douglas DC-8-54F (812)	Air Canada
	C–FTJO	Douglas DC-8-54F (815)	Air Canada
	C–FTJP	Douglas DC-8-54F (816)	Air Canada
	C–FTJQ	Douglas DC-8-54F (817)	Air Canada
	C–FTJR	Douglas DC-8-54F (818)	Air Canada
	C–FTJS	Douglas DC-8-54F (819)	Air Canada
	C–FTJT	Douglas DC-8-61 (860)	Air Canada
	C–FTJU	Douglas DC-8-61 (861)	Air Canada
	C–FTJV	Douglas DC-8-61 (862)	Air Canada
	C–FTJW	Douglas DC-8-61 (863)	Air Canada
	C–FTJX	Douglas DC-8-61 (864)	Air Canada
	C–FTJY	Douglas DC-8-61 (865)	Air Canada
	C–FTJZ	Douglas DC-8-61 (866)	Air Canada
	C–FTNI	L-1011-385 TriStar 100 (509)	Air Canada
	C–FTNJ	L-1011-385 TriStar 100 (510)	Air Canada
	C–FTNK	L-1011-385 TriStar 100 (511)	Air Canada
	C–FTNL	L-1011-385 TriStar 100 (512)	Air Canada
	C–FTOA	Boeing 747-133 (301)	Air Canada
	C–FTOB	Boeing 747-133 (302)	Air Canada
	C–FTOC	Boeing 747-133 (303)	Air Canada
	C–FTOD	Boeing 747-133 (304)	Air Canada
	C–FTOE	Boeing 747-133 (305)	Air Canada

Reg.	*Type*	*Owner or Operator*	*Notes*
C–GAGA	Boeing 747-233B (306)	Air Canada	
C–GAGB	Boeing 747-233B (307)	Air Canada	
C–	L-1011-385 TriStar 500	Air Canada	
C–	L-1011-385 TriStar 500	Air Canada	
C–	L-1011-385 TriStar 500	Air Canada	
C–	L-1011-385 TriStar 500	Air Canada	
C–	L-1011-385 TriStar 500	Air Canada	
C–	L-1011-385 TriStar 500	Air Canada	
C–GCPC	Douglas DC-10-30 (901)	CP Air *Empress of Amsterdam*	
C–GCPD	Douglas DC-10-30 (902)	CP Air	
C–GCPE	Douglas DC-10-30 (903)	CP Air *Empress of Ontario*	
C–GCPF	Douglas DC-10-30 (904)	CP Air *Empress of Buenos Aires*	
C–GCPG	Douglas DC-10-30	CP Air	
C–GCPH	Douglas DC-10-30	CP Air	
C–GHPW	L-100-30 Hercules (387)	Pacific Western	
C–GRYN	Boeing 707-338C	Ontario Worldair	
C–GRYO	Boeing 707-351C	Ontario Worldair	
C–GXRA	Boeing 747-211B	Wardair Canada *Herbert Hollick Kenyon*	
C–GXRB	Douglas DC-10-30	Wardair Canada *C. H. Punch Dickens*	
C–GXRC	Douglas DC-10-30	Wardair Canada *W. R. Wop May*	
C–GXRD	Boeing 747-211B	Wardair Canada	

NOTE: Airline fleet number carried on aircraft is shown in parenthesis.

CCCP (Russia)

All aircraft listed are operated by Aeroflot. The registrations are prefixed by CCCP in each case.

Reg.	*Type*	*Notes*	*Reg.*	*Type*	*Notes*
65001	Tu-134A		65050	Tu-134A	
65004	Tu-134A		65051	Tu-134A	
65005	Tu-134A		65052	Tu-134A	
65008	Tu-134A		65053	Tu-134A	
65009	Tu-134A		65055	Tu-134A	
65010	Tu-134A		65056	Tu-134A	
65011	Tu-134A		65058	Tu-134A	
65013	Tu-134A		65059	Tu-134A	
65014	Tu-134A		65060	Tu-134A	
65015	Tu-134A		65061	Tu-134A	
65017	Tu-134A		65062	Tu-134A	
65018	Tu-134A		65065	Tu-134A	
65020	Tu-134A		65066	Tu-134A	
65021	Tu-134A		65069	Tu-134A	
65023	Tu-134A		65070	Tu-134A	
65024	Tu-134A		65071	Tu-134A	
65025	Tu-134A		65072	Tu-134A	
65027	Tu-134A		65074	Tu-134A	
65028	Tu-134A		65075	Tu-134A	
65030	Tu-134A		65076	Tu-134A	
65031	Tu-134A		65077	Tu-134A	
65032	Tu-134A		65080	Tu-134A	
65034	Tu-134A		65082	Tu-134A	
65035	Tu-134A		65083	Tu-134A	
65036	Tu-134A		65085	Tu-134A	
65037	Tu-134A		65086	Tu-134A	
65038	Tu-134A		65089	Tu-134A	
65039	Tu-134A		65090	Tu-134A	
65040	Tu-134A		65095	Tu-134A	
65042	Tu-134A		65096	Tu-134A	
65043	Tu-134A		65098	Tu-134A	
65044	Tu-134A		65105	Tu-134A	
65045	Tu-134A		65107	Tu-134A	
65046	Tu-134A		65109	Tu-134A	
65047	Tu-134A		65113	Tu-134A	
65048	Tu-134A		65114	Tu-134A	
65049	Tu-134A		65118	Tu-134A	

Notes	Reg.	Type
	65119	Tu-134A
	65120	Tu-134A
	65132	Tu-134A
	65134	Tu-134A
	65140	Tu-134A
	65601	Tu-134
	65602	Tu-134
	65603	Tu-134
	65604	Tu-134
	65605	Tu-134
	65606	Tu-134
	65607	Tu-134
	65608	Tu-134
	65609	Tu-134
	65610	Tu-134
	65611	Tu-134
	65612	Tu-134
	65613	Tu-134
	65614	Tu-134
	65615	Tu-134
	65616	Tu-134
	65617	Tu-134
	65618	Tu-134
	65619	Tu-134
	65620	Tu-134
	65621	Tu-134
	65622	Tu-134
	65623	Tu-134
	65624	Tu-134A
	65625	Tu-134
	65626	Tu-134A
	65627	Tu-134
	65628	Tu-134
	65629	Tu-134
	65630	Tu-134
	65631	Tu-134
	65632	Tu-134
	65633	Tu-134
	65634	Tu-134
	65635	Tu-134
	65636	Tu-134
	65637	Tu-134
	65638	Tu-134
	65639	Tu-134
	65640	Tu-134
	65641	Tu-134
	65642	Tu-134
	65643	Tu-134
	65644	Tu-134A
	65645	Tu-134A
	65646	Tu-134A
	65647	Tu-134A
	65648	Tu-134A
	65649	Tu-134A
	65650	Tu-134A
	65651	Tu-134A
	65652	Tu-134A
	65653	Tu-134A
	65654	Tu-134A
	65655	Tu-134A
	65656	Tu-134A
	65657	Tu-134A
	65658	Tu-134A
	65659	Tu-134A
	65660	Tu-134A
	65661	Tu-134A
	65662	Tu-134A
	65663	Tu-134A
	65664	Tu-134A
	65665	Tu-134A
	65666	Tu-134A
	65667	Tu-134A
	65668	Tu-134A
	65669	Tu-134A
	65670	Tu-134A

Notes	Reg.	Type
	65671	Tu-134A
	65672	Tu-134A
	65673	Tu-134A
	65674	Tu-134A
	65675	Tu-134A
	65676	Tu-134A
	65677	Tu-134A
	65678	Tu-134A
	65679	Tu-134A
	65680	Tu-134A
	65683	Tu-134A
	65687	Tu-134A
	65690	Tu-134A
	65691	Tu-134A
	65718	Tu-134A
	65727	Tu-134A
	65728	Tu-134A
	65729	Tu-134A
	65730	Tu-134A
	65731	Tu-134A
	65732	Tu-134A
	65733	Tu-134A
	65734	Tu-134A
	65735	Tu-134A
	65739	Tu-134A
	65741	Tu-134A
	65742	Tu-134A
	65743	Tu-134A
	65744	Tu-134A
	65745	Tu-134A
	65746	Tu-134A
	65747	Tu-134A
	65748	Tu-134A
	65749	Tu-134A
	65753	Tu-134A
	65757	Tu-134A
	65761	Tu-134A
	65765	Tu-134A
	65769	Tu-134A
	65770	Tu-134A
	65771	Tu-134A
	65777	Tu-134A
	65780	Tu-134A
	65781	Tu-134A
	65783	Tu-134A
	65784	Tu-134A
	65785	Tu-134A
	65791	Tu-134A
	65792	Tu-134A
	65794	Tu-134A
	65801	Tu-134A
	65802	Tu-134A
	65804	Tu-134A
	65806	Tu-134A
	65810	Tu-134A
	65812	Tu-134A
	65815	Tu-134A
	65817	Tu-134A
	65818	Tu-134A
	65820	Tu-134A
	65821	Tu-134A
	65822	Tu-134A
	65823	Tu-134A
	65825	Tu-134A
	65828	Tu-134A
	65829	Tu-134A
	65830	Tu-134A
	65831	Tu-134A
	65832	Tu-134A
	65833	Tu-134A
	65834	Tu-134A
	65837	Tu-134A
	65839	Tu-134A
	65840	Tu-134A
	65841	Tu-134A

Reg.	Type	Notes
65843	Tu-134A	
65844	Tu-134A	
65845	Tu-134A	
65848	Tu-134A	
65851	Tu-134A	
65852	Tu-134A	
65853	Tu-134A	
65854	Tu-134A	
65857	Tu-134A	
65861	Tu-134A	
65862	Tu-134A	
65863	Tu-134A	
65864	Tu-134A	
65865	Tu-134A	
65866	Tu-134A	
65867	Tu-134A	
65868	Tu-134A	
65869	Tu-134A	
65870	Tu-134A	
65871	Tu-134A	
65872	Tu-134A	
65873	Tu-134A	
65874	Tu-134A	
65877	Tu-134A	
65878	Tu-134A	
65879	Tu-134A	
65880	Tu-134A	
65881	Tu-134A	
65882	Tu-134A	
65883	Tu-134A	
65884	Tu-134A	
65886	Tu-134A	
65888	Tu-134A	
65890	Tu-134A	
65891	Tu-134A	
65892	Tu-134A	
65893	Tu-134A	
65894	Tu-134A	
65895	Tu-134A	
65898	Tu-134A	
65899	Tu-134A	
65950	Tu-134A	
65951	Tu-134A	
65952	Tu-134A	
65953	Tu-134A	
65954	Tu-134A	
65955	Tu-134A	
65957	Tu-134A	
65960	Tu-134A	
65961	Tu-134A	
65962	Tu-134A	
65963	Tu-134A	
65964	Tu-134A	
65965	Tu-134A	
65967	Tu-134A	
65969	Tu-134A	
65970	Tu-134A	
65971	Tu-134A	
65972	Tu-134A	
65973	Tu-134A	
65974	Tu-134A	
65975	Tu-134A	
65976	Tu-134A	
76500	IL-76T	
76502	IL-76T	
76503	IL-76T	
76504	IL-76T	
76505	IL-76T	
76506	IL-76T	
76508	IL-76T	
76509	IL-76T	
76510	IL-76T	
76511	IL-76T	
76512	IL-76T	
76513	IL-76T	
76514	IL-76T	
76515	IL-76T	
76520	IL-76T	
85000	Tu-154	
85001	Tu-154	
85005	Tu-154	
85006	Tu-154	
85007	Tu-154	
85008	Tu-154	
85009	Tu-154	
85010	Tu-154	
85011	Tu-154	
85012	Tu-154	
85013	Tu-154	
85014	Tu-154	
85016	Tu-154	
85017	Tu-154	
85018	Tu-154	
85019	Tu-154	
85021	Tu-154	
85022	Tu-154	
85024	Tu-154	
85025	Tu-154	
85028	Tu-154	
85029	Tu-154	
85030	Tu-154	
85031	Tu-154	
85032	Tu-154	
85033	Tu-154	
85034	Tu-154	
85035	Tu-154	
85037	Tu-154	
85038	Tu-154	
85039	Tu-154	
85040	Tu-154	
85041	Tu-154	
85042	Tu-154	
85043	Tu-154	
85044	Tu-154	
85047	Tu-154	
85048	Tu-154	
85050	Tu-154	
85052	Tu-154	
85053	Tu-154	
85054	Tu-154	
85055	Tu-154	
85057	Tu-154	
85058	Tu-154A	
85060	Tu-154A	
85061	Tu-154A	
85062	Tu-154A	
85063	Tu-154A	
85064	Tu-154A	
85065	Tu-154A	
85066	Tu-154A	
85068	Tu-154A	
85069	Tu-154A	
85070	Tu-154A	
85071	Tu-154A	
85072	Tu-154A	
85074	Tu-154A	
85075	Tu-154B	
85076	Tu-154A	
85080	Tu-154A	
85081	Tu-154A	
85082	Tu-154A	
85083	Tu-154A	
85084	Tu-154A	
85085	Tu-154A	
85086	Tu-154A	
85088	Tu-154A	
85090	Tu-154A	
85091	Tu-154A	
85092	Tu-154A	
85093	Tu-154A	

Notes	Reg.	Type
	85094	Tu-154A
	85095	Tu-154A
	85096	Tu-154A
	85097	Tu-154A
	85098	Tu-154A
	85099	Tu-154A
	85100	Tu-154A
	85101	Tu-154A
	85102	Tu-154A
	85103	Tu-154A
	85104	Tu-154A
	85105	Tu-154A
	85106	Tu-154B
	85107	Tu-154A
	85108	Tu-154A
	85109	Tu-154A
	85111	Tu-154A
	85112	Tu-154A
	85113	Tu-154A
	85114	Tu-154A
	85115	Tu-154A
	85116	Tu-154A
	85117	Tu-154A
	85118	Tu-154B
	85119	Tu-154A
	85121	Tu-154B
	85122	Tu-154B
	85123	Tu-154B
	85124	Tu-154B
	85128	Tu-154B
	85129	Tu-154B
	85130	Tu-154B
	85131	Tu-154B
	85134	Tu-154B
	85136	Tu-154B
	85137	Tu-154B
	85138	Tu-154B
	85139	Tu-154B
	85140	Tu-154B
	85141	Tu-154B
	85142	Tu-154B
	85145	Tu-154B
	85146	Tu-154B
	85147	Tu-154B
	85148	Tu-154B
	85150	Tu-154B
	85152	Tu-154B
	85153	Tu-154B
	85154	Tu-154B
	85155	Tu-154B
	85156	Tu-154B
	85157	Tu-154B
	85160	Tu-154B
	85162	Tu-154B
	85163	Tu-154B
	85164	Tu-154B
	85165	Tu-154B
	85166	Tu-154B
	85167	Tu-154B
	85168	Tu-154B
	85169	Tu-154B
	85170	Tu-154B
	85171	Tu-154B
	85172	Tu-154B
	85174	Tu-154B
	85175	Tu-154B
	85176	Tu-154B
	85177	Tu-154B
	85178	Tu-154B
	85179	Tu-154B
	85180	Tu-154B
	85181	Tu-154B
	85182	Tu-154B
	85184	Tu-154B
	85185	Tu-154B

Notes	Reg.	Type
	85186	Tu-154B
	85187	Tu-154B
	85188	Tu-154B
	85189	Tu-154B
	85190	Tu-154B
	85192	Tu-154B
	85194	Tu-154B
	85196	Tu-154B
	85197	Tu-154B
	85198	Tu-154B
	85199	Tu-154B
	85200	Tu-154B
	85201	Tu-154B
	85202	Tu-154B
	85203	Tu-154B
	85204	Tu-154B
	85206	Tu-154B
	85207	Tu-154B
	85210	Tu-154B
	85211	Tu-154B
	85212	Tu-154B
	85214	Tu-154B
	85215	Tu-154B
	85216	Tu-154B
	85217	Tu-154B
	85218	Tu-154B
	85219	Tu-154B
	85220	Tu-154B
	85221	Tu-154B
	85222	Tu-154B
	85223	Tu-154B
	85228	Tu-154B
	85229	Tu-154B
	85230	Tu-154B
	85231	Tu-154B
	85232	Tu-154B
	85233	Tu-154B
	85234	Tu-154B
	85235	Tu-154B
	85236	Tu-154B
	85238	Tu-154B
	85240	Tu-154B
	85241	Tu-154B
	85242	Tu-154B
	85243	Tu-154B
	85244	Tu-154B
	85245	Tu-154B
	85247	Tu-154B
	85248	Tu-154B
	85249	Tu-154B
	85253	Tu-154B
	85254	Tu-154B
	85256	Tu-154B
	85257	Tu-154B
	85259	Tu-154B
	85260	Tu-154B
	85261	Tu-154B
	85263	Tu-154B
	85264	Tu-154B
	85265	Tu-154B
	85269	Tu-154B
	85271	Tu-154B
	85272	Tu-154B
	85274	Tu-154B
	85275	Tu-154B
	85276	Tu-154B
	85277	Tu-154B
	85278	Tu-154B
	85279	Tu-154B
	85280	Tu-154B
	85281	Tu-154B
	85282	Tu-154B
	85283	Tu-154B
	85284	Tu-154B
	85285	Tu-154B

Reg.	Type	Notes
85286	Tu-154B	
85287	Tu-154B	
85288	Tu-154B	
85290	Tu-154B	
85296	Tu-154B	
85298	Tu-154B	
85300	Tu-154B	
85301	Tu-154B	
85302	Tu-154B	
85303	Tu-154B	
85304	Tu-154B	
85305	Tu-154B	
85306	Tu-154B	
85311	Tu-154B	
85313	Tu-154B	
85314	Tu-154B	
85316	Tu-154B	
85323	Tu-154B	
85328	Tu-154B	
85330	Tu-154B	
85331	Tu-154B	
85334	Tu-154B	
85335	Tu-154B	
85336	Tu-154B	
85337	Tu-154B	
85339	Tu-154B	
85346	Tu-154B	
85347	Tu-154B	
85349	Tu-154B	
85350	Tu-154B	
85363	Tu-154B	
85364	Tu-154B	
85365	Tu-154B	
85366	Tu-154B	
85367	Tu-154B	
85368	Tu-154B	
85372	Tu-154B	
85374	Tu-154B	
85375	Tu-154B	
85376	Tu-154B	
85377	Tu-154B	
85378	Tu-154B	
85379	Tu-154B	
85381	Tu-154B	
85395	Tu-154B	
85396	Tu-154B	
85397	Tu-154B	
85398	Tu-154B	
85399	Tu-154B	
85400	Tu-154B	
85406	Tu-154B	
85407	Tu-154B	
85409	Tu-154B	
85410	Tu-154B	
85411	Tu-154B	
85412	Tu-154B	
85413	Tu-154B	
85423	Tu-154B	
85424	Tu-154B	
86450	IL-62	
86451	IL-62	
86452	IL-62M	
86453	IL-62M	
86454	IL-62M	
86455	IL-62M	
86456	IL-62M	
86457	IL-62M	
86458	IL-62M	
86459	IL-62M	
86460	IL-62	
86461	IL-62	
86462	IL-62M	
86463	IL-62M	
86464	IL-62M	
86465	IL-62M	
86469	IL-62M	
86470	IL-62M	
86471	IL-62M	
86472	IL-62M	
86473	IL-62M	
86474	IL-62M	
86475	IL-62M	
86476	IL-62M	
86477	IL-62M	
86478	IL-62M	
86479	IL-62M	
86480	IL-62M	
86481	IL-62M	
86482	IL-62M	
86483	IL-62M	
86484	IL-62M	
86485	IL-62M	
86486	IL-62M	
86487	IL-62M	
86488	IL-62M	
86489	IL-62M	
86490	IL-62M	
86491	IL-62M	
86497	IL-62M	
86498	IL-62M	
86499	IL-62M	
86500	IL-62M	
86501	IL-62M	
86502	IL-62M	
86504	IL-62M	
86507	IL-62M	
86509	IL-62M	
86510	IL-62M	
86511	IL-62M	
86605	IL-62	
86606	IL-62	
86607	IL-62M	
86608	IL-62	
86609	IL-62	
86610	IL-62	
86611	IL-62	
86612	IL-62	
86613	IL-62	
86614	IL-62M	
86615	IL-62	
86616	IL-62	
86617	IL-62	
86618	IL-62M	
86619	IL-62	
86620	IL-62M	
86621	IL-62M	
86622	IL-62M	
86623	IL-62M	
86624	IL-62	
86627	IL-62	
86648	IL-62	
86649	IL-62	
86650	IL-62	
86652	IL-62	
86653	IL-62	
86654	IL-62	
86655	IL-62	
86656	IL-62M	
86657	IL-62	
86658	IL-62M	
86659	IL-62	
86661	IL-62	
86662	IL-62	
86663	IL-62	
86664	IL-62	
86665	IL-62	
86666	IL-62	
86667	IL-62	
86668	IL-62	
86669	IL-62	

Notes	Reg.	Type
	86670	IL-62
	86671	IL-62
	86672	IL-62
	86673	IL-62M
	86674	IL-62
	86675	IL-62
	86676	IL-62
	86677	IL-62
	86678	IL-62
	86679	IL-62
	86680	IL-62
	86681	IL-62
	86682	IL-62
	86683	IL-62
	86684	IL-62
	86685	IL-62
	86686	IL-62
	86687	IL-62
	86688	IL-62
	86689	IL-62
	86690	IL-62
	86691	IL-62
	86692	IL-62M
	86693	IL-62M
	86694	IL-62
	86695	IL-62
	86696	IL-62
	86697	IL-62
	86698	IL-62
	86699	IL-62
	86700	IL-62M
	86701	IL-62M
	86702	IL-62M
	86703	IL-62
	86704	IL-62
	86705	IL-62M

CN (Morocco)

Notes	Reg.	Type	Owner or Operator
	CN–CCF	Boeing 727-2B6	Royal Air Maroc *Fes*
	CN–CCG	Boeing 727-2B6	Royal Air Maroc *l'Oiseau de la Providence*
	CN–CCH	Boeing 727-2B6	Royal Air Maroc *Marrakesh*
	CN–CCW	Boeing 727-2B6	Royal Air Maroc *Agadir*
	CN–RMO	Boeing 727-2B6	Royal Air Maroc
	CN–RMP	Boeing 727-2B6	Royal Air Maroc
	CN–RMQ	Boeing 727-2B6	Royal Air Maroc
	CN–RMR	Boeing 727-2B6	Royal Air Maroc

CS (Portugal)

	CS–TBA	Boeing 707-382B	TAP—Air Portugal *Santa Cruz*
	CS–TBB	Boeing 707-382B	TAP—Air Portugal *Santa Maria*
	CS–TBC	Boeing 707-382B	TAP—Air Portugal *Cidade de Luanda*
	CS–TBD	Boeing 707-382B	TAP—Air Portugal *Cidade de Lourenço Marques*
	CS–TBE	Boeing 707-382B	TAP—Air Portugal *Pedro Alvares Cabral*
	CS–TBF	Boeing 707-382B	TAP—Air Portugal *Vasco da Gama*
	CS–TBG	Boeing 707-382B	TAP—Air Portugal *Fernao de Magalhaes*
	CS–TBH	Boeing 707-399C	TAP—Air Portugal *Pedro Nunes*
	CS–TBI	Boeing 707-399C	TAP—Air Portugal *D. João de Castro*
	CS–TBJ	Boeing 707-373C	TAP—Air Portugal *Lisboa*
	CS–TBK	Boeing 727-82	TAP—Air Portugal *Açores*
	CS–TBL	Boeing 727-82	TAP—Air Portugal *Madeira*
	CS–TBM	Boeing 727-82	TAP—Air Portugal *Algarve*
	CS–TBN	Boeing 727-82QC	TAP—Air Portugal *Cidade do Porto*
	CS–TBO	Boeing 727-82QC	TAP—Air Portugal *Costa do Sol*
	CS–TBP	Boeing 727-82	TAP—Air Portugal *Cabo Verde*
	CS–TBQ	Boeing 727-172C	TAP—Air Portugal *Bissau*
	CS–TBS	Boeing 727-282	TAP—Air Portugal *Gago Coutinho*
	CS–TBT	Boeing 707-3F5C	TAP—Air Portugal *Humberto Delgado*
	CS–TBU	Boeing 707-3F5C	TAP—Air Portugal *Jaime Cortesão*

Reg.	Type	Owner or Operator	Notes
CS-TBV	Boeing 727-155C	TAP—Air Portugal	
CS-TBW	Boeing 727-282	TAP—Air Portugal *Coimbra*	
CS-TBX	Boeing 727-282	TAP—Air Portugal *Faro*	
CS-TJA	Boeing 747-282B	TAP—Air Portugal *Portugal*	
CS-TJB	Boeing 747-282B	TAP—Air Portugal *Brasil*	

D (German Federal Republic)

Hapag-Lloyd

Lufthansa

Condor Flugdienst

Reg.	Type	Owner or Operator	Notes
D-AAST	S.E.210 Caravelle 10-R	Aero Lloyd	
D-ABAK	S.E.210 Caravelle 10-R	Aero Lloyd	
D-ABAP	S.E.210 Caravelle 10-R	Special Air Transport	
D-ABAV	S.E.210 Caravelle 10-R	Special Air Transport	
D-ABAW	S.E.210 Caravelle 10-R	Special Air Transport	
D-ABBE	Boeing 737-230C	Lufthansa *City Jet Remscheid*	
D-ABCE	Boeing 737-230C	Lufthansa *City Jet Landshut*	
D-ABCI	Boeing 727-230	Lufthansa *Karlsruhe*	
D-ABDE	Boeing 737-230C	Lufthansa *City Jet Bamberg*	
D-ABDI	Boeing 727-230	Lufthansa *Lübeck*	
D-ABEA	Boeing 737-130	Lufthansa *City Jet Coburg*	
D-ABEC	Boeing 737-130	Lufthansa *City of Osnabrück*	
D-ABEF	Boeing 737-130	Lufthansa *City Jet Kempten*	
D-ABEG	Boeing 737-130	Lufthansa *City Jet Offenbach*	
D-ABEH	Boeing 737-130	Lufthansa *City Jet Solingen*	
D-ABEI	Boeing 737-130	Lufthansa *City Jet Oldenburg*	
D-ABEK	Boeing 737-130	Lufthansa *City Jet Konstanz*	
D-ABEL	Boeing 737-130	Lufthansa *City Jet Mülheim a.d.R.*	
D-ABEO	Boeing 737-130	Lufthansa *City Jet Göttingen*	
D-ABEP	Boeing 737-130	Lufthansa *City Jet Wilhelmshaven*	
D-ABEQ	Boeing 737-130	Lufthansa *City Jet Koblenz*	
D-ABET	Boeing 737-130	Lufthansa *City Jet Baden-Baden*	
D-ABEU	Boeing 737-130	Lufthansa *City Jet Heilbronn*	
D-ABEV	Boeing 737-130	Lufthansa *City Jet Marburg*	
D-ABEW	Boeing 737-130	Lufthansa *City Jet Bayreuth*	
D-ABEY	Boeing 737-130	Lufthansa *City Jet Worms*	
D-ABFA	Boeing 737-230	Lufthansa *Regensburg*	
D-ABFB	Boeing 737-230	Lufthansa *Flensburg*	
D-ABFC	Boeing 737-230	Lufthansa *Würzburg*	
D-ABFD	Boeing 737-230	Lufthansa *Bamberg*	
D-ABFE	Boeing 737-230C	Lufthansa *Trier*	
D-ABFF	Boeing 737-230	Lufthansa *Gelsenkirchen*	
D-ABFH	Boeing 737-230	Lufthansa Pforzheim	
D-ABFI	Boeing 727-230	Lufthansa *Münster*	
D-ABFK	Boeing 737-230	Lufthansa *Wuppertal*	
D-ABFL	Boeing 737-230	Lufthansa *Coburg*	
D-ABFM	Boeing 737-230	Lufthansa *Osnabrück*	
D-ABFN	Boeing 737-230	Lufthansa *Kempten*	
D-ABFP	Boeing 737-230	Lufthansa *Offenbach*	
D-ABFR	Boeing 737-230	Lufthansa *Solingen*	
D-ABFS	Boeing 737-230	Lufthansa *Oldenburg*	
D-ABFT	Boeing 737-230	Lufthansa (operated by Condor Flugdienst)	
D-ABFU	Boeing 737-230	Lufthansa *Mülheim a.d.R*	
D-ABFW	Boeing 737-230	Lufthansa *Wolfsberg*	
D-ABFX	Boeing 737-230	Lufthansa *Tübingen*	
D-ABFY	Boeing 737-230	Lufthansa *Göttingen*	
D-ABFZ	Boeing 737-230	Lufthansa *Wilhelmshaven*	

Notes	Reg.	Type	Owner or Operator
	D–ABGE	Boeing 737-230C	Lufthansa *Erlangen*
	D–ABGI	Boeing 727-230	Lufthansa *Leverkusen*
	D–ABHA	Boeing 737-230	Lufthansa *Koblenz*
	D–ABHB	Boeing 737-230	Lufthansa *Gosler*
	D–ABHC	Boeing 737-230	Lufthansa *Friedrichshafen*
	D–ABHD	Boeing 737-230	Lufthansa *Baden-Baden*
	D–ABHE	Boeing 737-230C	Lufthansa *Darmstadt*
	D–ABHF	Boeing 737-230	Lufthansa *Heilbronn*
	D–ABHH	Boeing 737-230	Lufthansa *Marburg*
	D–ABHI	Boeing 727-230	Lufthansa *Mönchengladbach*
	D–ABHK	Boeing 737-230	Lufthansa *Bayreuth*
	D–ABHL	Boeing 737-230	Lufthansa *Worms*
	D–ABHM	Boeing 737-230	Lufthansa *Landsnut*
	D–ABHN	Boeing 737-230	Lufthansa *Trier*
	D–ABHP	Boeing 737-230	Lufthansa *Erlangen*
	D–ABHR	Boeing 737-230	Lufthansa *Darmstadt*
	D–ABHS	Boeing 737-230	Lufthansa *Remscheid*
	D–ABIL	Boeing 727-30	Condor Flugdienst
	D–ABIP	Boeing 727-30	Condor Flugdienst
	D–ABIQ	Boeing 727-30	Condor Flugdienst
	D–ABIR	Boeing 727-30	Condor Flugdienst
	D–ABKA	Boeing 727-230	Lufthansa *Heidelberg*
	D–ABKB	Boeing 727-230	Lufthansa *Augsburg*
	D–ABKC	Boeing 727-230	Lufthansa *Braunschweig*
	D–ABKD	Boeing 727-230	Lufthansa *Freiburg*
	D–ABKE	Boeing 727-230	Lufthansa *Mannheim*
	D–ABKF	Boeing 727-230	Lufthansa *Saarbrücken*
	D–ABKG	Boeing 727-230	Lufthansa *Kassel*
	D–ABKH	Boeing 727-230	Lufthansa *Kiel*
	D–ABKI	Boeing 727-230	Lufthansa *Bremerhaven*
	D–ABKJ	Boeing 727-230	Lufthansa *Wiesbaden*
	D–ABKK	Boeing 727-230	Condor Flugdienst
	D–ABKL	Boeing 727-230	Condor Flugdienst
	D–ABKM	Boeing 727-230	Lufthansa *Hagen*
	D–ABKN	Boeing 727-230	Lufthansa *Ulm*
	D–ABKP	Boeing 727-230	Lufthansa *Krefeld*
	D–ABKQ	Boeing 727-230	Lufthansa *Mainz*
	D–ABKR	Boeing 727-230	Lufthansa *Bielefeld*
	D–ABKS	Boeing 727-230	Lufthansa *Oberhausen*
	D–ABKT	Boeing 727-230	Lufthansa *Aachen*
	D–ABLI	Boeing 727-230	Lufthansa *Ludwigshafen a.Rh.*
	D–ABMI	Boeing 727-230	Condor Flugdienst
	D–ABNI	Boeing 727-230	Condor Flugdienst
	D–ABPI	Boeing 727-230	Condor Flugdienst
	D–ABQI	Boeing 727-230	Lufthansa *Hildesheim*
	D–ABRI	Boeing 727-230	Lufthansa *Esslingen*
	D–ABSI	Boeing 727-230	Lufthansa *Hof*
	D–ABTI	Boeing 727-230	Condor Flugdienst
	D–ABUA	Boeing 707-330C	German Cargo
	D–ABUB	Boeing 707-330B	Lufthansa
	D–ABUD	Boeing 707-330B	Lufthansa
	D–ABUE	Boeing 707-330C	German Cargo
	D–ABUF	Boeing 707-330B	Condor Flugdienst
	D–ABUG	Boeing 707-330B	Condor Flugdienst
	D–ABUH	Boeing 707-330B	Lufthansa
	D–ABUI	Boeing 707-330C	German Cargo
	D–ABUJ	Boeing 707-330C	Lufthansa *Africa*
	D–ABUK	Boeing 707-330B	Lufthansa *Bachum*
	D–ABUL	Boeing 707-330B	Lufthansa *Duisburg*
	D–ABUM	Boeing 707-330B	Lufthansa
	D–ABUO	Boeing 707-330C	German Cargo
	D–ABVI	Boeing 727-230	Condor Flugdienst
	D–ABWI	Boeing 727-230	Condor Flugdienst
	D–ABYJ	Boeing 747-230B	Lufthansa *Hessen*
	D–ABYK	Boeing 747-230B	Lufthansa *Rheinland-Pfalz*
	D–ABYL	Boeing 747-230B	Lufthansa *Saarland*
	D–ABYM	Boeing 747-230B	Lufthansa *Schleswig-Holstein*
	D–ABYN	Boeing 747-230B	Lufthansa *Baden-Wurttemberg*
	D–ABYO	Boeing 747-230F	Lufthansa
	D–ABYP	Boeing 747-230B	Lufthansa *Niedersachsen*
	D–ABYQ	Boeing 747-230B	Lufthansa *Bremen*
	D–ABYR	Boeing 747-230B	Lufthansa *Nordrhein-Westfalen*
	D–ABYS	Boeing 747-230B	Lufthansa *Bayern*
	D–ABYT	Boeing 747-230B	Lufthansa *Hamburg*
	D–ABYU	Boeing 747-230F	Lufthansa

Reg.	Type	Owner or Operator	Notes
D–A	Boeing 747-230B	Lufthansa	
D–A	Boeing 747-230B	Lufthansa	
D–A	Boeing 747-230B	Lufthansa	
D–ACVK	S.E.210 Caravelle 10-R	Aero Lloyd	
D–ADAO	Douglas DC-10-30	Lufthansa *Düsseldorf*	
D–ADBO	Douglas DC-10-30	Lufthansa *Berlin*	
D–ADCO	Douglas DC-10-30	Lufthansa *Frankfurt*	
D–ADDO	Douglas DC-10-30	Lufthansa *Hamburg*	
D–ADFO	Douglas DC-10-30	Lufthansa *München*	
D–ADGO	Douglas DC-10-30	Lufthansa *Bonn*	
D–ADHO	Douglas DC-10-30	Lufthansa *Hannover*	
D–ADJO	Douglas DC-10-30	Lufthansa	
D–ADKO	Douglas DC-10-30	Lufthansa *Stuttgart*	
D–ADLO	Douglas DC-10-30	Lufthansa	
D–ADMO	Douglas DC-10-30	Lufthansa *Dortmund*	
D–ADPO	Douglas DC-10-30	Condor Flugdienst	
D–ADQO	Douglas DC-10-30	Condor Flugdienst	
D–ADSO	Douglas DC-10-30	Condor Flugdienst	
D–AERA	L-1011-385 TriStar 1	LTU	
D–AERE	L-1011-385 TriStar 1	LTU	
D–AERI	L-1011-385 TriStar 1	LTU	
D–AERL	L.1011-385 TriStar 500	LTU	
D–AERO	L-1011-385 TriStar 1	LTU	
D–AERU	L-1011-385 TriStar 100	LTU	
D–AERY	L.1011-385 TriStar 500	LTU	
D–AHLA	A.300B4 Airbus	Hapag-Lloyd	
D–AHLB	A.300B4 Airbus	Hapag-Lloyd	
D–AHLL	Boeing 727-81	Hapag-Lloyd	
D–AHLM	Boeing 727-81	Hapag-Lloyd	
D–AHLN	Boeing 727-81	Hapag-Lloyd	
D–AHLO	Boeing 727-29	Hapag-Lloyd	
D–AHLP	Boeing 727-14	Hapag-Lloyd	
D–AHLQ	Boeing 727-46	Hapag-Lloyd	
D–AHLR	Boeing 727-89	Hapag-Lloyd	
D–AHLS	Boeing 727-89	Hapag-Lloyd	
D–AHLT	Boeing 727-2K5	Hapag-Lloyd	
D–AHLU	Boeing 727-2K5	Hapag-Lloyd	
D–AHLV	Boeing 727-2K5	Hapag-Lloyd	
D–AIAA	A.300B2 Airbus	Lufthansa *Garmisch-Partenkirchen*	
D–AIAB	A.300B2 Airbus	Lufthansa *Rüdesheim*	
D–AIAC	A.300B2 Airbus	Lufthansa *Lüneburg*	
D–AIAD	A.300B2 Airbus	Lufthansa *Westerland/Sylt*	
D–AIAE	A.300B2 Airbus	Lufthansa *Neustadt an der Weinstrasse*	
D–AIBA	A.300B4 Airbus	Lufthansa *Kronberg im Taunus*	
D–AIBB	A.300B4 Airbus	Lufthansa *Freudenstadt/Schwarzwald*	
D–AIBC	A.300B4 Airbus	Lufthansa *Lindau-Bodenzee*	
D–AIBD	A.300B4 Airbus	Lufthansa *Erbach-Odenwald*	
D–AIBF	A.300B4 Airbus	Lufthansa *Kronberg im Taunus*	
D–ALFA	BAC One-Eleven 528	Hapag-Lloyd *Jakob Fugger*	
D–AMAM	BAC One-Eleven 515FB	Hapag-Lloyd	
D–AMAP	A.300B4 Airbus	Hapag-Lloyd	
D–AMAT	BAC One-Eleven 524FF	Hapag-Lloyd	
D–AMAX	A.300B4 Airbus	Hapag-Lloyd *Maximilian*	
D–AMAY	A.300B4 Airbus	Hapag-Lloyd *Ludwig I*	
D–AMAZ	A.300B4 Airbus	Hapag-Lloyd	
D–ANUE	BAC One-Eleven 528	Hapag-Lloyd *Albrecht Dürer*	
D–A	Boeing 737-200	Hapag-Lloyd	
D–A	Boeing 737-200	Hapag-Lloyd	
D–A	Boeing 737-200	Hapag-Lloyd	
D–A	Boeing 737-200	Hapag-Lloyd	
D–A	Boeing 737-200	Hapag-Lloyd	
D–A	Boeing 737-200	Hapag-Lloyd	
D–A	Boeing 737-2K9	Bavaria Fluggesellschaft	
D–A	Boeing 737-2K9	Bavaria Fluggesellschaft	
D–A	Boeing 737-2K9	Bavaria Fluggesellschaft	
D–A	Boeing 737-2K9	Bavaria Fluggesellschaft	
D–A	Boeing 737-2L9	Supair	
D–A	Boeing 737-2L9	Supair	
D–BAKA	F.27 Friendship Mk. 100	WDL	
D–BOOM	F.27J Friendship	Wirtschaftsflug Rhein-Main	

NOTE: Lufthansa is in the process of replacing its Boeing 737-130s with Series 230s.

DM (German Democratic Republic)

Notes	Reg.	Type	Owner or Operator
	DM–SCB	Tupolev Tu-134	Interflug
	DM–SCE	Tupolev Tu-134	Interflug
	DM–SCF	Tupolev Tu-134	Interflug
	DM–SCG	Tupolev Tu-134	Interflug
	DM–SCH	Tupolev Tu-134	Interflug
	DM–SCI	Tupolev Tu-134A	Interflug
	DM–SCK	Tupolev Tu-134A	Interflug
	DM–SCL	Tupolev Tu-134A	Interflug
	DM–SCN	Tupolev Tu-134A	Interflug
	DM–SCO	Tupolev Tu-134A	Interflug
	DM–SCP	Tupolev Tu-134A	Interflug
	DM–SCR	Tupolev Tu-134A	Interflug
	DM–SCS	Tupolev Tu-134A	Interflug
	DM–SCT	Tupolev Tu-134A	Interflug
	DM–SCU	Tupolev Tu-134A	Interflug
	DM–SCV	Tupolev Tu-134A	Interflug
	DM–SCW	Tupolev Tu-134A	Interflug
	DM–SCX	Tupolev Tu-134A	Interflug
	DM–SCY	Tupolev Tu-134A	Interflug
	DM–SCZ	Tupolev Tu-134	Interflug
	DM–SDE	Tupolev Tu-134A	Interflug
	DM–SDF	Tupolev Tu-134A	Interflug
	DM–SDG	Tupolev Tu-134A	Interflug
	DM–SDH	Tupolev Tu-134A	Interflug
	DM–SDI	Tupolev Tu-134A	Interflug
	DM–SDK	Tupolev Tu-134A	Interflug
	DM–SDL	Tupolev Tu-134A	Interflug
	DM–SDM	Tupolev Tu-134A	Interflug
	DM–SDN	Tupolev Tu-134A	Interflug
	DM–SDO	Tupolev Tu-134A	Interflug
	DM–SEB	Ilyushin IL-62	Interflug
	DM–SEC	Ilyushin IL-62	Interflug
	DM–SEF	Ilyushin IL-62	Interflug
	DM–SEG	Ilyushin IL-62	Interflug
	DM–SEH	Ilyushin IL-62	Interflug
	DM–SEI	Ilyushin IL-62	Interflug
	DM–SEK	Ilyushin IL-62M	Interflug

EC (Spain)

Transeuropa

Notes	Reg.	Type	Owner or Operator
	EC–ASN	Douglas DC-8-52	Aviaco *Murillo*
	EC–AUM	Douglas DC-8-52	Aviaco *Zurbaran*
	EC–BIB	S.E.210 Caravelle 10-R	Transeuropa
	EC–BIG	Douglas DC-9-32	Iberia *Villa de Madrid*
	EC–BIH	Douglas DC-9-32	Iberia *Ciudad de Barcelona*
	EC–BIJ	Douglas DC-9-32	Aviaco Iberia *Santa Cruz de Tenerife*
	EC–BIK	Douglas DC-9-32	Iberia *Las Palmas de Gran Canaria*
	EC–BIL	Douglas DC-9-32	Iberia *Ciudad de Zaragoza*
	EC–BIM	Douglas DC-9-32	Iberia *Ciudad de Santander*
	EC–BIN	Douglas DC-9-32	Iberia *Palma de Mallorca*
	EC–BIO	Douglas DC-9-32	Iberia *Villa de Bilbao*
	EC–BIP	Douglas DC-9-32	Aviaco *Santiago de Compostela*
	EC–BIQ	Douglas DC-9-32	Aviaco *Ciudad de Malaga*
	EC–BIR	Douglas DC-9-32	Iberia *Ciudad de Valencia*
	EC–BIS	Douglas DC-9-32	Iberia *Ciudad de Alicante*

Reg.	Type	Owner or Operator	Notes
EC-BIT	Douglas DC-9-32	Iberia *Ciudad de San Sebastian*	
EC-BIU	Douglas DC-9-32	Aviaco Iberia *Ciudad de Oviedo*	
EC-BJC	CV-990A Coronado	Spantax	
EC-BJD	CV-990A Coronado	Spantax	
EC-BMV	Douglas DC-8-55F	Aviaco *Pedro Berruguete*	
EC-BMY	Douglas DC-8-63	Iberia *Rosales*	
EC-BMZ	Douglas DC-8-63F	Iberia *Los Madrazo*	
EC-BPF	Douglas DC-9-32	Iberia *Ciudad de Almeria*	
EC-BPG	Douglas DC-9-32	Iberia *Ciudad de Vigo*	
EC-BPH	Douglas DC-9-32	Iberia *Ciudad de Gerona*	
EC-BQA	CV-990A Coronado	Spantax	
EC-BQQ	CV.990A Coronado	Spantax	
EC-BQS	Douglas DC-8-63	Iberia *Claudio Coello*	
EC-BQT	Douglas DC-9-32	Iberia *Villa de Murcia*	
EC-BQU	Douglas DC-9-32	Iberia *Ciudad de la Coruna*	
EC-BQV	Douglas DC-9-32	Iberia *Ciudad de Ibiza*	
EC-BQX	Douglas DC-9-32	Iberia *Ciudad de Valladolid*	
EC-BQY	Douglas DC-9-32	Iberia *Ciudad de Cordoba*	
EC-BQZ	Douglas DC-9-32	Iberia *Ciudad de Santa Cruz de la Palma*	
EC-BRQ	Boeing 747-256B	Iberia *Calderon de la Barca*	
EC-BRX	S.E.210 Caravelle 11-R	Transeuropa *Renacuajo I*	
EC-BSD	Douglas DC-8-63	Aviaco *Ribalta*	
EC-BSE	Douglas DC-8-63	Iberia *Alonso Cano*	
EC-BTE	CV-990A Coronado	Spantax	
EC-BXI	CV-990A Coronado	Spantax	
EC-BYD	Douglas DC-9-32	Iberia *Ciudad de Arrecife de Lanzarote*	
EC-BYE	Douglas DC-9-32	Iberia *Ciudad de Mahon*	
EC-BYF	Douglas DC-9-32	Iberia *Ciudad de Granada*	
EC-BYG	Douglas DC-9-32	Iberia *Ciudad de Pamplona*	
EC-BYH	Douglas DC-9-32	Aviaco *Ciudad de Cadiz*	
EC-BYI	Douglas DC-9-32	Iberia *Ciudad de Iberia Ciudad de Vitoria*	
EC-BYJ	Douglas DC-9-32	Iberia *Ciudad de Salamanca*	
EC-BYK	Douglas DC-9-33RC	Iberia *Ciudad de Badajoz*	
EC-BYL	Douglas DC-9-33RC	Iberia *Ciudad de Albacete*	
EC-BYM	Douglas DC-9-33RC	Iberia *Ciudad de Cangas de Onis*	
EC-BYN	Douglas DC-9-33RC	Iberia *Ciudad de Caceres*	
EC-BZO	CV-990A Coronado	Spantax	
EC-BZP	CV-990A Coronado	Spantax	
EC–CAI	Boeing 727-256	Iberia *Castilla la Nueva*	
EC–CAJ	Boeing 727-256	Iberia *Cataluna*	
EC–CAK	Boeing 727-256	Iberia *Aragon*	
EC–CBA	Boeing 727-256	Iberia *Vascongadas*	
EC–CBB	Boeing 727-256	Iberia *Valencia*	
EC–CBC	Boeing 727-256	Iberia *Navarra*	
EC–CBD	Boeing 727-256	Iberia *Murcia*	
EC–CBE	Boeing 727-256	Iberia *Leon*	
EC–CBF	Boeing 727-256	Iberia *Gran Canaria*	
EC–CBG	Boeing 727-256	Iberia *Extremadura*	
EC–CBH	Boeing 727-256	Iberia *Galicia*	
EC–CBI	Boeing 727-256	Iberia *Asturias*	
EC–CBJ	Boeing 727-256	Iberia *Andalucia*	
EC–CBK	Boeing 727-256	Iberia *Baleares*	
EC–CBL	Boeing 727-256	Iberia *Tenerife*	
EC–CBM	Boeing 727-256	Iberia *Castilla la Vieja*	
EC–CBO	Douglas DC-10-30	Iberia *Costa del Sol*	
EC–CBP	Douglas DC-10-30	Iberia *Costa Dorada*	
EC–CCF	Douglas DC-8-61CF	Spantax	
EC–CCG	Douglas DC-8-61CF	Spantax	
EC–CCN	Douglas DC-8-33	T.A.E.	
EC–CDC	Douglas DC-8-33	T.A.E.	
EC–CEZ	Douglas DC-10-30	Iberia *Costa del Azahar*	
EC–CFA	Boeing 727-256	Iberia *Jerez Xeres Sherry*	
EC–CFB	Boeing 727-256	Iberia *Rioja*	
EC–CFC	Boeing 727-256	Iberia *Tarragona*	
EC–CFD	Boeing 727-256	Iberia *Montilla Moriles*	
EC–CFE	Boeing 727-256	Iberia *Penedes*	
EC–CFF	Boeing 727-256	Iberia *Valdepenas*	
EC–CFG	Boeing 727-256	Iberia *La Mancha*	
EC–CFH	Boeing 727-256	Iberia *Priorato*	
EC–CFI	Boeing 727-256	Iberia *Carinena*	
EC–CFJ	Boeing 727-256	Iberia *Jumilla*	
EC–CFK	Boeing 727-256	Iberia *Rivero*	
EC–CGN	Douglas DC-9-32	Aviaco *Martin Alonso Pinzon*	
EC–CGO	Douglas DC-9-32	Aviaco *Pedro Alonso Nino*	
EC–CGP	Douglas DC-9-32	Aviaco *Juan Sebastian Elcano*	

Notes	Reg.	Type	Owner or Operator
	EC–CGQ	Douglas DC-9-32	Aviaco *Alonso de Ojeda*
	EC–CGR	Douglas DC-9-32	Aviaco *Francisco de Orellana*
	EC–CGS	Douglas DC-9-32	Aviaco *Vasco Nunez de Balboa*
	EC–CGY	Douglas DC-9-14	Spantax
	EC–CGZ	Douglas DC-9-14	Spantax
	EC-CID	Boeing 727-256	Iberia *Malaga*
	EC-CIE	Boeing 727-256	Iberia *Esparragosa*
	EC-CIZ	S.E.210 Caravelle 10-R	Transeuropa
	EC-CLB	Douglas DC-10-30	Iberia *Costa Blanca*
	EC-CLD	Douglas DC-9-32	Aviaco *Hernando de Soto*
	EC-CLE	Douglas DC-9-32	Aviaco *Juan Ponce de Leon*
	EC-CMS	S.E.210 Caravelle 10-B	T.A.E.
	EC–CNF	CV-990A Coronado	Spantax
	EC–CNG	CV-990A Coronado	Spantax
	EC–CNH	CV-990A Coronado	Spantax
	EC–CNJ	CV-990A Coronado	Spantax
	EC–CPI	S.E.210 Caravelle 10-R	Transeuropa
	EC–CQM	Douglas DC-8-54F	Aviaco
	EC–CSJ	Douglas DC-10-30	Iberia *Costa de la Luz*
	EC–CSK	Douglas DC-10-30	Iberia *Cornisa Cantabrica*
	EC–CTR	Douglas DC-9-34CF	Aviaco *Hernan Cortes*
	EC–CTS	Douglas DC-9-34CF	Aviaco *Francisco Pizarro*
	EC–CTT	Douglas DC-9-34CF	Aviaco *Pedro de Valdivia*
	EC–CTU	Douglas DC-9-34CF	Aviaco *Pedro de Alvarado*
	EC–CUM	S.E.210 Caravelle 10-B	T.A.E.
	EC–CYI	S.E.210 Caravelle 10-R	Transeuropa
	EC–CZE	Douglas DC-8-61	Spantax
	EC–DBE	Douglas DC-8-55F	Aviaco
	EC–DCC	Boeing 727–256	Iberia *Alvarino*
	EC–DCD	Boeing 727-256	Iberia *Chacoli*
	EC–DCE	Boeing 727-256	Iberia *Mentrida*
	EC–DCN	S.E.210 Caravelle 10-R	Transeuropa
	EC–DDU	Boeing 727-256	Iberia *Alhambra de Granada*
	EC–DDV	Boeing 727-256	Iberia *Acueducto de Segovia*
	EC–DDX	Boeing 727-256	Iberia *Monasterio de Poblet*
	EC–DDY	Boeing 727-256	Iberia *Cuevas de Altamira*
	EC–DDZ	Boeing 727-256	Iberia *Murallas de Avila*
	EC–DEA	Douglas DC-10-30	Iberia *Rias Gallegas*
	EC–DEG	Douglas DC-10-30CF	Spantax
	EC–DEM	Douglas DC-8-55F	Aviaco *Goya*
	EC–DFP	S.E.210 Caravelle 10-B	T.A.E.
	EC–DGB	Douglas DC-9-34	Aviaco *Castillo de Javier*
	EC–DGC	Douglas DC-9-34	Aviaco *Castillo de Sotomayor*
	EC–DGD	Douglas DC-9-34	Aviaco *Castillo de Arcos*
	EC–DGE	Douglas DC-9-34	Aviaco *Castillo de Bellver*
	EC–DHZ	Douglas DC-10-30	Iberia
	EC–DIA	Boeing 747-256B	Iberia
	EC–DIB	Boeing 747-256B	Iberia *Cervantes*
	EC–DIH	Douglas DC-8-55	T.A.E.
	EC–DIR	Douglas DC-9-14	Spantax
	EC–DLC	Boeing 747-256B	Iberia *Francisco de Quevedo*
	EC–DLD	Boeing 747-256B	Iberia *Lupe de Vega*

EI (Republic of Ireland)

Including complete current Irish Civil Register.

Notes	Reg.	Type	Owner or Operator
	EI–ADV	PA-12 Super Cruiser	R. E. Levis
	EI–AFF	*B.A. Swallow 2	J. McCarthy
	EI–AFK	*D.H.84 Dragon (EI–ABI)	Aer Lingus-Irish
	EI–AFN	*B.A. Swallow 2	J. McCarthy
	EI–AGB	*Miles M.38 Messenger 4	J. McLoughlin
	EI–AGD	*Taylorcraft Plus D	H. Wolf
	EI–AGE	*Miles M.38 Messenger 3	J. McLoughlin
	EI–AGJ	J/I Autocrat	W. G. Rafter
	EI–AGP	*D.H.82A Tiger Moth	D. W. R. Kennedy
	EI–AHA	*D.H.82A Tiger Moth	J. H. Maher

Reg.	Type	Owner or Operator	Notes
EI–AHR	*D.H.C.1 Chipmunk 22	C. Lane	
EI–AKM	Piper J-3C-65 Cub	Setanta Flying Group	
EI–ALH	Taylorcraft Plus D	N. Reilly	
EI–ALP	Avro 643 Cadet	J. C. O'Loughlin	
EI–ALU	Avro 631 Cadet	M. P. Cahill	
EI–AMD	M.S.880B Rallye Club	Morane Aircraft Ltd.	
EI–AMK	J/1 Autocrat	Irish Aero Club	
EI–AMO	J/1B Aiglet	R. Hassett	
EI–AMY	J/1N Alpha	Meath Flying Group Ltd.	
EI–AND	Cessna 175A	Jack Braithwaite (Ireland) Ltd.	
EI–ANE	BAC One-Eleven 208AL	Aer Lingus *St. Mel*	
EI–ANF	BAC One-Eleven 208AL	Aer Lingus *St. Malachy*	
EI–ANG	BAC One-Eleven 208AL	Aer Lingus *St. Declan*	
EI–ANH	BAC One-Eleven 208AL	Aer Lingus *St. Ronan*	
EI–ANO	Boeing 707-348C	Aer Lingus *St. Brigid*	
EI–ANT	Champion 7ECA Citabria	Setanta Flying Group	
EI–AOB	PA-28 Cherokee 140	Dr. H. J. R. Henderson	
EI–AOD	Cessna 182J Skylane	Oscar Delta Flying Training Co. Ltd.	
EI–AOK	Cessna F.172G	R. J. Cloughley & N. J. Simpson	
EI–AOO	Cessna 150E	R. Hassett	
EI–AOP	D.H.82A Tiger Moth	Dublin Tiger Group	
EI–AOS	Cessna 310B	Joyce Aviation Ltd.	
EI–APF	Cessna F.150F	Midland Flying Group Ltd.	
EI–APG	Boeing 707-348C	Aer Lingus *St. Senan*	
EI–APS	Schleicher ASK 14	G. W. Connolly & M. Slazenger	
EI–APT	Fokker D.VII/65 Replica	Blue Max Aviation Ltd.	
EI–APU	Fokker D.VII/65 Replica	Blue Max Aviation Ltd.	
EI–APV	Fokker D.VII/65 Replica	Blue Max Aviation Ltd.	
EI–APW	Fokker Dr.1 Replica	Blue Max Aviation Ltd.	
EI–ARA	SE.5A Replica	Blue Max Aviation Ltd.	
EI–ARB	SE.5A Replica	Blue Max Aviation Ltd.	
EI–ARC	Pfalz D.III Replica	Blue Max Aviation Ltd.	
EI–ARD	Pfalz D.III Replica	Blue Max Aviation Ltd.	
EI–ARE	Stampe SV.4C	Blue Max Aviation Ltd.	
EI–ARF	Caudron C.277 Luciole	Blue Max Aviation Ltd.	
EI–ARG	Morane-Saulnier M.S. 230	J. E. Hutchinson & ptnrs.	
EI–ARH	Currie Wot/S.E.5 Replica	Blue Max Aviation Ltd.	
EI–ARI	Currie Wot/S.E.5 Replica	Blue Max Aviation Ltd.	
EI–ARJ	Currie Wot/S.E.5 Replica	Blue Max Aviation Ltd.	
EI–ARK	Currie Wot/S.E.5 Replica	Blue Max Aviation Ltd.	
EI–ARL	Currie Wot/S.E.5 Replica	Blue Max Aviation Ltd.	
EI–ARM	Currie Wot/S.E.5 Replica	Blue Max Aviation Ltd.	
EI–ARN	Wren 460	Irish Parachute Club	
EI–ARW	Jodel D.R.1050	John B. Healy	
EI–ASA	Boeing 737-248	Aer Lingus *St. Jarlath*	
EI–ASB	Boeing 737-248	Aer Lingus *St. Albert*	
EI–ASC	Boeing 737-248C	Aer Lingus *St. Macartan*	
EI–ASD	Boeing 737-248C	Aer Lingus *St. Ide*	
EI–ASE	Boeing 737-248C	Aer Lingus *St. Fachtna*	
EI–ASF	Boeing 737-248	Aer Lingus *St. Nathy*	
EI–ASG	Boeing 737-248	Aer Lingus *St. Cormack*	
EI–ASH	Boeing 737-248	Aer Lingus *St. Eugene*	
EI–ASI	Boeing 747-148	Aer Lingus *St. Columcille*	
EI–ASL	Boeing 737-248C	Aer Lingus *St. Cillian*	
EI–ASO	Boeing 707-349C	Aer Lingus *St. Canice*	
EI–AST	Cessna F.150H	Liberty Flying Group	
EI–ASV	PA-28R Cherokee Arrow 180	F. E. A. Biggar	
EI–ATC	Cessna 310G	Iona National Airways	
EI–ATH	Cessna F.150J	Galway Flying Club	
EI–ATJ	B.121 Pup 1	Wexford Aero Club	
EI–ATK	PA-28 Cherokee 140	Mayo Flying Club Ltd.	
EI–ATS	M.S.880B Rallye Club	O. Bruton & G. Farrar	
EI–AUC	Cessna FA.150K Aerobat	Fifteen Flying Club Ltd.	
EI–AUD	M.S.880B Rallye Club	Kilkenny Flying Club Ltd.	
EI–AUE	M.S.880B Rallye Club	Slievenamon Air Ltd.	
EI–AUG	M.S.894 Rallye Minerva 220	Weston Ltd.	
EI–AUJ	M.S.880B Rallye Club	J. O'Conner & P. J. Ryan	
EI–AUM	J/1 Autocrat	J. G. Rafter	
EI–AUN	M.S.880B Rallye Club	Leinster Aero Club	
EI–AUO	Cessna FA.150K Aerobat	Kerry Aero Club	
EI–AUP	M.S.880B Rallye Club	Limerick Flying Club	
EI–AUR	Bölkow Bo 208C Junior	S. McCarthy	
EI–AUT	Forney F-1A Aircoupe	Joyce Aviation Ltd.	
EI–AUV	PA-23 Aztec 250	Shannon Executive Aviation	

Notes	Reg.	Type	Owner or Operator
	EI–AUY	Morane-Saulnier M.S.502	Historical Aircraft Preservation Group
	EI–AVB	Aeronca 7AC Champion	G. G. Bracken
	EI–AVC	Cessna F.337F	Iona National Airways
	EI–AVG	Beech A.55 Baron	C. O. Neale
	EI–AVL	J/5F Aiglet	George E. Flood
	EI–AVM	Cessna F.150L	Victor Mike Flying Group
	EI–AVN	Hughes 369HM	Helicopter Maintenance Ltd.
	EI–AVU	Stampe SV.4C	S. P. O'Carroll
	EI–AWA	Bell 206B JetRanger 2	Helicopter Maintenance Ltd.
	EI–AWE	Cessna F.150M	Third Flight Group
	EI–AWH	Cessna 210J	Southern Air Ltd.
	EI–AWJ	M.S.893A Rallye Commodore	Dr. W. J. Phelan
	EI–AWM	BN-2A Islander	Aer Arann
	EI–AWP	D.H.82A Tiger Moth	A. Lyons
	EI–AWR	Malmo MFI-9 Junior	W. M. Calder-Potts
	EI–AWU	M.S.880B Rallye Club	Longford Aviation Ltd.
	EI–AWW	Cessna 414	T. Farrington
	EI–AWY	Mitsubishi MU-2B-35	Helicopter Maintenance Ltd.
	EI–AYA	M.S.880B Rallye Club	Dundalk Aero Club Ltd.
	EI–AYB	GY-80 Horizon 180	Westwing Flying Group
	EI–AYD	AA-5 Traveler	Kenneth Walters
	EI–AYF	Cessna FRA.150L	Garda Flying Club
	EI–AYH	Cessna 172B	A. A. Kelly
	EI–AYI	M.S.880B Rallye Club	Irish Air Training Group
	EI–AYK	Cessna F.172M	P. J. Meade
	EI–AYL	A.109 Airedale	J. A. McWilliams & D. E. O'Sullivan
	EI–AYN	BN-2A Islander	Aer Arann
	EI–AYO	*Douglas DC-3A	Science Museum, Wroughton
	EI–AYR	Schleicher ASK-16	Kilkenny Airport Ltd.
	EI–AYS	PA-22 Colt 108	George Farrer & ptnrs.
	EI–AYT	M.S.894A Rallye Minerva	Dublin Gliding Club Ltd.
	EI–AYV	M.S.892A Rallye Commodore 150	P. Murtagh
	EI–AYW	PA-23 Aztec 250C	J. A. Chute Construction Co. Ltd.
	EI–AYY	Evans VP-1	Michael Donoghue
	EI–BAA	B.175 Britannia 307F	Aer Turas Teo *City of Dublin*
	EI–BAB	M.S.894E Rallye Minerva	P. Monaghan
	EI–BAF	Thunder Ax6-56 balloon	W. G. Woollett
	EI–BAG	Cessna 172A	Cork Skydiving Club Ltd.
	EI–BAJ	Stampe SV.4C	Dublin Tiger Group
	EI–BAM	Bell 212	Irish Helicopters Ltd.
	EI–BAN	Cameron O-65 balloon	C. M. Alexander
	EI–BAO	Cessna F.172G	Kingdom Air Ltd.
	EI–BAR	Thunder Ax8-105 balloon	Fr. D. Reid
	EI–BAS	Cessna F.172M	F. Fahey & B. Fitzgerald
	EI–BAT	Cessna F.150M	20th Air Training Co. Ltd.
	EI–BAU	Stampe SV.4C	S. P. O'Carroll
	EI–BAV	PA-22 Colt 108	J. P. Montcalm
	EI–BAY	Cameron Ax8-84 balloon	F. N. Lewis
	EI–BBB	R. Commander 112	W. J. O'Connor
	EI–BBC	PA-28 Cherokee 180C	The Cherokee Group
	EI–BBD	Evans VP-1	Volksplane Group
	EI–BBE	Champion 7FC Tri-Traveler	G. Treacy
	EI–BBG	M.S.880B Rallye Club	Weston Ltd.
	EI–BBH	B.175 Britannia 253F	Aer Turas Teo *City of Cork*
	EI–BBI	M.S.892 Rallye Commodore	Kilkenny Airport Ltd.
	EI–BBJ	M.S.880B Rallye Club	Weston Ltd.
	EI–BBK	A.109 Airedale	H. S. Igoe
	EI–BBL	R. 690A Commander	The Earl of Granard
	EI–BBM	Cameron O-65 balloon	Dublin Ballooning Club
	EI–BBN	Cessna F.150M	Iona National Airways
	EI–BBO	M.S.893E Rallye	W. Gavin
	EI–BBR	BN-2A-6 Islander	Aer Arann
	EI–BBS	PA-28 Cherokee 180	Patrick J. Whelan
	EI–BBT	F.8L Falco I Srs. 3	A. N. Johnston & D. M. Whelan
	EI–BBU	M.S.880B Rallye Club	Galway Aero Club Ltd.
	EI–BBV	Piper J-3C-65 Cub	Feidhlim Cronin
	EI–BBW	M.S.894A Rallye Minerva	T. P. Higgins
	EI–BCE	BN-2A-26 Islander	Aer Arann
	EI–BCF	Bensen B.8M	P. E. & E. O'Loughlin
	EI–BCH	M.S.892A Rallye Commodore 150	Bruton Aircraft Engineering Ltd.
	EI–BCJ	F.8L Falco I Srs. 3	D. Kelly
	EI–BCK	Cessna F.172K	Iona National Airways

Reg.	Type	Owner or Operator	Notes
EI-BCL	Cessna 182P	Iona National Airways	
EI-BCM	Piper J-3C-65 Cub	Kilmoon Flying Group	
EI-BCN	Piper J-3C-65 Cub	Snowflake Flying Group	
EI-BCO	Piper J-3C-65 Cub	J. Molloy	
EI-BCP	D.62B Condor	The Condor Group	
EI-BCR	Boeing 737-281	Aer Lingus *Sir Oliver Plunkett*	
EI-BCS	M.S.880B Rallye Club	J. J. Martyn & J. O'Neill	
EI-BCT	Cessna 411A	Air Surveys International	
EI-BCU	M.S.880B Rallye Club	Weston Ltd.	
EI-BCV	Cessna F.150M	Iona National Airways	
EI-BCW	M.S.880B Rallye Club	B. E. Lalor & L. Prendergast	
EI-BCY	Beech 200 Super King Air (232)	Minister of Defence	
EI-BDB	M.S.880B Rallye Club	Weston Ltd.	
EI-BDD	PA-34-200 Seneca	D. Gliksten	
EI-BDE	Colt Ax7-77A balloon	Colting Balloons	
EI-BDF	Enstrom F-28A	Enstrom Helicopters (Ireland)	
EI-BDG	Robin HR. 100/210	P. Parke	
EI-BDH	M.S.880B Rallye Club	Munster Wings Ltd.	
EI-BDI	MBB Bo 105C	Irish Helicopters Ltd.	
EI-BDK	M.S.880B Rallye Club	P. Colgan	
EI-BDL	Evans VP-2	J. Duggan	
EI-BDM	PA-23 Aztec 250D	Executive Air Services	
EI-BDO	Cessna F.152	Iona National Airways	
EI-BDP	Cessna 182P	Phoenix Engineering Ltd.	
EI-BDR	PA-28 Cherokee 180	Cork Flying Club	
EI-BDS	M.S.894A Rallye Minerva	Empire Enterprises	
EI-BDT	Dakota 4	Clyden Airways	
EI-BDU	Dakota 4	Clyden Airways	
EI-BDV	PA-32 Cherokee Six 260	Birr Flying Club	
EI-BDY	Boeing 737-2E1	Aer Lingus Teo *St. Erthne*	
EI-BEA	M.S.880B Rallye 100ST	Weston Ltd.	
EI-BEB	Boeing 737-248	Aer Lingus Teo	
EI-BEC	Boeing 737-248	Aer Lingus Teo	
EI-BED	Boeing 747-130	Aer Linte Eireann Teo	
EI-BEE	Boeing 737-281	Aer Lingus Teo *St. Cronin*	
EI-BEF	Boeing 737-281	Air Tara Ltd.	
EI-BEG	—	Aer Lingus Teo	
EI-BEH	—	Aer Lingus Teo	
EI-BEI	—	Aer Lingus Teo	
EI-BEJ	—	Aer Lingus Teo	
EI-BEK	—	Aer Lingus Teo	
EI-BEL	—	Aer Lingus Teo	
EI-BEM	—	Aer Lingus Teo	
EI-BEN	Piper J-3C-65 Cub	Capt. J. J. Sullivan	
EI-BEO	Cessna 310Q	Iona National Airways	
EI-BEP	M.S.892A Rallye Commodore	H. Lynch & J. O'Leary	
EI-BET	Cessna F.337G	Messrs. Donegan, Leonard & Meade	
EI-BEU	Auster J/4	G. H. Farrar	
EI-BEW	Beech D55 Baron	Marsh Nurseries	
EI-BEY	Naval N3N-3	Huntley & Huntley Ltd.	
EI-BFB	M.S.880B Rallye 100ST	Weston Ltd.	
EI-BFC	Boeing 737-2H4	Air Tara Ltd.	
EI-BFD	Scheibe SF.28A	Dublin Gliding Club	
EI-BFE	Cessna F.150G	Joyce Aviation Ltd.	
EI-BFF	Beech A.23 Musketeer	M. R. Cody	
EI-BFH	Bell 212	Irish Helicopters Ltd.	
EI-BFI	M.S.880B Rallye 100ST	J. O'Neill	
EI-BFJ	Beech A.200 Super King Air	Minister of Defence	
EI-BFK	Bell 206B JetRanger 3	M. V. O'Brien	
EI-BFM	M.S.893E Rallye 235GT	Michael Orr Leasing Ltd.	
EI-BFO	Piper J-3C-90 Cub	J. Molloy	
EI-BFP	M.S.880B Rallye 100ST	Weston Ltd.	
EI-BFR	MS.880B Rallye 100ST	Galway Flying Club	
EI-BFS	FRED Srs. 2	J. Farrant	
EI-BFT	Beech A200 Super King Air	Avair Ltd.	
EI-BFU	—	—	
EI-BFV	MS.880B Rallye 100ST	G. Treacy & ptnrs.	
EI-BFW	Beech A200 Super King Air	Avair Ltd.	
EI-BGA	MS.880B Rallye 100ST	T. Daly	
EI-BGB	MS.880B Rallye 100ST	M. Slattery & ptnrs.	
EI-BGC	MS.880B Rallye Club	B. Carpenter	
EI-BGD	MS.880B Rallye Club	S. O'Rourke & J. Lawlor	
EI-BGE	Cessna F.172M	J. I. Robinson	
EI-BGF	PA-28R Cherokee Arrow 180	Arrow Group	
EI-BGG	MS.892E Rallye Commodore	D. A. Weldon	

Notes	Reg.	Type	Owner or Operator
	EI–BGH	Cessna F.172N	Iona National Airways
	EI–BGI	Cessna F.152	Iona National Airways
	EI–BGK	Cessna P206D	Shannon Executive Aviation
	EI–BGL	R. Turbo Commander 690B	Flightline Ltd.
	EI–BGN	MS.880B Rallye Club	T. Guckian & E. Browne
	EI–BGO	Canadair CL-44D-4J	Aer Turas Teo
	EI–BGP	Cessna 414A	Iona National Airways
	EI–BGR	PA–31P Pressurised Navajo	Avair Ltd.
	EI–BGS	MS.893B Rallye 180GT	Mac Grids Ltd.
	EI–BGT	Colt 77A balloon	K. Haugh
	EI–BGU	MS.880B Rallye Club	M. F. Neary
	EI–BGV	AA-5 Traveler	Crowe & Fitzgerald Ltd.
	EI–BGW	H.S.125 Srs. 1B	Ledair (Ireland) Ltd.
	EI–BHA	Beech A200 Super King Air	Avair Ltd.
	EI–BHB	MS.887 Rallye 125	G. Drudy & J. Martyn
	EI–BHC	Cessna F.177RG	P. J. McGuire & B. Palfrey
	EI–BHD	MS.893E Rallye 180GT	Arvan Homes Ltd.
	EI–BHF	MS.892A Rallye Commodore 150	B. Mullen
	EI–BHG	Beech 200 Super King Air	Sunbird Air Charter
	EI–BHH	—	—
	EI–BHI	Bell 206B JetRanger 2	International Helicopter Sales Ltd.
	EI–BHJ	M.65 Gemini 3C	R. E. Winn
	EI–BHK	MS.880B Rallye Club	Bruton Aircraft Engineering Ltd.
	EI–BHL	Beech E90 King Air	Stewart Singlam Fabrics Ltd.
	EI–BHM	Cessna 337E	The Ross Flying Group
	EI–BHN	MS.893A Rallye Commodore	Bruton Aircraft Engineering Ltd.
	EI–BHO	Sikorsky S-61N	Irish Helicopters Ltd.
	EI–BHP	MS.893A Rallye Commodore	Wicklow Flying Group
	EI–BHT	Beech 77 Skipper	Avair Ltd.
	EI–BHU	Beech 77 Skipper	Avair Ltd.
	EI–BHV	Champion 7EC Traveler	M. MacDowell
	EI–BHW	Cessna F.150F	B. A. Carpenter
	EI–BHY	MS.892E Rallye Commodore	D. Killian
	EI–BIA	Cessna FA.152	Iona National Airways Ltd.
	EI–BIB	Cessna F.152	Galway Aero Club
	EI–BIC	Cessna F.172N	Wexford Aero Club
	EI–BID	PA-18 Super Cub 95	J. A. McWilliams
	EI–BIE	Cessna FA.152	Wexford Aero Club
	EI–BIF	MS.894 Rallye Minerva 235	Empire Enterprises
	EI–BIG	Zlin 526	P. von Lonkhuyzen
	EI–BIH	SOCATA TB.10 Tobago	A. D. D. Rogers
	EI–BIJ	AB-206B JetRanger	Irish Helicopters Ltd.
	EI–BIK	PA-18-150 Super Cub	Dublin Gliding Club
	EI–BIL	Beech A35 Bonanza	W. J. Phelan
	EI–BIM	MS.880B Rallye Club	D. Millar
	EI–BIN	Cessna F.172N	Iona National Airways Ltd.
	EI–BIO	Piper J-3C-65 Cub	Monasterevin Flying Club
	EI–BIP	Beech 200 Super King Air	Avair Ltd.
	EI–BIR	Cessna F.172M	M. Masterton
	EI–BIS	Robin R.1180TD	Abbeyshrule Aero Club
	EI–BIT	M.S.887 Rallye 125	P. Lacy & ptnrs.
	EI–BIU	Robin R.2112A	Bruton Aircraft Engineering Ltd.
	EI–BIV	Bellanca 8KCAB Citabria	Aerocrats Flying Group
	EI–BIW	MS.880B Rallye Club	B. A. Carpenter
	EI–BIY	Agusta-Bell 47G-3B1 (Soloy)	Irish Helicopters Ltd.
	EI–BJA	Cessna FRA.150L	Joyce Aviation Ltd.
	EI–BJB	Aeronca 7AC	Huntley & Huntley
	EI–BJC	Aeronca 7AC	Huntley & Huntley
	EI–BJD	Mooney M.20K	Athlone Flying Group
	EI–BJE	Boeing 737-275	Air Tara Ltd. (leased to Nigeria Airways)
	EI–BJF	AA-5 Traveler	B. O'Shea
	EI–BJG	Robin R.1180	N. Hanley
	EI–BJH	Nipper T.66 Srs. 3	B. A. Carpenter
	EI–BJI	Cessna FR.172E	Irish Parachute Club
	EI–BJJ	Aeronca 15AC Sedan	Huntley & Huntley
	EI–BJK	—	—
	EI–BJL	—	—
	EI–BJM	—	—
	EI–BJN	—	—
	EI–BJO	—	—
	EI–BJP	—	—
	EI–BJR	—	—

D–ABAP (Top) S.E.210 Caravelle 10-R of Special Air Transport

EI–BIE (Centre) Cessna FA.152/*A. S. Wright*

F–GCGH (Bottom) Fairchild Hiller FH.227B Friendship of Touraine Air Transport

Notes	Reg.	Type	Owner or Operator
	EI–BJS	—	—
	EI–BJT	—	—
	EI–BJU	—	—
	EI–BJV	—	—

EP (Iran)

Notes	Reg.	Type	Owner or Operator
	EP–IAA	Boeing 747SP-86	Iran Air *Fars*
	EP–IAB	Boeing 747SP-86	Iran Air *Kurdistan*
	EP–IAC	Boeing 747SP-86	Iran Air *Khuzestan*
	EP–IAD	Boeing 747SP-86	Iran Air
	EP–IAG	Boeing 747-286B	Iran Air *Azarabadegan*
	EP–IAH	Boeing 747-268B	Iran Air *Khorasan*
	EP–IAM	Boeing 747-186B	Iran Air
	EP–ICC	Boeing 747-2J9F	Iran Air
	EP–IRJ	Boeing 707-321B	Iran Air
	EP–IRK	Boeing 707-321C	Iran Air
	EP–IRL	Boeing 707-386C	Iran Air *Apadana*
	EP–IRM	Boeing 707-386C	Iran Air *Ekbatana*
	EP–IRN	Boeing 707-386C	Iran Air *Pasargad*

ET (Ethiopia)

Notes	Reg.	Type	Owner or Operator
	ET–AAH	Boeing 720-060B	Ethiopian Airlines *White Nile*
	ET–ABP	Boeing 720-060B	Ethiopian Airlines
	ET–ACQ	Boeing 707-379C	Ethiopian Airlines
	ET–AFA	Boeing 720-024B	Ethiopian Airlines
	ET–AFB	Boeing 720-024B	Ethiopian Airlines
	ET–AFK	Boeing 720-024B	Ethiopian Airlines

AIR FRANCE

F (France)

Notes	Reg.	Type	Owner or Operator
	F–BEIG	Douglas DC-3	Bretagne Air Services
	F–BHRD	S.E.210 Caravelle III	Air France *Guyenne*
	F–BHRF	S.E.210 Caravelle III	Air France *Auvergne*
	F–BHRH	S.E.210 Caravelle III	Air France *Bourgogne*

Reg.	Type	Owner or Operator	Notes
F–BHRI	S.E.210 Caravelle III	Air France *Bretagne*	
F–BHRK	S.E.210 Caravelle III	Air France *Corse*	
F–BHRQ	S.E.210 Caravelle III	Air Inter	
F–BHRR	S.E.210 Caravelle III	Air Inter	
F–BHRS	S.E.210 Caravelle III	Air Inter	
F–BHRU	S.E.210 Caravelle III	Air France *Poitou*	
F–BHRV	S.E.210 Caravelle III	Air France *Provence*	
F–BHRY	S.E.210 Caravelle III	Air France *Touraine*	
F–BHRZ	S.E.210 Caravelle III	Air Inter	
F–BHSV	Boeing 707-328B	Air France	
F–BHSX	Boeing 707-328B	Air France	
F–BHSY	Boeing 707-328B	Air France	
F–BJCM	Boeing 707-355C (Cargo)	Air France	
F–BJTE	S.E.210 Caravelle III	Air Charter International	
F–BJTG	S.E.210 Caravelle III	Air Charter International *Roussillon*	
F–BJTH	S.E.210 Caravelle III	Air Charter International *Franche-Comté*	
F–BJTJ	S.E.210 Caravelle III	Air Charter International *Bourbonnais*	
F–BJTL	S.E.210 Caravelle III	Air France *Aunis et Saintonge*	
F–BJTO	S.E.210 Caravelle III	Air Charter International *Pays Basque*	
F–BJTP	S.E.210 Caravelle III	Air France *Comtat Venaissin*	
F–BJTQ	S.E.210 Caravelle III	Air France *Champagne*	
F–BJTR	S.E.210 Caravelle III	Air France *Principauté de Monaco*	
F–BJTS	S.E.210 Caravelle III	Air France *Vercors*	
F–BLCA	Boeing 707-328B	Air France	
F–BLCC	Boeing 707-328C (Cargo)	Air France	
F–BLCD	Boeing 707-328B	Air France	
F–BLCE	Boeing 707-328B	Air France	
F–BLCF	Boeing 707-328C (Cargo)	Air France	
F–BLCG	Boeing 707-328C (Cargo)	Air France *Chateau du Lude*	
F–BLCH	Boeing 707-328C	Air France	
F–BLCI	Boeing 707-328C	Air France	
F–BLCK	Boeing 707-328C	Air France	
F–BLCL	Boeing 707-328C	Air France	
F–BLKF	S.E.210 Caravelle III	Air Charter International	
F–BLLB	Boeing 707-328B	Air France	
F–BLOY	HPR-7 Herald 210	Europe Aero Service	
F–BNKA	S.E.210 Caravelle III	Air Inter	
F–BNKB	S.E.210 Caravelle III	Air Inter	
F–BNKC	S.E.210 Caravelle III	Air Inter	
F–BNKD	S.E.210 Caravelle III	Air Inter	
F–BNKE	S.E.210 Caravelle III	Air Inter	
F–BNKF	S.E.210 Caravelle III	Air Inter	
F–BNKG	S.E.210 Caravelle III	Air Inter	
F–BNKH	S.E.210 Caravelle III	Air Inter	
F–BNKJ	S.E.210 Caravelle III	Air Inter	
F–BNKK	S.E.210 Caravelle III	Air Inter	
F–BNKL	S.E.210 Caravelle III	Air Inter	
F–BNOG	S.E.210 Caravelle 12B	Air Inter	
F–BOHA	S.E.210 Caravelle III	Air France *Comté de Nice*	
F–BOIZ	HPR-7 Herald 210	Europe Aero Service	
F–BOJA	Boeing 727-228	Air France	
F–BOJB	Boeing 727-228	Air France	
F–BOJC	Boeing 727-228	Air France	
F–BOJD	Boeing 727-228	Air France	
F–BOJE	Boeing 727-228	Air France	
F–BPJG	Boeing 727-228	Air France	
F–BPJH	Boeing 727-228	Air France	
F–BPJI	Boeing 727-228	Air France	
F–BPJJ	Boeing 727-228	Air France	
F–BPJK	Boeing 727-228	Air France	
F–BPJL	Boeing 727-228	Air France	
F–BPJM	Boeing 727-228	Air France	
F–BPJN	Boeing 727-228	Air France	
F–BPJO	Boeing 727-228	Air France	
F–BPJP	Boeing 727-228	Air France	
F–BPJQ	Boeing 727-228	Air France	
F–BPJR	Boeing 727-228	Air France	
F–BPJS	Boeing 727-228	Air France	
F–BPJT	Boeing 727-228	Air France	
F–BPJU	Boeing 727-214	Air Charter International	
F–BPJV	Boeing 727-214	Air Charter International	
F–BPNA	F.27 Friendship Mk. 500	Air Inter	

Notes	*Reg.*	*Type*	*Owner or Operator*
	F–BPNB	F.27 Friendship Mk. 500	Air Inter
	F–BPNC	F.27 Friendship Mk. 500	Air Inter
	F–BPND	F.27 Friendship Mk. 500	Air Inter
	F–BPNE	F.27 Friendship Mk. 500	Air Inter
	F–BPNG	F.27 Friendship Mk. 500	Air Inter
	F–BPNH	F.27 Friendship Mk. 500	Air Inter
	F–BPNI	F.27 Friendship Mk. 500	Air Inter
	F–BPNJ	F.27 Friendship Mk. 500	Air Inter
	F–BPPA	Aero Spacelines Guppy-201	Aeromaritime
	F–BPUA	F.27 Friendship Mk. 500	Air France
	F–BPUB	F.27 Friendship Mk. 500	Air France
	F–BPUC	F.27 Friendship Mk. 500	Air France
	F–BPUD	F.27 Friendship Mk. 500	Air France
	F–BPUE	F.27 Friendship Mk. 500	Air France
	F–BPUF	F.27 Friendship Mk. 500	Air France
	F–BPUG	F.27 Friendship Mk. 500	Air France
	F–BPUH	F.27 Friendship Mk. 500	Air France
	F–BPUI	F.27 Friendship Mk. 500	Air France
	F–BPUJ	F.27 Friendship Mk. 500	Air France
	F–BPUK	F.27 Friendship Mk. 500	Air France
	F–BPUL	F.27 Friendship Mk. 500	Air France
	F–BPVA	Boeing 747-128	Air France
	F–BPVB	Boeing 747-128	Air France
	F–BPVC	Boeing 747-128	Air France
	F–BPVD	Boeing 747-128	Air France
	F–BPVE	Boeing 747-128	Air France
	F–BPVF	Boeing 747-128	Air France
	F–BPVG	Boeing 747-128	Air France
	F–BPVH	Boeing 747-128	Air France
	F–BPVL	Boeing 747-128	Air France
	F–BPVP	Boeing 747-128	Air France
	F–BPVR	Boeing 747-228F	Air France
	F–BPVS	Boeing 747-228B	Air France
	F–BPVT	Boeing 747-228B	Air France
	F–BPVV	Boeing 747-228F	Air France
	F–BPVX	Boeing 747-228B	Air France
	F–BPVY	Boeing 747-228B	Air France
	F–BPVZ	Boeing 747-228F	Air France
	F–BRGU	S.E.210 Caravelle VI-N	Minerve
	F–BRNI	Beech 70 Queen Air	Lucas Air Transport
	F–BSIF	F.27 Friendship Mk. 200	T.A.T.
	F–BSRR	S.E.210 Caravelle III	Air Inter
	F–BSUM	F.27 Friendship Mk. 500	Air France
	F–BSUN	F.27 Friendship Mk. 500	Air France
	F–BSUO	F.27 Friendship Mk. 500	Air France
	F–BTGV	Aero Spacelines Guppy-201	Aeromaritime
	F–BTOA	S.E.210 Caravelle 12	Air Inter
	F–BTOB	S.E.210 Caravelle 12	Air Inter
	F–BTOC	S.E.210 Caravelle 12	Air Inter
	F–BTOD	S.E.210 Caravelle 12	Air Inter
	F–BTOE	S.E.210 Caravelle 12	Air Inter
	F–BTOV	V.952 Vanguard	Europe Aero Service
	F–BTSC	Concorde 101	Air France
	F–BTSD	Concorde 101	Air France
	F–BTTA	Mercure 100	Air Inter
	F–BTTB	Mercure 100	Air Inter
	F–BTTC	Mercure 100	Air Inter
	F–BTTD	Mercure 100	Air Inter
	F–BTTE	Mercure 100	Air Inter
	F–BTTF	Mercure 100	Air Inter
	F–BTTG	Mercure 100	Air Inter
	F–BTTH	Mercure 100	Air Inter
	F–BTTI	Mercure 100	Air Inter
	F–BTTJ	Mercure 100	Air Inter
	F–BUAE	A.300B2 Airbus	Air Inter
	F–BUAF	A.300B2 Airbus	Air Inter
	F–BUAG	A.300B2 Airbus	Air Inter
	F–BUAH	A.300B2 Airbus	Air Inter
	F–BUAI	A.300B2 Airbus	Air Inter
	F–BUAJ	A.300B2 Airbus	Air Inter
	F–BUAK	A.300B2 Airbus	Air Inter
	F–BUFH	S.E.210 Caravelle III	Aerotour
	F–BUTE	F-28 Fellowship 1000	T.A.T.
	F–BUTI	F-28 Fellowship 1000	Air Alpes (Air France)
	F–BUZC	S.E.210 Caravelle VI-R	Minerve

Reg.	Type	Owner or Operator	Notes
F–BVFA	Concorde 101	Air France	
F–BVFB	Concorde 101	Air France	
F–BVFC	Concorde 101	Air France	
F–BVFD	Concorde 101	Air France	
F–BVGA	A.300B2 Airbus	Air France	
F–BVGB	A.300B2 Airbus	Air France	
F–BVGC	A.300B2 Airbus	Air France	
F–BVGD	A.300B2 Airbus	Air France	
F–BVGE	A.300B2 Airbus	Air France	
F–BVGF	A.300B2 Airbus	Air France	
F–BVGG	A.300B4 Airbus	Air France	
F–BVGH	A.300B4 Airbus	Air France	
F–BVGI	A.300B4 Airbus	Air France	
F–BVGJ	A.300B4 Airbus	Air France	
F–BVGK	A.300B4 Airbus	Air France	
F–BVGL	A.300B4 Airbus	Air France	
F–BVGM	A.300B4 Airbus	Air France	
F–BVGN	A.300B4 Airbus	Air France	
F–	A.300B4 Airbus	Air France	
F–	A.300B4 Airbus	Air France	
F–	A.300B4 Airbus	Air France	
F–	A.300B4 Airbus	Air France	
F–	A.300B4 Airbus	Air France	
F–BVOX	Partenavia P.68B	Bretagne Air Services	
F–BVPL	SN-601 Corvette	Air Languedoc	
F–BVPU	S.E.210 Caravelle VI-N	Aerotour	
F–BVPZ	S.E.210 Caravelle VI-N	Aerotour	
F–BVRZ	V.952 Vanguard	Europe Aero Service	
F–BVSF	S.E.210 Caravelle VI-N	Aerotex	
F–BVTA	F.27 Friendship Mk. 200	T.A.T.	
F–BVUY	V.952 Vanguard	Europe Aero Service	
F–BXOO	S.E.210 Caravelle VI-N	Europe Aero Service *Languedoc*	
F–BYAA	F.27 Friendship Mk. 400	Air Alpes	
F–BYAB	F.27 Friendship Mk. 600	Air Alpes	
F–BYAI	S.E.210 Caravelle VI-N	Aerotour	
F–BYAO	F.27 Friendship Mk. 100	Uni-Air	
F–BYAP	F.27 Friendship Mk. 100	Uni-Air	
F–BYAR	F.27 Friendship Mk. 600	Air Alpes	
F–BYAT	S.E.210 Caravelle VI-N	Aerotour	
F–BYCA	S.E.210 Caravelle VI-N	Europe Aero Service *Alsace*	
F–BYCD	S.E.210 Caravelle VI-N	Europe Aero Service	
F–BYCN	Boeing 707-321C (Cargo)	Air France	
F–BYCO	Boeing 707-321C (Cargo)	Air France	
F–BYCP	Boeing 707-321C (Cargo)	Air France	
F–BYCU	Douglas DC-3	Bretagne Air Services	
F–BYCX	Douglas DC-3	Trans Europe-Air	
F–BYCY	S.E.210 Caravelle VI-N	Aerotour	
F–GAPA	S.E.210 Caravelle VI-R	Minerve	
F–GATP	S.E.210 Caravelle 10B	Minerve	
F–GATZ	S.E.210 Caravelle VI-N	Minerve	
F–GBBR	F.28 Fellowship 1000	T.A.T.	
F–GBBS	F.28 Fellowship 1000	T.A.T.	
F–GBBT	F.28 Fellowship 1000	T.A.T.	
F–GBBX	F.28 Fellowship 1000	Air Alsace	
F–GBDE	F.27 Friendship Mk. 400	Air Alpes	
F–GBEA	A.300B2 Airbus	Air France	
F–GBEB	A.300B2 Airbus	Air France	
F–GBEC	A.300B2 Airbus	Air France	
F–GBGA	EMB-110P2 Bandeirante	Brit. Air	
F–GBGI	F.27 Friendship Mk. 600	Air Alpes	
F–GBLE	EMB-110P2 Bandeirante	Brit. Air	
F–GBLY	Partenavia P.68B	Bretagne Air Services	
F–GBMG	EMB-110P2 Bandeirante	Brit. Air	
F–GBMI	S.E.210 Caravelle VI-N	Europe Aero Service	
F–GBMJ	S.E.210 Caravelle VI-N	Europe Aero Service *Valentinois*	
F–GBMK	S.E.210 Caravelle VI-N	Europe Aero Service *Rousillon*	
F–GBRQ	FH.227B Friendship	T.A.T.	
F–GBRU	FH.227B Friendship	T.A.T.	
F–GBRZ	FH.227B Friendship	T.A.T.	
F–GCBA	Boeing 747-228B	Air France	
F–GCBC	Boeing 747-228B	Air France	
F–GCDA	Boeing 727-228	Air France	
F–GCDB	Boeing 727-228	Air France	
F–GCDC	Boeing 727-228	Air France	
F–GCDD	Boeing 727-228	Air France	

Notes	Reg.	Type	Owner or Operator
	F–GCDE	Boeing 727-228	Air France
	F–GCFC	FH.227B Friendship	T.A.T.
	F–GCFD	FH.227B Friendship	T.A.T.
	F–GCGH	FH.227B Friendship	T.A.T.
	F–GCGQ	Boeing 727-227	Europe Aero Service *Normandie*
	F–GCGR	Boeing 737-2L9	Aerotour
	F–GCGS	Boeing 737-2L9	Aerotour
	F–GCJL	Boeing 737-222	Euralair
	F–GCJO	FH.227B Friendship	T.A.T.
	F–GCJT	S.E.210 Caravelle 10B	Europe Aero Service
	F–GCJV	F.27 Friendship Mk. 400	Air Alpes
	F–GCLA	EMB-110P2 Bandeirante	Lucas Air Transport
	F–GCLL	Boeing 737-222	Euralair
	F–GCLM	FH.227B Friendship	T.A.T.
	F–GCLN	FH.227B Friendship	T.A.T.
	F–GCLO	FH.227B Friendship	T.A.T.
	F–GCLP	FH.227B Friendship	T.A.T.
	F–GCLQ	FH.227B Friendship	T.A.T.
	F–GCMR	F.27 Friendship Mk. 200	Air Alsace
	F–GCPA	F.27 Friendship Mk. 100	Air Alpes
	F–GCPS	FH.227B Friendship	T.A.T.
	F–GCSL	Boeing 737-222	Euralair
	F–	Boeing 727-227	Europe Aero Service
	F–	Boeing 727-227	Europe Aero Service
	F–	Boeing 727-227	Europe Aero Service

NOTE: Air France also operates eight more Boeing 747s which retain the US registrations N1252E, N1289E, N18815, N28366, N28899, N28903, N40116 and N63305. Air Alsace operates two Fellowships G–JCWW and G–WWJC on behalf of Air France.

HA (Hungary)

MALÉV

	HA–LBE	Tupolev Tu-134	Malev
	HA–LBF	Tupolev Tu-134	Malev
	HA–LBG	Tupolev Tu-134	Malev
	HA–LBH	Tupolev Tu-134	Malev
	HA–LBI	Tupolev Tu-134A	Malev
	HA–LBK	Tupolev Tu-134A	Malev
	HA–LCA	Tupolev Tu-154B	Malev
	HA–LCB	Tupolev Tu-154B	Malev
	HA–LCE	Tupolev Tu-154B	Malev
	HA–LCF	Tupolev Tu-154B	Malev
	HA–LCG	Tupolev Tu-154B	Malev
	HA–LCH	Tupolev Tu-154B	Malev
	HA–LCL	Tupolev Tu-154	Malev
	HA–LCM	Tupolev Tu-154B	Malev
	HA–LCN	Tupolev Tu-154B	Malev
	HA–MAA	Ilyushin IL-62M	Malev
	HA–YSA	Tupolev Tu-154B	Malev

HB (Switzerland)

BALAIR

	HB–IBS	Douglas DC-6A	Balair
	HB–ICI	S.E.210 Caravelle 10-R	CTA
	HB–ICN	S.E.210 Caravelle 10-R	CTA Ville de Genève
	HB–ICO	S.E.210 Caravelle 10-R	CTA
	HB–ICQ	S.E.210 Caravelle 10-R	CTA

Reg.	Type	Owner or Operator	Notes
HB–IDF	Douglas DC-8-62	Swissair *Schwyz*	
HB–IDG	Douglas DC-8-62	Swissair *Neuchatel*	
HB–IDH	Douglas DC-8-62CF	Balair	
HB–IDI	Douglas DC-8-62	Swissair *Solothurn*	
HB–IDK	Douglas DC-8-62CF	Swissair *Matterhorn*	
HB–IDL	Douglas DC-8-62	Swissair *Aargau*	
HB–IDO	Douglas DC-9-30	Swissair *Cointrin*	
HB–IDP	Douglas DC-9-30	Swissair *Basel-Land*	
HB–IDR	Douglas DC-9-30	Swissair *Baden*	
HB–IDT	Douglas DC-9-34	Balair	
HB–IDZ	Douglas DC-8-63PF	Balair	
HB–IFG	Douglas DC-9-30	Swissair *Valais*	
HB–IFH	Douglas DC-9-30	Swissair *Opfikon*	
HB–IFI	Douglas DC-9-30	Swissair *Zug*	
HB–IFK	Douglas DC-9-30	Swissair *Kloten*	
HB–IFL	Douglas DC-9-30	Swissair *Appenzell I. Rh.*	
HB–IFN	Douglas DC-9-30	Swissair *Obwalden*	
HB–IFO	Douglas DC-9-30	Swissair *Appenzell A. Rh.*	
HB–IFP	Douglas DC-9-30	Swissair *Glarus*	
HB–IFR	Douglas DC-9-30	Swissair *Ticino*	
HB–IFS	Douglas DC-9-30	Swissair *Winterthur*	
HB–IFT	Douglas DC-9-30	Swissair *Rümlang*	
HB–IFU	Douglas DC-9-30	Swissair *Chur*	
HB–IFV	Douglas DC-9-30	Swissair *Bülach*	
HB–IFW	Douglas DC-9-30F	Swissair *Payerne*	
HB–IFX	Douglas DC-9-30	Swissair *Lausanne*	
HB–IFY	Douglas DC-9-30	Swissair *Bellinzana*	
HB–IFZ	Douglas DC-9-30	Balair	
HB–IGA	Boeing 747-257B	Swissair *Geneve*	
HB–IGB	Boeing 747-257B	Swissair *Zurich*	
HB–IHA	Douglas DC-10-30	Swissair *St. Gallen*	
HB–IHB	Douglas DC-10-30	Swissair *Schaffhausen*	
HB–IHC	Douglas DC-10-30	Swissair *Luzern*	
HB–IHD	Douglas DC-10-30	Swissair *Bern*	
HB–IHE	Douglas DC-10-30	Swissair *Vaud*	
HB–IHF	Douglas DC-10-30	Swissair *Nidwalden*	
HB–IHG	Douglas DC-10-30	Swissair *Grisons*	
HB–IHH	Douglas DC-10-30	Swissair *Basel-Stadt*	
HB–IHI	Douglas DC-10-30	Swissair *Fribourg*	
HB–IHK	Douglas DC-10-30	Balair	
HB–IHL	Douglas DC-10-30	Swissair	
HB–IHM	Douglas DC-10-30	Swissair	
HB–IKB	Douglas DC-9-32	Alisarda (Italy)	
HB–IKC	Douglas DC-9-32	Alisarda (Italy)	
HB–INA	Douglas DC-9-81	Swissair	
HB–INB	Douglas DC-9-81	Swissair	
HB–INC	Douglas DC-9-81	Swissair	
HB–IND	Douglas DC-9-81	Swissair	
HB–INE	Douglas DC-9-81	Swissair	
HB–INF	Douglas DC-9-81	Swissair	
HB–ING	Douglas DC-9-81	Swissair	
HB–INH	Douglas DC-9-81	Swissair	
HB–INI	Douglas DC-9-81	Swissair	
HB–INK	Douglas DC-9-81	Swissair	
HB–INL	Douglas DC-9-81	Swissair	
HB–INM	Douglas DC-9-81	Swissair	
HB–INN	Douglas DC-9-81	Swissair	
HB–INO	Douglas DC-9-81	Swissair	
HB–INP	Douglas DC-9-81	Swissair	
HB–ISK	Douglas DC-9-51	Swissair *Höri*	
HB–ISL	Douglas DC-9-51	Swissair *Köniz*	
HB–ISM	Douglas DC-9-51	Swissair *Wettingen*	
HB–ISN	Douglas DC-9-51	Swissair *Sion*	
HB–ISO	Douglas DC-9-51	Swissair *Bienne*	
HB–ISP	Douglas DC-9-51	Swissair *Lugano*	
HB–ISR	Douglas DC-9-51	Swissair *Locarno*	
HB–ISS	Douglas DC-9-51	Swissair *Dietikon*	
HB–IST	Douglas DC-9-51	Swissair *Aarau*	
HB–ISU	Douglas DC-9-51	Swissair *Bachenbülach*	
HB–ISV	Douglas DC-9-51	Swissair *Winkel*	
HB–ISW	Douglas DC-9-51	Swissair *Dubendorf*	

NOTE: Swissair is in the process of replacing its DC-9-30s with Series 81s.

HK (Colombia)

Notes	Reg.	Type	Owner or Operator
	HK-1402	Boeing 707-359B	Avianca *Sucre*
	HK-1410	Boeing 707-359B	Avianca *Bolivar*
	HK-1718X	Boeing 707-321C	Avianca
	HK-1849	Boeing 707-321C	Avianca
	HK-2000X	Boeing 747-124	Avianca *Eldorado*
	HK-2015	Boeing 707-321B	Avianca
	HK-2016	Boeing 707-321B	Avianca
	HK-2070X	Boeing 707-321B	Avianca
	HK-2300	Boeing 747-259B	Avianca

HS (Thailand)

Notes	Reg.	Type	Owner or Operator
	HS-TGA	Boeing 747-2D7B	Thai Airways International *Visuthakasatriya*
	HS-TGB	Boeing 747-2D7B	Thai Airways International *Sirisobhakya*
	HS-TGC	Boeing 747-2D7B	Thai Airways International *Dararasmi*
	HS-TGD	Douglas DC-10-30	Thai Airways International *Phimara*
	HS-TGE	Douglas DC-10-30	Thai Airways International *Sriwanna*
	HS-TGF	Boeing 747-2D7B	Thai Airways International
	HS-TGG	Boeing 747-2D7B	Thai Airways International

HZ (Saudi Arabia)

Notes	Reg.	Type	Owner or Operator
	HZ-AHA	L.1011-385 TriStar 200	SAUDIA—Saudi Arabian Airlines
	HZ-AHB	L.1011-385 TriStar 200	SAUDIA—Saudi Arabian Airlines
	HZ-AHC	L.1011-385 TriStar 200	SAUDIA—Saudi Arabian Airlines
	HZ-AHD	L.1011-385 TriStar 200	SAUDIA—Saudi Arabian Airlines
	HZ-AHE	L.1011-385 TriStar 200	SAUDIA—Saudi Arabian Airlines
	HZ-AHF	L.1011-385 TriStar 200	SAUDIA—Saudi Arabian Airlines
	HZ-AHG	L.1011-385 TriStar 200	SAUDIA—Saudi Arabian Airlines
	HZ-AHH	L.1011-385 TriStar 200	SAUDIA—Saudi Arabian Airlines
	HZ-AHI	L.1011-385 TriStar 200	SAUDIA—Saudi Arabian Airlines
	HZ-AHJ	L.1011-385 TriStar 200	SAUDIA—Saudi Arabian Airlines
	HZ-AHL	L.1011-385 TriStar 200	SAUDIA—Saudi Arabian Airlines
	HZ-AHM	L.1011-385 TriStar 200	SAUDIA—Saudi Arabian Airlines
	HZ-AHN	L.1011-385 TriStar 200	SAUDIA—Saudi Arabian Airlines
	HZ-AHO	L.1011-385 TriStar 200	SAUDIA—Saudi Arabian Airlines
	HZ-AHP	L.1011-385 TriStar 200	SAUDIA—Saudi Arabian Airlines
	HZ-AHR	L.1011-385 TriStar 200	SAUDIA—Saudi Arabian Airlines

NOTE: SAUDIA also operate on lease DC-8s, N800EV, N865F, N868F, N911CL, N8632 and TF-LLC.

I (Italy)

Alitalia

Reg.	Type	Owner or Operator	Notes
I–ATIA	Douglas DC-9-32	Aero Trasporti Italiani (ATI)	
I–ATIE	Douglas DC-9-32	Aero Trasporti Italiani (ATI)	
I–ATIH	Douglas DC-9-32	Aero Trasporti Italiani (ATI)	
I–ATIJ	Douglas DC-9-32	Aero Trasporti Italiani (ATI)	
I–ATIK	Douglas DC-9-32	Aero Trasporti Italiani (ATI)	
I–ATIO	Douglas DC-9-32	Aero Trasporti Italiani (ATI)	
I–ATIQ	Douglas DC-9-32	Aero Trasporti Italiani (ATI)	
I–ATIU	Douglas DC-9-32	Aero Trasporti Italiani (ATI)	
I–ATIW	Douglas DC-9-32	Aero Trasporti Italiani (ATI)	
I–ATIX	Douglas DC-9-32	Aero Trasporti Italiani (ATI)	
I–ATIY	Douglas DC-9-32	Aero Trasporti Italiani (ATI)	
I–ATJA	Douglas DC-9-32	Aero Trasporti Italiani (ATI)	
I–ATJB	Douglas DC-9-32	Aero Trasporti Italiani (ATI)	
I–BUSB	A.300B4 Airbus	Alitalia *Tiziano*	
I–BUSC	A.300B4 Airbus	Alitalia	
I–BUSD	A.300B4 Airbus	Alitalia *Caravaggio*	
I–BUSF	A.300B4 Airbus	Alitalia	
I–	A.300B4 Airbus	Alitalia	
I–	A.300B4 Airbus	Alitalia	
I–	A.300B4 Airbus	Alitalia	
I–	A.300B4 Airbus	Alitalia	
I–DEMA	Boeing 747-143A	Alitalia *Neil Alden Armstrong*	
I–DEMB	Boeing 747-243B	Alitalia *Carlo del Prete*	
I–DEMC	Boeing 747-243B	Alitalia	
I–DEMD	Boeing 747-243B	Alitalia	
I–DEME	Boeing 747-143A	Alitalia *Arturo Ferrarin*	
I–DEMF	Boeing 747-243B	Alitalia	
I–	Boeing 747-243B	Alitalia	
I–	Boeing 747-243B	Alitalia	
I–	Boeing 747-243B	Alitalia	
I–	Boeing 747-243B	Alitalia	
I–DEMO	Boeing 747-243B	Alitalia *Francesco de Pinedo*	
I–DEMU	Boeing 747-243B	Alitalia *Geo Chavez*	
I–DIBC	Douglas DC-9-32	Alitalia *Isola di Lampedusa*	
I–DIBD	Douglas DC-9-32	Alitalia *Isola di Montecristo*	
I–DIBJ	Douglas DC-9-32	Alitalia *Isola della Capraia*	
I–DIBK	Douglas DC-9-32F	Alitalia *Ercole*	
I–DIBN	Douglas DC-9-32	Alitalia *Isola della Palmaria*	
I–DIBO	Douglas DC-9-32	Alitalia *Isola di Procida*	
I–DIBQ	Douglas DC-9-32	Alitalia *Isola di Pianosa*	
I–DIKA	Douglas DC-9-32	Alitalia *Isola di Capri*	
I–DIKC	Douglas DC-9-32	Alitalia *Isola di Ponza*	
I–DIKD	Douglas DC-9-32	Alitalia *Isola del Giglio*	
I–DIKE	Douglas DC-9-32	Alitalia *Isola d'Elba*	
I–DIKF	Douglas DC-9-32F	Alitalia *Atlante*	
I–DIKI	Douglas DC-9-32	Alitalia *Isola di Murano*	
I–DIKJ	Douglas DC-9-32	Alitalia *Isola di Lipari*	
I–DIKL	Douglas DC-9-32	Alitalia *Isola di Panarea*	
I–DIKM	Douglas DC-9-32	Alitalia *Isola di Tavolara*	
I–DIKN	Douglas DC-9-32	Alitalia *Isola di Nisida*	
I–DIKO	Douglas DC-9-32	Alitalia *Isola di Pantelleria*	
I–DIKP	Douglas DC-9-32	Aero Trasporti Italiani (ATI)	
I–DIKR	Douglas DC-9-32	Alitalia *Isola di Torcello*	
I–DIKS	Douglas DC-9-32	Alitalia *Isola di Filicudi*	
I–DIKT	Douglas DC-9-32	Alitalia *Isola di Ustica*	
I–DIKU	Douglas DC-9-32	Alitalia *Isola d'Ischia*	
I–DIKV	Douglas DC-9-32	Alitalia *Isola di Vulcano*	
I–DIKW	Douglas DC-9-32	Alitalia *Isola di Giannutri*	
I–DIKY	Douglas DC-9-32	Aero Trasporti Italiani (ATI)	
I–DIKZ	Douglas DC-9-32	Alitalia *Isola di Linosa*	
I–DIRA	Boeing 727-243	Alitalia *Citta di Gubbio*	
I–DIRB	Boeing 727-243	Alitalia *Citta di Siracusa*	
I–DIRC	Boeing 727-243	Alitalia *Citta di Aosta*	
I–DIRD	Boeing 727-243	Alitalia *Citta di Bergamo*	
I–DIRF	Boeing 727-243	Alitalia *Citta di Lecce*	

Notes	Reg.	Type	Owner or Operator
	I–DIRG	Boeing 727-243	Alitalia *Citta di Urbino*
	I–DIRI	Boeing 727-243	Alitalia *Citta di Siena*
	I–DIRJ	Boeing 727-243	Alitalia *Citta di Verona*
	I–DIRL	Boeing 727-243	Alitalia *Citta di Viterbo*
	I–DIRM	Boeing 727-243	Alitalia *Citta di Genova*
	I–DIRN	Boeing 727-243	Alitalia *Citta di Aquilea*
	I–DIRO	Boeing 727-243	Alitalia *Citta di Amalfi*
	I–DIRP	Boeing 727-243	Alitalia *Citta di Ivrea*
	I–DIRQ	Boeing 727-243	Alitalia *Citta di Sassari*
	I–DIRR	Boeing 727-243	Alitalia *Citta di Trento*
	I–DIRS	Boeing 727-243	Alitalia *Citta di Sulmona*
	I–DIRU	Boeing 727-243	Alitalia *Citta di Ravenna*
	I–DIWC	Douglas DC-8-62F	Alitalia *Titano*
	I–DIWH	Douglas DC-8-62	Alitalia *Giovanni Pierluigi da Palestrina*
	I–DIWJ	Douglas DC-8-62	Alitalia *Antonio Vivaldi*
	I–DIWQ	Douglas DC-8-62F	Alitalia *Ciclope*
	I–DIWV	Douglas DC-8-62	Alitalia *Gioacchino Rossini*
	I–DIWW	Douglas DC-8-62	Alitalia *Arcangelo Corelli*
	I–DIWX	Douglas DC-8-62	Alitalia *Luigi Cherubini*
	I–DIWY	Douglas DC-8-62	Alitalia *Vincenzo Bellini*
	I–DIZA	Douglas DC-9-32	Alitalia *Isola di Palmarola*
	I–DIZB	Douglas DC-9-32	Aero Trasporti Italiani (ATI)
	I–DIZC	Douglas DC-9-32	Aero Trasporti Italiani (ATI)
	I–DIZE	Douglas DC-9-32	Aero Trasporti Italiani (ATI)
	I–DIZF	Douglas DC-9-32	Alitalia *Isola di Spargi*
	I–DIZI	Douglas DC-9-32	Aero Trasporti Italiani (ATI)
	I–DIZO	Douglas DC-9-32	Alitalia *Isola di Tino*
	I–DIZU	Douglas DC-9-32	Aero Trasporti Italiani (ATI)
	I–DYNA	Douglas DC-10-30	Alitalia *Galileo Galilei*
	I–DYNB	Douglas DC-10-30	Alitalia *Giotto di Bondone*
	I–DYNC	Douglas DC-10-30	Alitalia *Luigi Pirandello*
	I–DYND	Douglas DC-10-30	Alitalia *Enrico Fermi*
	I–DYNE	Douglas DC-10-30	Alitalia *Dante Alghieri*
	I–DYNI	Douglas DC-10-30	Alitalia *Michelangelo Buonarrotti*
	I–DYNO	Douglas DC-10-30	Alitalia *Benvenuto Cellini*
	I–DYNU	Douglas DC-10-30	Alitalia *Guglielmo Marconi*
	I–SARJ	Douglas DC-9-14	Alisarda
	I–SARV	Douglas DC-9-14	Alisarda
	I–TIDB	F.28 Fellowship 1000	Itavia
	I–TIDI	F.28 Fellowship 1000	Itavia
	I–TIDU	F.28 Fellowship 1000	Itavia
	I–TIGA	Douglas DC-9-11	Itavia
	I–TIGB	Douglas DC-9-15	Itavia
	I–TIGE	Douglas DC-9-15	Itavia
		Douglas DC-9-15	Itavia

NOTE: Itavia also operate three DC-9s which retain US registrations N934F, N7465B and N94454. Alisarda uses two DC-9s, which retain their Swiss registrations HB–IKB and HB–IKC.

JA (Japan)

Notes	Reg.	Type	Owner or Operator
	JA8031	Douglas DC-8-62	Japan Air Lines
	JA8032	Douglas DC-8-62	Japan Air Lines
	JA8033	Douglas DC-8-62	Japan Air Lines
	JA8034	Douglas DC-8-62	Japan Air Lines
	JA8035	Douglas DC-8-62	Japan Air Lines
	JA8036	Douglas DC-8-62AF	Japan Air Lines
	JA8037	Douglas DC-8-62	Japan Air Lines
	JA8044	Douglas DC-8-62AF	Japan Air Lines
	JA8052	Douglas DC-8-62	Japan Air Lines

Reg.	Type	Owner or Operator	Notes
JA8053	Douglas DC-8-62	Japan Air Lines	
JA8055	Douglas DC-8-62AF	Japan Air Lines	
JA8056	Douglas DC-8-62AF	Japan Air Lines	
JA8101	Boeing 747-146	Japan Air Lines	
JA8102	Boeing 747-146	Japan Air Lines	
JA8103	Boeing 747-146	Japan Air Lines	
JA8104	Boeing 747-246B	Japan Air Lines	
JA8105	Boeing 747-246B	Japan Air Lines	
JA8106	Boeing 747-246B	Japan Air Lines	
JA8107	Boeing 747-146A	Japan Air Lines	
JA8108	Boeing 747-246B	Japan Air Lines	
JA8110	Boeing 747-246B	Japan Air Lines	
JA8111	Boeing 747-246B	Japan Air Lines	
JA8112	Boeing 747-146A	Japan Air Lines	
JA8113	Boeing 747-246B	Japan Air Lines	
JA8114	Boeing 747-246B	Japan Air Lines	
JA8115	Boeing 747-146A	Japan Air Lines	
JA8116	Boeing 747-146A	Japan Air Lines	
JA8122	Boeing 747-246B	Japan Air Lines	
JA8123	Boeing 747-246F	Japan Air Lines	
JA8125	Boeing 747-246B	Japan Air Lines	
JA8127	Boeing 747-246B	Japan Air Lines	
JA8128	Boeing 747-146A	Japan Air Lines	
JA8129	Boeing 747-246B	Japan Air Lines	
JA8130	Boeing 747-246B	Japan Air Lines	
JA8131	Boeing 747-246B	Japan Air Lines	
JA8132	Boeing 747-246F	Japan Air Lines	
JA8140	Boeing 747-246B	Japan Air Lines	
JA8141	Boeing 747-246B	Japan Air Lines	
JA8142	Boeing 747-146B	Japan Air Lines	
JA8143	Boeing 747-146B	Japan Air Lines	
JA8144	Boeing 747-246F	Japan Air Lines	
JA8149	Boeing 747-246B	Japan Air Lines	
JA8150	Boeing 747-246B	Japan Air Lines	
JA8151	Boeing 747-246F	Japan Air Lines	

JY (Jordan)

Reg.	Type	Owner or Operator	Notes
JY–ADP	Boeing 707-383C	Alia—The Royal Jordanian Airline *City of Amman*	
JY–ADT	Boeing 720-030B	Alia—The Royal Jordanian Airline *City of Madaba*	
JY–AEB	Boeing 707-384C	Alia—The Royal Jordanian Airline *City of Jerash*	
JY–AEC	Boeing 707-384C	Alia—The Royal Jordanian Airline *City of Umquais*	
JY–AED	Boeing 707-321C (Cargo)	Alia—The Royal Jordanian Airline *City of Al-Karameh*	
JY–AES	Boeing 707-321C	Alia—The Royal Jordanian Airline	
JY–AFA	Boeing 747-2D3B	Alia—The Royal Jordanian Airline *Prince Ali*	
JY–AFB	Boeing 747-2D3B	Alia—The Royal Jordanian Airline *Princess Haya*	
JY–AFQ	Boeing 707-344C	Alia—The Royal Jordanian Airline	
JY–AFR	Boeing 707-344C	Alia—The Royal Jordanian Airline	
JY–AGA	L.1011-385 TriStar 500	Alia—The Royal Jordanian Airline	
JY–AGB	L.1011-385 TriStar 500	Alia—The Royal Jordanian Airline	
JY–AGC	L.1011-385 TriStar 500	Alia—The Royal Jordanian Airline	
JY–	Boeing 747-2D3B	Alia—The Royal Jordanian Airline	

LN (Norway)

Notes	*Reg.*	*Type*	*Owner or Operator*
	LN–BWG	Convair 580	Nor-Fly
	LN–FOG	L-188AF Electra	Fred Olsen Airtransport
	LN–FOH	L-188AF Electra	Fred Olsen Airtransport
	LN–FOI	L-188CF Electra	Fred Olsen Airtransport
	LN–KLK	Convair 440	Nor-Fly
	LN–MAM	Convair 440	Nor-Fly
	LN–MAP	Convair 440	Nor-Fly
	LN–MOF	Douglas DC-8-63	Scanair *Bue Viking*
	LN–MOG	Douglas DC-8-62	SAS *Odd Viking*
	LN–MOW	Douglas DC-8-62	SAS *Roald Viking*
	LN–NPB	Boeing 737-2R4C	Air Executive
	LN–NPH	F.27 Friendship MK. 300	Air Executive
	LN–NPI	F.27 Friendship Mk. 100	Air Executive
	LN–RCA	A.300B2 Airbus	SAS *Snorry Viking*
	LN–RKA	Douglas DC-10-30	SAS *Olav Viking*
	LN–RKB	Douglas DC-10-30	SAS *Haakon Viking*
	LN–RLA	Douglas DC-9-41	SAS *Are Viking*
	LN–RLB	Douglas DC-9-41	SAS *Arne Viking*
	LN–RLC	Douglas DC-9-41	SAS *Gunnar Viking*
	LN–RLD	Douglas DC-9-41	SAS *Torleif Viking*
	LN–RLH	Douglas DC-9-41	SAS *Einar Viking*
	LN–RLJ	Douglas DC-9-41	SAS *Stein Viking*
	LN–RLK	Douglas DC-9-41	SAS *Erling Viking*
	LN–RLL	Douglas DC-9-21	SAS *Guttorm Viking*
	LN–RLN	Douglas DC-9-41	SAS *Halldor Viking*
	LN–RLO	Douglas DC-9-21	SAS *Gunder Viking*
	LN–RLP	Douglas DC-9-41	SAS *Froste Viking*
	LN–RLS	Douglas DC-9-41	SAS *Asmund Viking*
	LN–RLT	Douglas DC-9-41	SAS *Audun Viking*
	LN–RLU	Douglas DC-9-41	SAS *Eivind Viking*
	LN–RLW	Douglas DC-9-33AF	SAS *Rand Viking*
	LN–RLX	Douglas DC-9-41	SAS *Sote Viking*
	LN–RLZ	Douglas DC-9-41	SAS *Bodvar Viking*
	LN–RNA	Boeing 747-283B	SAS *Magnus Viking*
	LN–RNB	Boeing 747-283B	SAS
	LN–SUA	Boeing 737-205C	Braathens SAFE *Halvdan Svarte*
	LN–SUB	Boeing 737-205	Braathens SAFE *Magnus Den Gode*
	LN–SUC	F.28 Fellowship 1000	Braathens SAFE *Olav Kyrre*
	LN–SUD	Boeing 737-205	Braathens SAFE *Olav Tryggvason*
	LN–SUE	F.27 Friendship Mk. 100	Air Executive
	LN–SUF	F.27 Friendship Mk. 100	Air Executive
	LN–SUG	Boeing 737-205	Braathens SAFE *Harald Harfagre*
	LN–SUH	Boeing 737-205	Braathens SAFE *Sigurd Jorsalfar*
	LN–SUI	Boeing 737-205	Braathens SAFE *Haakon den Gode*
	LN–SUK	Boeing 737-205	Braathens SAFE *Olav Haraldsson*
	LN–SUL	F.27 Friendship Mk. 100	Air Executive
	LN–SUM	Boeing 737-205	Braathens SAFE *Magnus Lagabøter*
	LN–SUN	F.28 Fellowship 1000	Braathens SAFE *Haakon Sverresson*
	LN–SUO	F.28 Fellowship 1000	Braathens SAFE *Magnus Barfot*
	LN–SUP	Boeing 737-205	Braathens SAFE *Haakon IV*
	LN–SUS	Boeing 737-205	Braathens SAFE *Haakon V*
	LN–SUT	Boeing 737-205	Braathens SAFE *Oystein Magnusson*
	LN–SUX	F.28 Fellowship 1000	Braathens SAFE *Harald Hardrade*

LV (Argentina)

Reg.	Type	Owner or Operator	Notes
LV–ISA	Boeing 707-387B	Aerolineas Argentinas *Antartida Argentina*	
LV–ISB	Boeing 707-387B	Aerolineas Argentinas *Almilan*	
LV–ISC	Boeing 707-387B	Aerolineas Argentinas *Betelgeuse*	
LV–ISD	Boeing 707-387B	Aerolineas Argentinas *Procion*	
LV–JGP	Boeing 707-387C	Aerolineas Argentinas *Achernar*	
LV–JGR	Boeing 707-387C	Aerolineas Argentinas *Canopus*	
LV–LGO	Boeing 707-372C	Aerolineas Argentinas	
LV–LGP	Boeing 707-372C	Aerolineas Argentinas	
LV–LZD	Boeing 747-287B	Aerolineas Argentinas	
LV–MLO	Boeing 747-287B	Aerolineas Argentinas	
LV–MLP	Boeing 747-287B	Aerolineas Argentinas	
LV–MLR	Boeing 747-287B	Aerolineas Argentinas	
LV–	Boeing 747-287B	Aerolineas Argentinas	
LV–MSG	Boeing 707-321C	Transporte Aereo Rioplatense	
LV–MZE	Boeing 707-338C	Transporte Aereo Rioplatense	

LX (Luxembourg)

Reg.	Type	Owner or Operator	Notes
LX–ACV	Douglas DC-8-63CF	Air India Cargo	
LX–DCV	Boeing 747-2R7F	Cargolux *City of Luxembourg*	
LX–ECV	Boeing 747-2R7F	Cargolux	
LX–LGA	F.27 Friendship Mk. 100	Luxair *Prince Henri*	
LX–LGB	F.27 Friendship Mk. 100	Luxair *Prince Jean*	
LX–LGD	F.27 Freindship Mk. 400	Luxair *Princesse Margaretha*	
LX–LGH	Boeing 737-2C9	Luxair *Prince Guillaume*	
LX–LGI	Boeing 737-2C9	Luxair *Princesse Marie-Astrid*	

LZ (Bulgaria)

Reg.	Type	Owner or Operator	Notes
LZ–BTA	Tupolev Tu-154	Balkan Bulgarian Airlines	
LZ–BTC	Tupolev Tu-154	Balkan Bulgarian Airlines	
LZ–BTD	Tupolev Tu-154B	Balkan Bulgarian Airlines	
LZ–BTE	Tupolev Tu-154B	Balkan Bulgarian Airlines	
LZ–BTF	Tupolev Tu-154A	Balkan Bulgarian Airlines	
LZ–BTG	Tupolev Tu-154A	Balkan Bulgarian Airlines	
LZ–BTJ	Tupolev Tu-154B	Balkan Bulgarian Airlines	
LZ–BTK	Tupolev Tu-154B	Balkan Bulgarian Airlines	
LZ–BTL	Tupolev Tu-154B	Balkan Bulgarian Airlines	
LZ–BTM	Tupolev Tu-154B	Balkan Bulgarian Airlines	
LZ–BTO	Tupolev Tu-154B	Balkan Bulgarian Airlines	
LZ–BTP	Tupolev Tu-154B	Balkan Bulgarian Airlines	
LZ–BTR	Tupolev Tu-154B	Balkan Bulgarian Airlines	
LZ–BTS	Tupolev Tu-154B	Balkan Bulgarian Airlines	
LZ–TUA	Tupolev Tu-134	Balkan Bulgarian Airlines	
LZ–TUC	Tupolev Tu-134	Balkan Bulgarian Airlines	
LZ–TUD	Tupolev Tu-134	Balkan Bulgarian Airlines	
LZ–TUE	Tupolev Tu-134	Balkan Bulgarian Airlines	
LZ–TUF	Tupolev Tu-134	Balkan Bulgarian Airlines	
LZ–TUK	Tupolev Tu-134A	Balkan Bulgarian Airlines	
LZ–TUL	Tupolev Tu-134A	Balkan Bulgarian Airlines	

Notes	Reg.	Type	Owner or Operator
	LZ–TUM	Tupolev Tu-134A	Balkan Bulgarian Airlines
	LZ–TUN	Tupolev Tu-134A	Balkan Bulgarian Airlines
	LZ–TUO	Tupolev Tu-134	Balkan Bulgarian Airlines
	LZ–TUP	Tupolev Tu-134A	Balkan Bulgarian Airlines
	LZ–TUR	Tupolev Tu-134A	Balkan Bulgarian Airlines
	LZ–TUS	Tupolev Tu-134A	Balkan Bulgarian Airlines
	LZ–TUZ	Tupolev Tu-134A	Balkan Bulgarian Airlines

N (USA)

BI BRANIFF INTERNATIONAL

Notes	Reg.	Type	Owner or Operator
	N10ST	L-100-30 Hercules	Transamerica Airlines
	N11ST	L-100-30 Hercules	Transamerica Airlines
	N12ST	L-100-30 Hercules	Transamerica Airlines
	N15ST	L-100-30 Hercules	Transamerica Airlines
	N16ST	L-100-30 Hercules	Transamerica Airlines
	N17ST	L-100-30 Hercules	Transamerica Airlines
	N18ST	L-100-30 Hercules	Transamerica Airlines
	N19ST	L-100-30 Hercules	Transamerica Airlines
	N20ST	L-100-30 Hercules	Transamerica Airlines
	N21ST	L-100-30 Hercules	Transamerica Airlines
	N23ST	L-100-30 Hercules	Transamerica Airlines
	N24ST	L-100-30 Hercules	Transamerica Airlines
	N101AK	L-100-30 Hercules	Alaska International Air
	N101TV	Douglas DC-10-30CF	Transamerica Airlines
	N102TV	Douglas DC-10-30CF	Transamerica Airlines
	N103TV	Douglas DC-10-30CF	Transamerica Airlines
	N103WA	Douglas DC-10-30CF	World Airways
	N104AK	L-100-30 Hercules	Alaska International Air
	N104WA	Douglas DC-10-30CF	World Airways
	N105WA	Douglas DC-10-30CF	World Airways
	N106AK	L-100-30 Hercules	Alaska International Air
	N106WA	Douglas DC-10-30CF	World Airways
	N107AK	L-100-30 Hercules	Alaska International Air

Reg.	Type	Owner or Operator	Notes
N107WA	Douglas DC-10-30CF	World Airways	
N108AK	L-100-30 Hercules	Alaska International Air	
N108WA	Douglas DC-10-30CF	World Airways	
N109WA	Douglas DC-10-30CF	World Airways	
N112WA	Douglas DC-10-30CF	World Airways	
N113WA	Douglas DC-10-30CF	World Airways	
N114WA	Douglas DC-10-30CF	World Airways	
N133TW	Boeing 747-146	Trans World Airlines	
N319PA	Boeing 727-21	Pan Am *Clipper Ponce de Leon*	
N323PA	Boeing 727-21	Pan Am *Clipper Star of Peace*	
N325PA	Boeing 727-21	Pan Am *Clipper Syren*	
N327PA	Boeing 727-21	Pan Am *Clipper Meteor*	
N329PA	Boeing 727-21	Pan Am *Clipper ICC Berlin*	
N339PA	Boeing 727-21C	Pan Am *Clipper Dawn*	
N340PA	Boeing 727-21C	Pan Am *Clipper Talisman*	
N355PA	Boeing 727-21	Pan Am *Clipper Andrew Jackson*	
N356PA	Boeing 727-21	Pan Am *Clipper Belle of the Sky*	
N357PA	Boeing 727-21	Pan Am *Clipper Betsy Ross*	
N358PA	Boeing 727-21	Pan Am *Clipper David Crockett*	
N359PA	Boeing 727-21	Pan Am *Clipper Orpheus*	
N360PA	Boeing 727-21	Pan Am *Clipper Golden Rule*	
N361PA	Boeing 727-51	Pan Am	
N362PA	Boeing 727-51	Pan Am *Clipper Frankfurt*	
N363PA	Boeing 727-121	Pan Am	
N364PA	Boeing 727-121	Pan Am	
N365PA	Boeing 727-121	Pan Am	
N366PA	Boeing 727-121	Pan Am	
N447T	Conroy CL-44-O	*Skymonster*	
N496PA	Boeing 707-321B	Pan Am *Clipper Northern Eagle*	
N501PA	L-1011-385 TriStar 500	Pan Am *Clipper Eagle*	
N503PA	L-1011-385 TriStar 500	Pan Am *Clipper Flying Eagle*	
N504PA	L-1011-385 TriStar 500	Pan Am *Clipper National Eagle*	
N505PA	L-1011-385 TriStar 500	Pan Am *Clipper Eagle Wing*	
N507PA	L-1011-385 TriStar 500	Pan Am *Clipper Northern Eagle*	
N508PA	L-1011-385 TriStar 500	Pan Am *Clipper Bald Eagle*	
N509PA	L-1011-385 TriStar 500	Pan Am	
N510PA	L-1011-385 TriStar 500	Pan Am	
N511PA	L-1011-385 TriStar 500	Pan Am	
N512PA	L-1011-385 TriStar 500	Pan Am	
N513PA	L-1011-385 TriStar 500	Pan Am	
N514PA	L-1011-385 TriStar 500	Pan Am	
N530PA	Boeing 747SP-21	Pan Am *Clipper Mayflower*	
N531PA	Boeing 747SP-21	Pan Am *Clipper Freedom*	
N532PA	Boeing 747SP-21	Pan Am *Clipper Constitution*	
N533PA	Boeing 747SP-21	Pan Am *Clipper New Horizons*	
N534PA	Boeing 747SP-21	Pan Am *Clipper Great Republic*	
N535PA	Boeing 747-273C	Pan Am *Clipper Mercury*	
N536PA	Boeing 747SP-21	Pan Am *Clipper Lindbergh*	
N537PA	Boeing 747SP-21	Pan Am *Clipper High Flyer*	
N538PA	Boeing 747SP-21	Pan Am *Clipper Fleetwing*	
N539PA	Boeing 747SP-21	Pan Am *Clipper Liberty Bell*	
N540PA	Boeing 747SP-21	Pan Am *Clipper White Falcon*	
N601BN	Boeing 747-127	Braniff International	
N601US	Boeing 747-151	Northwest Orient	
N602BN	Boeing 747-227B	Braniff International	
N602US	Boeing 747-151	Northwest Orient	
N603US	Boeing 747-151	Northwest Orient	
N604US	Boeing 747-151	Northwest Orient	
N605BN	Boeing 747-227B	Braniff International	
N605US	Boeing 747-151	Northwest Orient	
N606US	Boeing 747-151	Northwest Orient	
N607US	Boeing 747-151	Northwest Orient	
N608US	Boeing 747-151	Northwest Orient	
N609BN	Boeing 747-227B	Braniff International	
N609US	Boeing 747-151	Northwest Orient	
N610BN	Boeing 747-130	Braniff International	
N610US	Boeing 747-151	Northwest Orient	
N611BN	Boeing 747-230B	Braniff International	
N611US	Boeing 747-251B	Northwest Orient	
N612BN	Boeing 747-227B	Braniff International	
N612US	Boeing 747-251B	Northwest Orient	
N613US	Boeing 747-251B	Northwest Orient	
N614US	Boeing 747-251B	Northwest Orient	
N615US	Boeing 747-251B	Northwest Orient	
N616US	Boeing 747-251F	Northwest Orient	

Notes	Reg.	Type	Owner or Operator
	N617US	Boeing 747-251F	Northwest Orient
	N618US	Boeing 747-251F	Northwest Orient
	N619US	Boeing 747-251F	Northwest Orient
	N620US	Boeing 747-135	Northwest Orient
	N621US	Boeing 747-135	Northwest Orient
	N622US	Boeing 747-251B	Northwest Orient
	N623US	Boeing 747-251B	Northwest Orient
	N624US	Boeing 747-251B	Northwest Orient
	N625US	Boeing 747-251B	Northwest Orient
	N626US	Boeing 747-251B	Northwest Orient
	N627US	Boeing 747-251B	Northwest Orient
	N628US	Boeing 747-251B	Northwest Orient
	N629US	Boeing 747-251F	Northwest Orient
	N652PA	Boeing 747-121	Pan Am *Clipper Pacific Trader*
	N653PA	Boeing 747-121	Pan Am *Clipper Unity*
	N654PA	Boeing 747-121F	Pan Am *Clipper White Wing*
	N655PA	Boeing 747-121	Pan Am *Clipper Wild Fire*
	N656PA	Boeing 747-121	Pan Am *Clipper Live Yankee*
	N657PA	Boeing 747-121	Pan Am *Clipper Arctic*
	N658PA	Boeing 747-121F	Pan Am *Clipper Fortune*
	N659PA	Boeing 747-121	Pan Am *Clipper Plymouth Rock*
	N702SW	Boeing 747-245F	Flying Tiger Line
	N703SW	Boeing 747-245F	Flying Tiger Line
	N705SW	Boeing 747-245F	Flying Tiger Line
	N706SW	Boeing 747-245F	Flying Tiger Line
	N731PA	Boeing 747-121	Pan Am *Clipper Bostonian*
	N732PA	Boeing 747-121	Pan Am *Clipper Storm King*
	N733PA	Boeing 747-121	Pan Am *Clipper Washington*
	N734PA	Boeing 747-121	Pan Am *Clipper Flying Cloud*
	N735PA	Boeing 747-121	Pan Am *Clipper Young America*
	N737PA	Boeing 747-121	Pan Am *Clipper Red Jacket*
	N738PA	Boeing 747-121	Pan Am *Clipper Defender*
	N739PA	Boeing 747-121	Pan Am *Clipper Morning Light*
	N740PA	Boeing 747-121	Pan Am *Clipper Rival*
	N741PA	Boeing 747-121	Pan Am *Clipper Kit Carson*
	N741PR	Boeing 747-2F6B	Philippine Airlines
	N741TV	Boeing 747-271C	Transamerica Airlines
	N742PA	Boeing 747-121	Pan Am *Clipper Rainbow*
	N742PR	Boeing 747-2F6B	Philippine Airlines
	N742TV	Boeing 747-271C	Transamerica Airlines
	N743PA	Boeing 747-121	Pan Am *Clipper Derby*
	N743PR	Boeing 747-2F6B	Philippine Airlines
	N743TV	Boeing 747-271C	Transamerica Airlines
	N744PA	Boeing 747-121	Pan Am *Clipper Star of the Union*
	N744TV	Boeing 747-271C	Transamerica Airlines
	N745TV	Boeing 747-271C	Transamerica Airlines
	N747PA	Boeing 747-121	Pan Am *Clipper America*
	N748PA	Boeing 747-121	Pan Am *Clipper Hornet*
	N748TA	Boeing 747-212B	Flying Tiger Line
	N748WA	Boeing 747-273C	World Airways (leased to Flying Tiger Line)
	N749PA	Boeing 747-121	Pan Am *Clipper Intrepid*
	N749WA	Boeing 747-273C	Braniff International
	N750PA	Boeing 747-121	Pan Am *Clipper Rambler*
	N750WA	Boeing 747-124F	World Airways
	N751DA	L-1011-385 TriStar 500	Delta Air Lines
	N751PA	Boeing 747-121	Pan Am *Clipper Midnight Sun*
	N752DA	L-1011-385 TriStar 500	Delta Air Lines
	N753DA	L-1011-385 TriStar 500	Delta Air Lines
	N753PA	Boeing 747-121	Pan Am *Clipper Westwind*
	N754PA	Boeing 747-121	Pan Am *Clipper Ocean Rover*
	N755PA	Boeing 747-121	Pan Am *Clipper Sovereign of the Seas*
	N760TW	Boeing 707-331B	Trans World Airlines
	N762TW	Boeing 707-331	Air Tanzania *Mogorongo Crater*
	N770PA	Boeing 747-121	Pan Am *Clipper Express*
	N771PA	Boeing 747-121F	Pan Am *Clipper Donald McKay*
	N773TW	Boeing 707-331B	Trans World Airlines
	N774TW	Boeing 707-331B	Trans World Airlines
	N775TW	Boeing 707-331B	Trans World Airlines
	N778TW	Boeing 707-331B	Trans World Airlines
	N779TW	Boeing 707-331B	Trans World Airlines
	N780TW	Boeing 707-331B	Trans World Airlines
	N781FT	Douglas DC-8-63CF	Flying Tiger Line
	N783FT	Douglas DC-8-63AF	Flying Tiger Line

I–BUSB (Top) A.300B4 Airbus of Alitalia

N741PR (Centre) Boeing 747-2F6B of Philippine Airlines/*A. S. Wright*

N103TV (Bottom) Douglas DC-10-30CF of Transamerica airlines/*A. S. Wright*

Notes	Reg.	Type	Owner or Operator
	N784FT	Douglas DC-8-63AF	Flying Tiger Line
	N786FT	Douglas DC-8-63AF	Flying Tiger Line
	N787FT	Douglas DC-8-63AF	Flying Tiger Line
	N788FT	Douglas DC-8-63AF	Flying Tiger Line
	N788TW	Boeing 707-331C	Trans World Airlines
	N790FT	Douglas DC-8-63AF	Flying Tiger Line
	N791FT	Douglas DC-8-63CF	Flying Tiger Line
	N792FT	Douglas DC-8-63CF	Flying Tiger Line
	N793FT	Douglas DC-8-63CF	Flying Tiger Line
	N793TW	Boeing 707-331B	Trans World Airlines
	N795FT	Douglas DC-8-63CF	Flying Tiger Line
	N796FT	Douglas DC-8-63CF	Flying Tiger Line
	N797FT	Douglas DC-8-63CF	Flying Tiger Line
	N798FT	Douglas DC-8-63CF	Flying Tiger Line
	N800FT	Boeing 747-123F	Flying Tiger Line
	N800EV	Douglas DC-8-52	SAUDIA-Saudi Arabian Airlines
	N801EV	Douglas DC-8-52	Evergreen International Airlines
	N801FT	Boeing 747-123F	Flying Tiger Line
	N801WA	Douglas DC-8-63CF	World Airways
	N802FT	Boeing 747-123F	Flying Tiger Line
	N803FT	Boeing 747-132F	Flying Tiger Line
	N804EV	Douglas DC-8-52	Evergreen International Airlines
	N804FT	Boeing 747-132F	Flying Tiger Line
	N804WA	Douglas DC-8-63CF	World Airways
	N805FT	Boeing 747-132F	Flying Tiger Line
	N805WA	Douglas DC-8-63CF	World Airways
	N806FT	Boeing 747-249F	Flying Tiger Line *Robert W. Prescott*
	N806WA	Douglas DC-8-63CF	World Airways
	N807FT	Boeing 747-249F	Flying Tiger Line *Thomas Haywood*
	N808FT	Boeing 747-249F	Flying Tiger Line *William E. Bartlett*
	N809FT	Boeing 747-249F	Flying Tiger Line
	N810EV	Douglas DC-8-61CF	Evergreen International Airlines
	N810FT	Boeing 747-249F	Flying Tiger Line
	N811FT	Boeing 747-245F	Flying Tiger Line
	N814FT	Boeing 747-245F	Flying Tiger Line
	N815FT	Boeing 747-245F	Flying Tiger Line
	N816FT	Boeing 747-245F	Flying Tiger Line
	N860FT	Douglas DC-8-61CF	Flying Tiger Line
	N861FT	Douglas DC-8-61CF	Flying Tiger Line
	N862FT	Douglas DC-8-61CF	Flying Tiger Line
	N863FT	Douglas DC-8-61CF	Flying Tiger Line
	N864F	Douglas DC-8-63CF	Capitol International Airways
	N864FT	Douglas DC-8-61CF	Flying Tiger Line
	N865F	Douglas DC-8-63CF	SAUDIA—Saudi Arabian Airlines
	N867FT	Douglas DC-8-61CF	Flying Tiger Line
	N868F	Douglas DC-8-61CF	SAUDIA—Saudi Arabian Airlines
	N870TV	Douglas DC-8-63CF	Transamerica Airlines
	N871TV	Douglas DC-8-63CF	Transamerica Airlines
	N872TV	Douglas DC-8-63CF	Transamerica Airlines
	N880PA	Boeing 707-321B	Pan Am *Clipper Emerald Isle*
	N881PA	Boeing 707-321B	Pan Am *Clipper Reindeer*
	N884PA	Boeing 707-321B	Pan Am *Clipper Nightingale*
	N885PA	Boeing 707-321B	Pan Am *Clipper Northern Light*
	N886PA	Boeing 707-321B	Pan Am *Clipper Sea Lark*
	N887PA	Boeing 707-321B	Pan Am *Clipper Flora Temple*
	N901PA	Boeing 747-123F	Pan Am *Clipper Carrier Dove*
	N901WA	Douglas DC-10-10	Western Airlines
	N902PA	Boeing 747-132	Pan Am *Clipper Mandarin*
	N902WA	Douglas DC-10-10	Western Airlines
	N904PA	Boeing 747-221F	Pan Am *Clipper Industry*
	N904WA	Douglas DC-10-10	Western Airlines
	N905PA	Boeing 747-221F	Pan Am *Clipper Resolute*
	N905WA	Douglas DC-10-10	Western Airlines
	N906WA	Douglas DC-10-10	Western Airlines
	N907CL	Douglas DC-8-63CF	Capitol International Airways
	N907WA	Douglas DC-10-10	Western Airlines
	N908WA	Douglas DC-10-10	Western Airlines
	N909WA	Douglas DC-10-10	Western Airlines
	N910CL	Douglas DC-8-63CF	Capitol International Airways
	N911CL	Douglas DC-8-61	SAUDIA—Saudi Arabian Airways
	N912CL	Douglas DC-8-61	Capitol International Airways
	N912WA	Douglas DC-10-10	Western Airlines
	N913WA	Douglas DC-10-10	Western Airlines
	N914WA	Douglas DC-10-10	Western Airlines

Reg.	Type	Owner or Operator	Notes
N915WA	Douglas DC-10-10	Western Airlines	
N934F	Douglas DC-9-32CF	Itavia (Italy)	
N1035F	Douglas DC-10-30CF	Air Florida	
N1252E	Boeing 747-228B	Air France	
N1289E	Boeing 747-228B	Air France	
N1295E	Boeing 747-206B	KLM *Ganges*	
N1298E	Boeing 747-206B	KLM *Indus*	
N1301E	Boeing 747-206B	KLM *Admiral Richard E. Byrd*	
N1607E	Boeing 747-227B	Braniff International	
N4864T	Douglas DC-8-63CF	Transamerica Airlines	
N4865T	Douglas DC-8-63CF	Transamerica Airlines	
N4866T	Douglas DC-8-63CF	Transamerica Airlines	
N4867T	Douglas DC-8-63CF	Transamerica Airlines	
N4868T	Douglas DC-8-63CF	Transamerica Airlines	
N4869T	Douglas DC-8-63CF	Transamerica Airlines	
N6161A	Douglas DC-8-63CF	Airlift International	
N6162A	Douglas DC-8-63CF	Airlift International	
N6163A	Douglas DC-8-63CF	Airlift International	
N7465B	Douglas DC-9-33CF	Itavia (Italy)	
N7984S	L-100-20 Hercules	Southern Air Transport	
N8632	Douglas DC-8-63F	SAUDIA—Saudi Arabian Airlines	
N8636	Douglas DC-8-63F	Evergreen International Airlines	
N8642	Douglas DC-8-63F	Air India Cargo	
N8705T	Boeing 707-331B	Trans World Airlines	
N8725T	Boeing 707-331B	Trans World Airlines	
N8729	Boeing 707-331B	Trans World Airlines	
N8730	Boeing 707-331B	Trans World Airlines	
N8733	Boeing 707-331B	Trans World Airlines	
N8735	Boeing 707-331B	Trans World Airlines	
N8736	Boeing 707-331B	Trans World Airlines	
N8737	Boeing 707-331B	Trans World Airlines	
N8738	Boeing 707-331B	Trans World Airlines	
N8762	Douglas DC-8-61	Capitol International Airways	
N8763	Douglas DC-8-61	Capitol International Airways	
N8764	Douglas DC-8-61	Capitol International Airways	
N8765	Douglas DC-8-61	Capitol International Airways	
N8766	Douglas DC-8-61	Capitol International Airways	
N9232R	L-100-30 Hercules	Southern Air Transport	
N9266R	L-100-20 Hercules	Southern Air Transport	
N9666	Boeing 747-123	Braniff International	
N18702	Boeing 707-331B	Trans World Airlines	
N18703	Boeing 707-331B	Trans World Airlines	
N18704	Boeing 707-331B	Trans World Airlines	
N18706	Boeing 707-331B	Trans World Airlines	
N18707	Boeing 707-331B	Trans World Airlines	
N18708	Boeing 707-331B	Trans World Airlines	
N18709	Boeing 707-331B	Trans World Airlines	
N18710	Boeing 707-331B	Trans World Airlines	
N18711	Boeing 707-331B	Trans World Airlines	
N18712	Boeing 707-331B	Trans World Airlines	
N18713	Boeing 707-331B	Trans World Airlines	
N18815	Boeing 747-228F	Air France	
N28366	Boeing 747-128	Air France	
N28724	Boeing 707-331B	Trans World Airlines	
N28726	Boeing 707-331B	Trans World Airlines	
N28727	Boeing 707-331B	Trans World Airlines	
N28899	Boeing 747-128	Air France	
N28903	Boeing 747-128	Air France	
N31029	L.1011-385 TriStar 100	Trans World Airlines	
N31030	L.1011-385 TriStar 100	Trans World Airlines	
N31031	L.1011-385 TriStar 100	Trans World Airlines	
N31032	L.1011-385 TriStar 100	Trans World Airlines	
N31033	L.1011-385 TriStar 100	Trans World Airlines	
N40116	Boeing 747-128	Air France	
N41020	L.1011-385 TriStar 100	Trans World Airlines	
N53110	Boeing 747-131	Trans World Airlines	
N53116	Boeing 747-131	Trans World Airlines	
N63305	Boeing 747-128	Air France	
N81025	L.1011-385 TriStar 100	Trans World Airlines	
N81026	L.1011-385 TriStar 100	Trans World Airlines	
N81027	L.1011-385 TriStar 100	Trans World Airlines	
N81028	L.1011-385 TriStar 100	Trans World Airlines	
N93104	Boeing 747-131	Trans World Airlines	
N93105	Boeing 747-131	Trans World Airlines	

Notes	Reg.	Type	Owner or Operator
	N93106	Boeing 747-131	Trans World Airlines
	N93107	Boeing 747-131	Trans World Airlines
	N93108	Boeing 747-131	Trans World Airlines
	N93109	Boeing 747-131	Trans World Airlines
	N93115	Boeing 747-131	Trans World Airlines
	N93117	Boeing 747-131	Trans World Airlines
	N93119	Boeing 747-131	Trans World Airlines
	N94454	Douglas DC-9-33RC	Itavia (Italy)

NOTE: The fleet names of some of the Pan Am aircraft will change during 1981.

OD (Lebanon)

Notes	Reg.	Type	Owner or Operator
	OD–AFB	Boeing 707-3B4C	Middle East Airlines
	OD–AFD	Boeing 707-3B4C	Middle East Airlines
	OD–AFE	Boeing 707-3B4C	Middle East Airlines
	OD–AFL	Boeing 720-023B	Middle East Airlines
	OD–AFM	Boeing 720-023B	Middle East Airlines
	OD–AFN	Boeing 720-023B	Middle East Airlines
	OD–AFO	Boeing 720-023B	Middle East Airlines
	OD–AFP	Boeing 720-023B	Middle East Airlines
	OD–AFQ	Boeing 720-023B	Middle East Airlines
	OD–AFR	Boeing 720-023B	Middle East Airlines
	OD–AFS	Boeing 720-023B	Middle East Airlines
	OD–AFU	Boeing 720-023B	Middle East Airlines
	OD–AFW	Boeing 720-023B	Middle East Airlines
	OD–AFY	Boeing 707-327C	Trans Mediterranean Airways
	OD–AFZ	Boeing 720-023B	Middle East Airlines
	OD–AGB	Boeing 720-023B	Middle East Airlines
	OD–AGD	Boeing 707-323C	Trans Mediterranean Airways
	OD–AGF	Boeing 720-047B	Middle East Airlines
	OD–AGG	Boeing 720-047B	Middle East Airlines
	OD–AGH	Boeing 747-2B4B	Middle East Airlines (leased to SAUDIA)
	OD–AGI	Boeing 747-2B4B	Middle East Airlines (leased to SAUDIA)
	OD–AGJ	Boeing 747-2B4B	Middle East Airlines (leased to SAUDIA)
	OD–AGN	Boeing 707-323C	Trans Mediterranean Airways
	OD–AGO	Boeing 707-321C	Trans Mediterranean Airways
	OD–AGP	Boeing 707-321C	Trans Mediterranean Airways
	OD–AGQ	Boeing 720-047	Middle East Airlines
	OD–AGR	Boeing 720-047	Middle East Airlines
	OD–AGS	Boeing 707-331C	Trans Mediterranean Airways
	OD–AGT	Boeing 707-331C	Trans Mediterranean Airways
	OD–AGU	Boeing 707-	Middle East Airlines
	OD–AGW	Boeing 707-327C	Trans Mediterranean Airways
	OD–AGX	Boeing 707-327C	Trans Mediterranean Airways
	OD–AGY	Boeing 707-327C	Trans Mediterranean Airways

OE (Austria)

MONTANA AUSTRIA

Notes	Reg.	Type	Owner or Operator
	OE–HLA	F.27 Friendship Mk.100	Lauda Air
	OE–IDA	Boeing 707-396C	Montana Flugbetrieb
	OE–ILB	F.27 Friendship Mk.600	Lauda Air
	OE–INA	Boeing 707-138B	Montana Flugbetrieb
	OE–IRA	Boeing 707-138B	Montana Flugbetrieb
	OE–LDA	Douglas DC-9-32	Austrian Airlines *Niederösterreich*
	OE–LDB	Douglas DC-9-32	Austrian Airlines *Burgenland*
	OE–LDC	Douglas DC-9-32	Austrian Airlines *Kärnten*

Reg.	Type	Owner or Operator	Notes
OE–LDD	Douglas DC-9-32	Austrian Airlines *Steiermark*	
OE–LDE	Douglas DC-9-32	Austrian Airlines *Oberösterreich*	
OE–LDF	Douglas DC-9-32	Austrian Airlines *Salzburg*	
OE–LDG	Douglas DC-9-32	Austrian Airlines *Tirol*	
OE–LDH	Douglas DC-9-32	Austrian Airlines *Vorarlberg*	
OE–LDI	Douglas DC-9-32	Austrian Airlines *Bregenz*	
OE–LDK	Douglas DC-9-51	Austrian Airlines *Graz*	
OE–LDL	Douglas DC-9-51	Austrian Airlines *Linz*	
OE–LDM	Douglas DC-9-51	Austrian Airlines *Klagenfurt*	
OE–LDN	Douglas DC-9-51	Austrian Airlines *Innsbruck*	
OE–LDO	Douglas DC-9-51	Austrian Airlines *Eisenstädt*	
OE–LDP	Douglas DC-9-81	Austrian Airlines *Wien*	
OE–LDR	Douglas DC-9-81	Austrian Airlines *Niederösterreich*	
OE–LDS	Douglas DC-9-81	Austrian Airlines *Burgenland*	
OE–LDT	Douglas DC-9-81	Austrian Airlines *Kärnten*	
OE–LDU	Douglas DC-9-81	Austrian Airlines *Steirmark*	
OE–LDV	Douglas DC-9-81	Austrian Airlines *Oberösterreich*	
OE–LDW	Douglas DC-9-81	Austrian Airlines *Salzburg*	
OE–LDX	Douglas DC-9-81	Austrian Airlines *Tirol*	
OE–LDY	Douglas DC-9-81	Austrian Airlines *Vorarlberg*	

OH (Finland)

Reg.	Type	Owner or Operator	Notes
OH–KDA	Douglas DC-6B (swing-tail freighter)	Kar-Air	
OH–KDM	Douglas DC-8-51	Kar Air	
OH–LFT	Douglas DC-8-62CF	Finnair *Paavo Nurmi*	
OH–LFY	Douglas DC-8-62CF	Finnair *J. K. Paasikivi*	
OH–LFZ	Douglas DC-8-62	Finnair *Jean Sibelius*	
OH–LHA	Douglas DC-10-30	Finnair *Iso Antti*	
OH–LHB	Douglas DC-10-30	Finnair	
OH–LSA	S.E.210 Caravelle 10-B	Finnair *Helsinki*	
OH–LSB	S.E.210 Caravelle 10-B	Finnair *Tampere*	
OH–LSC	S.E.210 Caravelle 10-B	Finnair *Turku*	
OH–LSD	S.E.210 Caravelle 10-B	Finnair *Oulu*	
OH–LSE	S.E.210 Caravelle 10-B	Finnair *Lahti*	
OH–LSF	S.E.210 Caravelle 10-B	Finnair *Pori*	
OH–LSG	S.E.210 Caravelle 10-B	Finnair *Jyväskylä*	
OH–LSH	S.E.210 Caravelle 10-B	Finnair *Kuopio*	
OH–LYA	Douglas DC-9-14	Finnair	
OH–LYB	Douglas DC-9-14	Finnair	
OH–LYC	Douglas DC-9-14	Finnair	
OH–LYD	Douglas DC-9-14	Finnair	
OH–LYE	Douglas DC-9-14	Finnair	
OH–LYG	Douglas DC-9-14	Finnair	
OH–LYH	Douglas DC-9-15MC	Finnair	
OH–LYI	Douglas DC-9-15MC	Finnair	
OH–LYK	Douglas DC-9-15	Finnair	
OH–LYN	Douglas DC-9-51	Finnair	
OH–LYO	Douglas DC-9-51	Finnair	
OH–LYP	Douglas DC-9-51	Finnair	
OH–LYR	Douglas DC-9-51	Finnair	
OH–LYS	Douglas DC-9-51	Finnair	
OH–LYT	Douglas DC-9-51	Finnair	
OH–LYU	Douglas DC-9-51	Finnair	
OH–LYV	Douglas DC-9-51	Finnair	
OH–LYW	Douglas DC-9-51	Finnair	
OH–LYX	Douglas DC-9-51	Finnair	
OH–LYY	Douglas DC-9-51	Finnair	
OH–LYZ	Douglas DC-9-51	Finnair	

OK (Czechoslovakia)

Notes	Reg.	Type	Owner or Operator
	OK–ABD	Ilyushin IL-62	Ceskoslovenske Aerolinie *Kosice*
	OK–AFA	Tupolev Tu-134A	Ceskoslovenske Aerolinie
	OK–AFB	Tupolev Tu-134A	Ceskoslovenske Aerolinie
	OK–CFC	Tupolev Tu-134A	Ceskoslovenske Aerolinie
	OK–CFE	Tupolev Tu-134A	Ceskoslovenske Aerolinie
	OK–CFF	Tupolev Tu-134A	Ceskoslovenske Aerolinie
	OK–CFG	Tupolev Tu-134A	Ceskoslovenske Aerolinie
	OK–CFH	Tupolev Tu-134A	Ceskoslovenske Aerolinie
	OK–DBE	Ilyushin IL-62	Ceskoslovenske Aerolinie *Brno*
	OK–DFI	Tupolev Tu-134A	Ceskoslovenske Aerolinie
	OK–EBG	Ilyushin IL-62	Ceskoslovenske Aerolinie *Banska Bystrica*
	OK–EFJ	Tupolev Tu-134A	Ceskoslovenske Aerolinie
	OK–EFK	Tupolev Tu-134A	Ceskoslovenske Aerolinie
	OK–GBH	Ilyushin IL-62	Ceskoslovenske Aerolinie *Usti Nad Labem*
	OK–HFL	Tupolev Tu-134A	Ceskoslovenske Aerolinie
	OK–HFM	Tupolev Tu-134A	Ceskoslovenske Aerolinie
	OK–IFN	Tupolev Tu-134A	Ceskoslovenske Aerolinie
	OK–JBI	Ilyushin IL-62M	Ceskoslovenske Aerolinie *Pizen*
	OK–JBJ	Ilyushin IL-62M	Ceskoslovenske Aerolinie *Hradec Kralové*
	OK–KBK	Ilyushin IL-62M	Ceskoslovenske Aerolinie *Ceske Budejovice*
	OK–YBA	Ilyushin IL-62	Ceskoslovenske Aerolinie *Praha*
	OK–YBB	Ilyushin IL-62	Ceskoslovenske Aerolinie *Bratislava*
	OK–ZBC	Ilyushin IL-62	Ceskoslovenske Aerolinie *Ostrava*

OO (Belgium)

Notes	Reg.	Type	Owner or Operator
	OO–ABB	Boeing 737-2P6	Air Belguim
	OO–DTA	FH-227B Friendship	Delta Air Transport
	OO–DTB	FH-227B Friendship	Delta Air Transport
	OO–DTC	FH-227B Friendship	Delta Air Transport
	OO–DTD	FH-227B Friendship	Delta Air Transport
	OO–JPI	Swearingen SA226TC Metro II	Sabena
	OO–JPK	Swearingen SA226TC Metro II	Sabena
	OO–SBQ	Boeing 737-229	Sobelair
	OO–SBS	Boeing 737-229	Sobelair
	OO–SBT	Boeing 737-229	Sobelair
	OO–SBU	Boeing 707-373C	Sobelair
	OO–SBW	Boeing 707-344	Sobelair
	OO–SDA	Boeing 737-229	Sabena
	OO–SDB	Boeing 737-229	Sabena
	OO–SDC	Boeing 737-229	Sabena
	OO–SDD	Boeing 737-229	Sabena
	OO–SDE	Boeing 737-229	Sabena
	OO–SDF	Boeing 737-229	Sabena
	OO–SDG	Boeing 737-229	Sabena
	OO–SDJ	Boeing 737-229C	Sabena
	OO–SDK	Boeing 737-229C	Sabena
	OO–SDL	Boeing 737-229	Sabena

Reg.	Type	Owner or Operator	Notes
OO–SDM	Boeing 737-229	Sabena	
OO–SDN	Boeing 737-229	Sabena	
OO–SDO	Boeing 737-229	Sabena	
OO–SDP	Boeing 737-229C	Sabena	
OO–SDR	Boeing 737-229C	Sabena	
OO–SGA	Boeing 747-129	Sabena	
OO–SGB	Boeing 747-129	Sabena	
OO–SJA	Boeing 707-329	Sobelair	
OO–SJD	Boeing 707-329	Sobelair	
OO–SJJ	Boeing 707-329C	Sabena	
OO–SJL	Boeing 707-329C	Sabena	
OO–SJM	Boeing 707-329C	Sabena	
OO–SJN	Boeing 707-329C	Sabena	
OO–SJO	Boeing 707-329C	Sabena	
OO–SLA	Douglas DC-10-30CF	Sabena	
OO–SLB	Douglas DC-10-30CF	Sabena	
OO–SLC	Douglas DC-10-30CF	Sabena	
OO–SLD	Douglas DC-10-30CF	Sabena	
OO–SLE	Douglas DC-10-30CF	Sabena	
OO–TEC	Boeing 707-131	Trans European Airways *Oostende*	
OO–TED	Boeing 707-131	Trans European Airways *Rena*	
OO–TEF	A.300B1 Airbus	Trans European Airways *Aline*	
OO–TEH	Boeing 737-2M8	Trans European Airways *Marcus Johannes*	
OO–TEK	Boeing 737-2Q9	Trans European Airways *Liège*	
OO–TEL	Boeing 737-2M8	Trans European Airways *Antwerpen*	
OO–TEM	Boeing 737-2Q8	Trans European Airways	
OO–TEN	Boeing 737-2M8	Trans European Airways	
OO–TEO	Boeing 737-2M8	Trans European Airways	
OO–TE	Boeing 737-2M8	Trans European Airways	
OO–WAY	Beech 99	Sabena	
OO–WAZ	Beech 99	Sabena	
OO–YCM	Boeing 707-320C	Young Cargo	

OY (Denmark)

Reg.	Type	Owner or Operator	Notes
OY–APJ	Boeing 737-2L9	Maersk Air	
OY–APK	Boeing 737-2L9	Maersk Air	
OY–APL	Boeing 737-2L9	Maersk Air	
OY–APN	Boeing 737-2L9	Maersk Air	
OY–APO	Boeing 737-2L9	Maersk Air	
OY–APP	Boeing 737-2L9	Maersk Air	
OY–APS	Boeing 737-2L9	Maersk Air	
OY–APU	Boeing 720-051B	Maersk Air	
OY–APV	Boeing 720-051B	Maersk Air	
OY–APW	Boeing 720-051B	Maersk Air	
OY–APZ	Boeing 720-051B	Maersk Air	
OY–DSK	Boeing 720-025	Conair	
OY–DSL	Boeing 720-025	Conair	
OY–DSM	Boeing 720-025	Conair	
OY–DSP	Boeing 720-025	Conair	
OY–KAA	A.300B2 Airbus	SAS	
OY–KDA	Douglas DC-10-30	SAS *Gorm Viking*	
OY–KGA	Douglas DC-9-41	SAS *Heming Viking*	
OY–KGB	Douglas DC-9-41	SAS *Toste Viking*	
OY–KGC	Douglas DC-9-41	SAS *Helge Viking*	
OY–KGD	Douglas DC-9-21	SAS *Ubbe Viking*	
OY–KGE	Douglas DC-9-21	SAS *Orvar Viking*	
OY–KGF	Douglas DC-9-21	SAS *Rolf Viking*	

Notes	Reg.	Type	Owner or Operator
	OY–KGG	Douglas DC-9-41	SAS *Sune Viking*
	OY–KGH	Douglas DC-9-41	SAS *Eiliv Viking*
	OY–KGI	Douglas DC-9-41	SAS *Bent Viking*
	OY–KGK	Douglas DC-9-41	SAS *Ebbe Viking*
	OY–KGL	Douglas DC-9-41	SAS *Angantyr Viking*
	OY–KGM	Douglas DC-9-41	SAS *Arnfinn Viking*
	OY–KGN	Douglas DC-9-41	SAS *Gram Viking*
	OY–KGO	Douglas DC-9-41	SAS *Holte Viking*
	OY–KGP	Douglas DC-9-41	SAS *Torbern Viking*
	OY–KGR	Douglas DC-9-41	SAS *Holger Viking*
	OY–KGS	Douglas DC-9-41	SAS *Hall Viking*
	OY–KHA	Boeing 747-283B	SAS *Ivar Viking*
	OY–KHB	Boeing 747-283B	SAS
	OY–KTF	Douglas DC-8-63	SAS *Frode Viking*
	OY–KTG	Douglas DC-8-63PF	SAS *Torodd Viking*
	OY–SAA	S.E.210 Caravelle 12	Sterling Airways
	OY–SAC	S.E.210 Caravelle 12	Sterling Airways
	OY–SAD	S.E.210 Caravelle 12	Sterling Airways
	OY–SAE	S.E.210 Caravelle 12	Sterling Airways
	OY–SAF	S.E.210 Caravelle 12	Sterling Airways
	OY–SAG	S.E.210 Caravelle 12	Sterling Airways
	OY–SAS	Boeing 727-2J4	Sterling Airways
	OY–SAT	Boeing 727-2J4	Sterling Airways
	OY–SAU	Boeing 727-2J4	Sterling Airways
	OY–SBE	Boeing 727-2J4	Sterling Airways
	OY–SBF	Boeing 727-2J4	Sterling Airways
	OY–SBG	Boeing 727-2J4	Sterling Airways
	OY–STH	S.E.210 Caravelle 10B	Sterling Airways
	OY–STI	S.E.210 Caravelle 10B	Sterling Airways
	OY–STM	S.E.210 Caravelle 10B	Sterling Airways

PH (Netherlands)

Notes	Reg.	Type	Owner or Operator
	PH–ARO	F.27 Friendship Mk 400	N.L.M.
	PH–BUA	Boeing 747-206B	K.L.M. *The Mississippi*
	PH–BUB	Boeing 747-206B	K.L.M. *The Danube*
	PH–BUC	Boeing 747-206B	K.L.M. *The Amazon*
	PH–BUD	Boeing 747-206B	K.L.M. *The Nile*
	PH–BUE	Boeing 747-206B	K.L.M. *Rio de la Plata*
	PH–BUG	Boeing 747-206B	K.L.M. *The Orinoco*
	PH–BUH	Boeing 747-206B	K.L.M. *Dr Albert Plesman*
	PH–BUI	Boeing 747-206B	K.L.M. *Wilbur Wright*
	PH–BUK	Boeing 747-206B	K.L.M. *Louis Blériot*
	PH–BUL	Boeing 747-206B	K.L.M. *Charles Lindbergh*
	PH–BUM	Boeing 747-206B	K.L.M. *Sir Charles E. Kingsford-Smith*
	PH–BUN	Boeing 747-206B	K.L.M. *Anthony H. G. Fokker*
	PH–BUO	Boeing 747-206B	K.L.M. *Missouri*
	PH–	Boeing 747-206B	K.L.M.
	PH–	Boeing 747-206B	K.L.M.
	PH–	Boeing 747-206B	K.L.M.
	PH–CHB	F.28 Fellowship 4000	N.L.M. *Birmingham*
	PH–CHD	F.28 Fellowship 4000	N.L.M. *Maastricht*
	PH–CHF	F.28 Fellowship 4000	N.L.M. *Island of Guernsey*
	PH–CHI	F.28 Fellowship 4000	N.L.M. *Eindhoven*
	PH–DCT	Douglas DC-8-55F	K.L.M. *Baron Pierre de Coubertin*
	PH–DCU	Douglas DC-8-55F	K.L.M. *Sir Winston Churchill*
	PH–DCZ	Douglas DC-8-55F	K.L.M. *Hans Christiaan Andersen*
	PH–DEA	Douglas DC-8-63	K.L.M. *Amerigo Vespucci*
	PH–DEB	Douglas DC-8-63	K.L.M. *Christopher Columbus*
	PH–DEC	Douglas DC-8-63	K.L.M. *Marco Polo*

Reg.	Type	Owner or Operator	Notes
PH–DED	Douglas DC-8-63	K.L.M. *Leifur Eiriksson*	
PH–DEE	Douglas DC-8-63	K.L.M. *Abel Tasman*	
PH–DEF	Douglas DC-8-63	K.L.M. *Henry Hudson*	
PH–DEG	Douglas DC-8-63	K.L.M. *Jan van Reibeeck*	
PH–DEH	Douglas DC-8-63	K.L.M. *Vasco da Gama*	
PH–DEK	Douglas DC-8-63	K.L.M. *David Livingstone*	
PH–DEL	Douglas DC-8-63	K.L.M. *Fernao de Magalhaes*	
PH–DEM	Douglas DC-8-63	K.L.M. *James Cook* (on lease to Surinam Airways)	
PH–DNB	Douglas DC-9-15	K.L.M. *City of Brussels*	
PH–DNC	Douglas DC-9-15	K.L.M. *City of Luxembourg*	
PH–DNG	Douglas DC-9-32	K.L.M. *City of Rotterdam*	
PH–DNH	Douglas DC-9-32	K.L.M. *City of Zurich*	
PH–DNI	Douglas DC-9-32	K.L.M. *City of Istanbul*	
PH–DNK	Douglas DC-9-32	K.L.M. *City of Copenhagen*	
PH–DNL	Douglas DC-9-32	K.L.M. *City of London*	
PH–DNM	Douglas DC-9-33RC	K.L.M. *City of Madrid*	
PH–DNN	Douglas DC-9-33RC	K.L.M. *City of Vienna*	
PH–DNO	Douglas DC-9-33RC	K.L.M. *City of Oslo*	
PH–DNP	Douglas DC-9-33RC	K.L.M. *City of Athens*	
PH–DNR	Douglas DC-9-33RC	K.L.M. *City of Stockholm*	
PH–DNS	Douglas DC-9-32	K.L.M. *City of Arnhem*	
PH–DNT	Douglas DC-9-32	K.L.M. *City of Lisbon*	
PH–DNV	Douglas DC-9-32	K.L.M. *City of Warsaw*	
PH–DNW	Douglas DC-9-32	K.L.M. *City of Moscow*	
PH–DNY	Douglas DC-9-33RC	K.L.M. *City of Paris*	
PH–DNZ	Douglas DC-9-33RC	K.L.M. *City of Rome*	
PH–DOA	Douglas DC-9-32	K.L.M. *City of Utrecht*	
PH–DOB	Douglas DC-9-32	K.L.M. *City of Santa Monica*	
PH–DTA	Douglas DC-10-30	K.L.M. *Johann Sebastian Bach*	
PH–DTB	Douglas DC-10-30	K.L.M. *Ludwig van Beethoven*	
PH–DTC	Douglas DC-10-30	K.L.M. *Frédéric François Chopin*	
PH–DTD	Douglas DC-10-30	K.L.M. *Maurice Ravel*	
PH–DTE	Douglas DC-10-30	K.L.M. *Wolfgang Amadeus Mozart*	
PH–DTI	Douglas DC-10-30	K.L.M. (leased to PAL)	
PH–DTK	Douglas DC-10-30	K.L.M. (leased to PAL)	
PH–DTL	Douglas DC-10-30	K.L.M. *Edvard Hagerup Grieg*	
PH–KFD	F.27 Friendship Mk. 200	N.L.M. *Jan Moll*	
PH–KFE	F.27 Friendship Mk. 400	N.L.M. *Jan Dellaert*	
PH–KFG	F.27 Friendship Mk. 200	N.L.M. *Koos Abspoel*	
PH–KFH	F.27 Friendship Mk. 200	N.L.M. *Koene Dirk Parmentier*	
PH–KFI	F.27 Friendship Mk. 500	N.L.M.	
PH–KFK	F.27 Friendship Mk. 500	N.L.M.	
PH–KFL	F.27 Friendship Mk. 500	N.L.M.	
PH–MAO	Douglas DC-9-33RC	Martinair *Desiderius Erasmus*	
PH–MAR	Douglas DC-9-33RC	Martinair *Jean Monnet*	
PH–MAT	F.28 Fellowship 1000	Martinair *Prinses Margriet*	
PH–MAX	Douglas DC-9-32	Martinair *Europe*	
PH–MBG	Douglas DC-10-30CF	Martinair *Kohoutek*	
PH–MBN	Douglas DC-10-30CF	Martinair *Anthony Ruys*	
PH–MBP	Douglas DC-10-30CF	Martinair *Hong Kong*	
PH–MBT	Douglas DC-10-30CF	Martinair	
PH–SAD	F.27 Friendship Mk. 200	N.L.M. *Evert van Dijk*	
PH–TVA	Boeing 707-123B	Transavia *Provincie Zeeland*	
PH–TVC	Boeing 737-2K2C	Transavia *Richard Gordon*	
PH–TVE	Boeing 737-2K2C	Transavia *Alan Bean*	
PH–TVH	Boeing 737-222	Transavia *Neil Armstrong*	
PH–TVI	Boeing 737-222	Transavia *Michael Collins*	
PH–TVP	Boeing 737-2K2	Transavia	
PH–TVR	Boeing 737-2K2	Transavia	
PH–TVS	Boeing 737-2K2	Transavia	

NOTE: KLM also operate Boeing 747-206Bs N1295E, N1298E and N1301E.

PK (Indonesia)

Reg.	Type	Owner or Operator	Notes
PK–GIA	Douglas DC-10-30	Garuda Indonesian Airways *Irian Jaya*	
PK–GIB	Douglas DC-10-30	Garuda Indonesian Airways *Bali*	
PK–GIC	Douglas DC-10-30	Garuda Indonesian Airways *Java*	
PK–GID	Douglas DC-10-30	Garuda Indonesian Airways *Sumatra*	
PK–GIE	Douglas DC-10-30	Garuda Indonesian Airways	

Notes	Reg.	Type	Owner or Operator
	PK–GIF	Douglas DC-10-30	Garuda Indonesian Airways *Sulawesi*
	PK–GSA	Boeing 747-2U3B	Garuda Indonesian Airways *City of Jakarta*
	PK–GSB	Boeing 747-2U3B	Garuda Indonesian Airways *City of Bandung*
	PK–GSC	Boeing 747-2U3B	Garuda Indonesian Airways *City of Medan*
	PK–GSD	Boeing 747-2U3B	Garuda Indonesian Airways *City of Surabaya*

PP (Brazil)

Notes	Reg.	Type	Owner or Operator
	PP–VJH	Boeing 707-320C	VARIG
	PP–VJK	Boeing 707-379C	VARIG
	PP–VJS	Boeing 707-341C	VARIG
	PP–VJT	Boeing 707-341C	VARIG
	PP–VJX	Boeing 707-345C	VARIG
	PP–VJY	Boeing 707-345C	VARIG
	PP–VLI	Boeing 707-385C	VARIG
	PP–VLK	Boeing 707-324C	VARIG
	PP–VLL	Boeing 707-324C	VARIG
	PP–VLM	Boeing 707-324C	VARIG
	PP–VLN	Boeing 707-324C	VARIG
	PP–VLO	Boeing 707-324C	VARIG
	PP–VLP	Boeing 707-323C	VARIG
	PP–VMA	Douglas DC-10-30	VARIG
	PP–VMB	Douglas DC-10-30	VARIG
	PP–VMD	Douglas DC-10-30	VARIG
	PP–VMQ	Douglas DC-10-30	VARIG

RP (Philippines)

NOTE: Philippine Airlines operate three Boeing 747s which retain their US registrations N741PR, N742PR and N743PR.

S2 (Bangladesh)

Bangladesh Biman

Notes	Reg.	Type	Owner or Operator
	S2–ABN	Boeing 707-351C	Bangladesh Biman *City of Shah Jalal*
	S2–ACA	Boeing 707-351C	Bangladesh Biman
	S2–ACE	Boeing 707-351C	Bangladesh Biman *City of Tokyo*
	S2–ACG	Boeing 707-349C	Bangladesh Biman

SE (Sweden)

Notes	Reg.	Type	Owner or Operator
	SE–DAK	Douglas DC-9-41	SAS *Ragnvald Viking*
	SE–DAL	Douglas DC-9-41	SAS *Algot Viking*
	SE–DAM	Douglas DC-9-41	SAS *Starkad Viking*
	SE–DAN	Douglas DC-9-41	SAS *Alf Viking*
	SE–DAO	Douglas DC-9-41	SAS *Asgaut Viking*

Reg.	Type	Owner or Operator	Notes
SE–DAP	Douglas DC-9-41	SAS *Torgils Viking*	
SE–DAR	Douglas DC-9-41	SAS *Agnar Viking*	
SE–DAS	Douglas DC-9-41	SAS *Garder Viking*	
SE–DAT	Douglas DC-9-41	SAS *Gissur Viking*	
SE–DAU	Douglas DC-9-41	SAS *Hadding Viking*	
SE–DAW	Douglas DC-9-41	SAS *Gotrik Viking*	
SE–DAX	Douglas DC-9-41	SAS *Helsing Viking*	
SE–DBG	Douglas DC-8-62	Scanair *Jorund Viking*	
SE–DBI	Douglas DC-8-62CF	Scanair *Ylva Viking*	
SE–DBK	Douglas DC-8-63	SAS *Edmund Viking*	
SE–DBL	Douglas DC-8-63	SAS *Tord Viking*	
SE–DBM	Douglas DC-9-41	SAS *Ossur Viking*	
SE–DBN	Douglas DC-9-33AF	SAS *Sigtrygge Viking*	
SE–DBO	Douglas DC-9-21	SAS *Siger Viking*	
SE–DBP	Douglas DC-9-21	SAS *Rane Viking*	
SE–DBR	Douglas DC-9-21	SAS *Skate Viking*	
SE–DBS	Douglas DC-9-21	SAS *Svipdag Viking*	
SE–DBT	Douglas DC-9-41	SAS *Agne Viking*	
SE–DBU	Douglas DC-9-41	SAS *Hjalmar Viking*	
SE–DBW	Douglas DC-9-41	SAS *Adils Viking*	
SE–DBX	Douglas DC-9-41	SAS *Arnljot Viking*	
SE–DDA	Boeing 727-134	Scanair/Transair Sweden *Midnight Sun*	
SE–DDB	Boeing 727-134	Scanair/Transair Sweden *Northern Light*	
SE–DDC	Boeing 727-134C	Scanair/Transair Sweden *Polar Circle*	
SE–DDL	Boeing 747-283B	SAS *Huge Viking*	
SE–DDP	Douglas DC-9-41	SAS *Brun Viking*	
SE–DDR	Douglas DC-9-41	SAS *Atle Viking*	
SE–DDS	Douglas DC-9-41	SAS *Alrik Viking*	
SE–DDT	Douglas DC-9-41	SAS *Amund Viking*	
SE–DDU	Douglas DC-8-62	Scanair *Knud Viking*	
SE–DFD	Douglas DC-10-30	SAS *Dag Viking*	
SE–DFE	Douglas DC-10-30	SAS *Sverker Viking*	
SE–DFK	A.300B2 Airbus	SAS *Sven Viking*	
SE–DFL	A.300B2 Airbus	SAS	
SE–DFZ	Boeing 747-283B	SAS *Knut Viking*	
SE–DGA	F.28 Fellowship 1000	Linjeflyg	
SE–DGB	F.28 Fellowship 1000	Linjeflyg	
SE–DGC	F.28 Fellowship 1000	Linjeflyg	
SE–DGD	F.28 Fellowship 4000	Linjeflyg	
SE–DGE	F.28 Fellowship 4000	Linjeflyg	
SE–DGF	F.28 Fellowship 4000	Linjeflyg	
SE–DGG	F.28 Fellowship 4000	Linjeflyg	
SE–DGH	F.28 Fellowship 4000	Linjeflyg	
SE–DGI	F.28 Fellowship 4000	Linjeflyg	
SE–DGK	F.28 Fellowship 4000	Linjeflyg	
SE–DGL	F.28 Fellowship 4000	Linjeflyg	
SE–DGM	F.28 Fellowship 4000	Linjeflyg	
SE–DGN	F.28 Fellowship 4000	Linjeflyg	

SP (Poland)

Reg.	Type	Owner or Operator	Notes
SP–LAB	Ilyushin IL-62	Polskie Linie Lotnicze (LOT) *Tadeusz Kosciuszko*	
SP–LAC	Ilyushin IL-62	Polskie Linie Lotnicze (LOT) *Fryderyk Chopin*	
SP–LAD	Ilyushin IL-62	Polskie Linie Lotnicze (LOT) *Kazimierz Pulaski*	
SP–LAE	Ilyushin IL-62	Polskie Linie Lotnicze (LOT) *Henryk Sienkiewicz*	
SP–LAF	Ilyushin IL-62	Polskie Linie Lotnicze (LOT) *Adam Michiewicz*	
SP–LAG	Ilyushin IL-62	Polskie Linie Lotnicze (LOT) *Maria Sklodowska-Curie*	
SP–LBA	Ilyushin IL-62M	Polskie Linie Lotnicze (LOT) *Juliusz Sowacki*	

Notes	Reg.	Type	Owner or Operator
	SP–LBB	Ilyushin IL-62M	Polskie Linie Lotnicze (LOT)
	SP–	Ilyushin IL-62M	Polskie Linie Lotnicze (LOT)
	SP–	Ilyushin IL-62M	Polskie Linie Lotnicze (LOT)
	SP–LGA	Tupolev Tu-134	Polskie Linie Lotnicze (LOT)
	SP–LGC	Tupolev Tu-134	Polskie Linie Lotnicze (LOT)
	SP–LGD	Tupolev Tu-134	Polskie Linie Lotnicze (LOT)
	SP–LGE	Tupolev Tu-134	Polskie Linie Lotnicze (LOT)
	SP–LHA	Tupolev Tu-134A	Polskie Linie Lotnicze (LOT)
	SP–LHB	Tupolev Tu-134A	Polskie Linie Lotnicze (LOT)
	SP–LHC	Tupolev Tu-134A	Polskie Linie Lotnicze (LOT)
	SP–LHD	Tupolev Tu-134A	Polskie Linie Lotnicze (LOT)
	SP–LHE	Tupolev Tu-134A	Polskie Linie Lotnicze (LOT)
	SP–LHF	Tupolev Tu-134A	Polskie Linie Lotnicze (LOT)
	SP–LHG	Tupolev Tu-134A	Polskie Linie Lotnicze (LOT)
	SP–LSA	Ilyushin IL-18V (Cargo)	Polskie Linie Lotnicze (LOT)
	SP–LSB	Ilyushin IL-18V	Polskie Linie Lotnicze (LOT)
	SP–LSC	Ilyushin IL-18V (Cargo)	Polskie Linie Lotnicze (LOT)
	SP–LSD	Ilyushin IL-18V	Polskie Linie Lotnicze (LOT)
	SP–LSE	Ilyushin IL-18V	Polskie Linie Lotnicze (LOT)
	SP–LSF	Ilyushin IL-18E	Polskie Linie Lotnicze (LOT)
	SP–LSG	Ilyushin IL-18E (Cargo)	Polskie Linie Lotnicze (LOT)
	SP–LSH	Ilyushin IL-18V	Polskie Linie Lotnicze (LOT)
	SP–LSI	Ilyushin IL-18D	Polskie Linie Lotnicze (LOT)

ST (Sudan)

Notes	Reg.	Type	Owner or Operator
	ST–AFA	Boeing 707-3J8C	Sudan Airways
	ST–AFB	Boeing 707-3J8C	Sudan Airways

SU (Egypt)

مصر للطيران

EGYPTAIR

Notes	Reg.	Type	Owner or Operator
	SU–AOU	Boeing 707-366C	EgyptAir *Khopho*
	SU–APD	Boeing 707-366C	EgyptAir *Khafrah*
	SU–APE	Boeing 707-366C	EgyptAir *Mankara*
	SU–AVX	Boeing 707-366C	EgyptAir *Tutankhamun*
	SU–AVY	Boeing 707-366C	EgyptAir *Akhenaton*
	SU–AVZ	Boeing 707-366C	EgyptAir *Mena*
	SU–AXK	Boeing 707-366C	EgyptAir *Seti I*
	SU–BBA	Boeing 707-338C	Air Cargo Egypt

SX (Greece)

Notes	Reg.	Type	Owner or Operator
	SX–BEB	A.300B4 Airbus	Olympic Airways *Odysseus*
	SX–BEC	A.300B4 Airbus	Olympic Airways *Achilles*
	SX–BED	A.300B4 Airbus	Olympic Airways *Telemachos*
	SX–BEE	A.300B4 Airbus	Olympic Airways *Nestor*
	SX–BEF	A.300B4 Airbus	Olympic Airways *Ajax*
	SX–DBC	Boeing 707-384C	Olympic Airways *City of Knossos*
	SX–DBD	Boeing 707-384C	Olympic Airways *City of Sparta*
	SX–DBE	Boeing 707-384B	Olympic Airways *City of Pella*

OY–SAU (Top) Boeing 727-2J4 of Sterling Airways/*A. S. Wright*

PH–CHI (Centre) Fokker F.28 Fellowship 4000 of NLM/*S. G. Richards*

SP–LSF (Bottom) Ilyushin IL-18E of Polskie Linie Lotnicze (LOT)

Notes	Reg.	Type	Owner or Operator
	SX–DBF	Boeing 707-384B	Olympic Airways *City of Mycenae*
	SX–DBG	Boeing 720-051B	Olympic Airways *Axios River*
	SX–DBH	Boeing 720-051B	Olympic Airways *Acheloos River*
	SX–DBI	Boeing 720-051B	Olympic Airways *Pinios River*
	SX–DBK	Boeing 720-051B	Olympic Airways *Strimon River*
	SX–DBL	Boeing 720-051B	Olympic Airways *Evros River*
	SX–DBM	Boeing 720-051B	Olympic Airways *Aliakmon River*
	SX–DBN	Boeing 720-051B	Olympic Airways *Nestos River*
	SX–DBO	Boeing 707-351C	Olympic Airways *City of Lindos*
	SX–DBP	Boeing 707-351C	Olympic Airways *City of Thebes*
	SX–OAA	Boeing 747-284B	Olympic Airways *Olympic Zeus*
	SX–OAB	Boeing 747-284B	Olympic Airways *Olympic Eagle*

TC (Turkey)

Notes	Reg.	Type	Owner or Operator
	TC–JAU	Douglas DC-10-10	Türk Hava Yollari (THY) *Istanbul*
	TC–JAY	Douglas DC-10-10	Türk Hava Yollari (THY) *Izmir*
	TC–JBF	Boeing 727-2F2	Türk Hava Yollari (THY) *Adana*
	TC–JBG	Boeing 727-2F2	Türk Hava Yollari (THY) *Ankara*
	TC–JBJ	Boeing 727-2F2	Türk Hava Yollari (THY) *Diyarbakir*
	TC–JBM	Boeing 727-2F2	Türk Hava Yollari (THY) *Menderes*
	TC–JBR	Boeing 727-2F2	Türk Hava Yollari (THY) *Afyon*
	TC–JBS	Boeing 707-321B	Türk Hava Yollari (THY) *Basak*
	TC–JBT	Boeing 707-321B	Türk Hava Yollari (THY) *Baris*
	TC–JBU	Boeing 707-321B	Türk Hava Yollari (THY) *Yurdum*
	TC–JBV	Douglas DC-8-21	Bursa Hava Yollari
	TC–JBY	Douglas DC-8-21	Bursa Hava Yollari
	TC–JBZ	Douglas DC-8-21	Bursa Hava Yollari

TF (Iceland)

Notes	Reg.	Type	Owner or Operator
	TF–BCV	Douglas DC-8-63CF	Cargolux
	TF–CCV	Douglas DC-8-63CF	Cargolux *City of Echternach*
	TF–FLB	Douglas DC-8-63CF	Icelandair
	TF–FLC	Douglas DC-8-63CF	Icelandair (leased to SAUDIA)
	TF–FLE	Douglas DC-8-63CF	Icelandair
	TF–FLF	Douglas DC-8-63CF	Icelandair
	TF–FLG	Boeing 727-185C	Icelandair
	TF–FLH	Boeing 727-108C	Icelandair
	TF–FLI	Boeing 727-208	Icelandair
	TF–IUB	Douglas DC-6A	Iscargo
	TF–ISC	L-188 Electra	Iscargo
	TF–VLB	Boeing 720-047B	Eagle Air
	TF–VLG	Boeing 707-347C	Eagle Air

TS (Tunisia)

الخطوط الجوية التونسية

TUNIS AIR

Notes	Reg.	Type	Owner or Operator
	TS–JHN	Boeing 727-2H3	Tunis-Air *Carthago*
	TS–JHO	Boeing 727-2H3	Tunis-Air *Jerba*
	TS–JHP	Boeing 727-2H3	Tunis-Air *Monastir*

Reg.	Type	Owner or Operator	Notes
TS–JHQ	Boeing 727-2H3	Tunis-Air *Tozeur-Nefta*	
TS–JHR	Boeing 727-2H3	Tunis-Air *Bizerte*	
TS–JHS	Boeing 727-2H3	Tunis-Air *Kairouan*	
TS–JHT	Boeing 727-2H3	Tunis-Air *Sidi Bousaid*	
TS–JHU	Boeing 727-2H3	Tunis-Air *Hannibal*	
TS–JHV	Boeing 727-2H3	Tunis-Air *Jugurtha*	
TS–JHW	Boeing 727-2H3	Tunis-Air *Ibn Khaldoun*	

VH (Australia)

VH–EBA	Boeing 747-238B	Qantas Airways *City of Canberra*	
VH–EBB	Boeing 747-238B	Qantas Airways *City of Melbourne*	
VH–EBC	Boeing 747-238B	Qantas Airways *City of Sydney*	
VH–EBD	Boeing 747-238B	Qantas Airways *City of Perth*	
VH–EBE	Boeing 747-238B	Qantas Airways *City of Brisbane*	
VH–EBF	Boeing 747-238B	Qantas Airways *City of Adelaide*	
VH–EBG	Boeing 747-238B	Qantas Airways *City of Hobart*	
VH–EBH	Boeing 747-238B	Qantas Airways *City of Newcastle*	
VH–EBI	Boeing 747-238B	Qantas Airways *City of Darwin*	
VH–EBJ	Boeing 747-238B	Qantas Airways *City of Geelong*	
VH–EBK	Boeing 747-238B	Qantas Airways *City of Wollongong*	
VH–EBL	Boeing 747-238B	Qantas Airways *City of Townsville*	
VH–EBM	Boeing 747-238B	Qantas Airways *City of Parramatta*	
VH–EBN	Boeing 747-238B	Qantas Airways *City of Albury*	
VH–EBO	Boeing 747-238B	Qantas Airways *City of Elizabeth*	
VH–EBP	Boeing 747-238B	Qantas Airways *City of Freemantle*	
VH–EBQ	Boeing 747-238B	Qantas Airways	
VH–EBR	Boeing 747-238B	Qantas Airways	
VH–EBS	Boeing 747-238B	Qantas Airways	
VH–EBT	Boeing 747-238B	Qantas Airways	
VH–ECA	Boeing 747-238B	Qantas Airways *City of Sale*	
VH–ECB	Boeing 747-238B	Qantas Airways *City of Swan Hill*	
VH–ECC	Boeing 747-238B	Qantas Airways	

VP–W (Zimbabwe)

VP–WGA	Boeing 707-344C	Air Zimbabwe	

VH–R (Hong Kong)

VR–HIA	Boeing 747-267B	Cathay Pacific Airways	
VR–HIB	Boeing 747-267B	Cathay Pacific Airways	
VR–HKG	Boeing 747-267B	Cathay Pacific Airways	
VR–	Boeing 747-267B	Cathay Pacific Airways	

VT (India)

VT–EBE	Boeing 747-237B	Air India *Emperor Shahjehan*	
VT–EBN	Boeing 747-237B	Air-India *Emperor Rajendra Chola*	
VT–EBO	Boeing 747-237B	Air-India *Emperor Nikramaditya*	
VT–EDU	Boeing 747-237B	Air-India *Emperor Akbar*	

Notes	Reg.	Type	Owner or Operator
	VT–EFJ	Boeing 747-237B	Air-India *Emperor Chandragupta*
	VT–EFO	Boeing 747-237B	Air-India *Emperor Kanishka*
	VT–EFU	Boeing 747-237B	Air-India *Emperor Krishna Deva*
	VT–EGA	Boeing 747-237B	Air-India *Samudra Gupto*
	VT–EGB	Boeing 747-237B	Air-India Mahendra Varman
	VT–EGC	Boeing 747-237B	Air-India *Harsha Vardhuma*

NOTE: Air India Cargo operates a Douglas DC-8-63F which retains its US registration N8642.

YA (Afghanistan)

Notes	Reg.	Type	Owner or Operator
	YA–LAS	Douglas DC-10-30	Ariana

YI (Iraq)

Notes	Reg.	Type	Owner or Operator
	YI–AGE	Boeing 707-370C	Iraqi Airways
	YI–AGF	Boeing 707-370C	Iraqi Airways
	YI–AGG	Boeing 707-370C	Iraqi Airways
	YI–AGN	Boeing 747-270C	Iraqi Airways
	YI–AGO	Boeing 747-270C	Iraqi Airways

YK (Syria)

Notes	Reg.	Type	Owner or Operator
	YK–AHA	Boeing 747SP-94	Syrian Arab Airlines *16 Novembre*
	YK–AHB	Boeing 747SP-94	Syrian Arab Airlines *Arab Solidarity*

YR (Romania)

Notes	Reg.	Type	Owner or Operator
	YR–ABA	Boeing 707-3K1C	Tarom
	YR–ABC	Boeing 707-3K1C	Tarom
	YR–ABM	Boeing 707-321C	Tarom
	YR–ABN	Boeing 707-321C	Tarom
	YR–BCB	BAC One-Eleven 424	Tarom
	YR–BCE	BAC One-Eleven 424	Tarom
	YR–BCG	BAC One-Eleven 401	Tarom
	YR–BCH	BAC One-Eleven 402	Tarom
	YR–BCI	BAC One-Eleven 525FT	Tarom
	YR–BCJ	BAC One-Eleven 525FT	Tarom
	YR–BCK	BAC One-Eleven 525FT	Tarom
	YR–BCL	BAC One-Eleven 525FT	Tarom
	YR–BCM	BAC One-Eleven 525FT	Tarom
	YR–	BAC One-Eleven 525FT	Tarom
	YR–	BAC One-Eleven 525FT	Tarom

Reg.	Type	Owner or Operator	Notes
YR–	BAC One-Eleven 487	Tarom	
YR–IRA	Ilyushin IL-62	Tarom	
YR–IRB	Ilyushin IL-62	Tarom	
YR–IRC	Ilyushin IL-62	Tarom	
YR–IRD	Ilyushin IL-62M	Tarom	
YR–IRE	Ilyushin IL-62M	Tarom	
YR–TPA	Tupolev Tu-154B	Tarom	
YR–TPB	Tupolev Tu-154B	Tarom	
YR–TPC	Tupolev Tu-154B	Tarom	
YR–TPD	Tupolev Tu-154B	Tarom	
YR–TPE	Tupolev Tu-154B	Tarom	
YR–TPF	Tupolev Tu-154B	Tarom	
YR–TPG	Tupolev Tu-154B	Tarom	
YR–TPI	Tupolev Tu-154B	Tarom	
YR–TPJ	Tupolev Tu-154B	Tarom	
YR–TPK	Tupolev Tu-154B	Tarom	

AVIOGENEX

YU (Yugoslavia)

INEX ADRIA

Reg.	Type	Owner or Operator	Notes
YU–AGE	Boeing 707-340C	Jugoslovenski Aerotransport	
YU–AGG	Boeing 707-340C	Jugoslovenski Aerotransport	
YU–AGI	Boeing 707-351C	Jugoslovenski Aerotransport	
YU–AGJ	Boeing 707-351C	Jugoslovenski Aerotransport	
YU–AHJ	Douglas DC-9-32	Inex Adria Airways *Ljubljana*	
YU–AHL	Douglas DC-9-32	Jugoslovenski Aerotransport	
YU–AHM	Douglas DC-9-32	Jugoslovenski Aerotransport *Tivat*	
YU–AHN	Douglas DC-9-32	Jugoslovenski Aerotransport	
YU–AHO	Douglas DC-9-32	Jugoslovenski Aerotransport	
YU–AHP	Douglas DC-9-32	Jugoslovenski Aerotransport	
YU–AHU	Douglas DC-9-32	Jugoslovenski Aerotransport	
YU–AHV	Douglas DC-9-32	Jugoslovenski Aerotransport	
YU–AHW	Douglas DC-9-33CF	Inex Adria Airways *Sarajevo*	
YU–AHX	Tupolev Tu-134A	Aviogenex *Beograd*	
YU–AHY	Tupolev Tu-134A	Aviogenex *Zagreb*	
YU–AJA	Tupolev Tu-134A	Aviogenex *Titograd*	
YU–AJB	Douglas DC-9-32	Inex Adria Airways	
YU–AJD	Tupolev Tu-134A	Aviogenex *Skopje*	
YU–AJF	Douglas DC-9-32	Inex Adria Airways	
YU–AJH	Douglas DC-9-32	Jugoslovenski Aerotransport	
YU–AJI	Douglas DC-9-32	Jugoslovenski Aerotransport	
YU–AJJ	Douglas DC-9-32	Jugoslovenski Aerotransport	
YU–AJK	Douglas DC-9-32	Jugoslovenski Aerotransport	
YU–AJL	Douglas DC-9-32	Jugoslovenski Aerotransport	
YU–AJM	Douglas DC-9-32	Jugoslovenski Aerotransport	
YU–AJP	Douglas DC-9-33CF	Inex Adria Airways	
YU–AJT	Douglas DC-9-51	Inex Adria Airways	
YU–AJU	Douglas DC-9-51	Inex Adria Airways *Maribor*	
YU–AJV	Tupolev Tu-134A	Aviogenex *Mostar*	
YU–AJW	Tupolev Tu-134A	Aviogenex *Pristina*	
YU–AJX	Douglas DC-9-32	Inex Adria Airways	
YU–AKA	Boeing 727-2H9	Jugoslovenski Aerotransport	
YU–AKB	Boeing 727-2H9	Jugoslovenski Aerotransport	
YU–AKE	Boeing 727-2H9	Jugoslovenski Aerotransport	
YU–AKF	Boeing 727-2H9	Jugoslovenski Aerotransport	
YU–AKG	Boeing 727-2H9	Jugoslovenski Aerotransport	
YU–AKI	Boeing 727-2H9	Jugoslovenski Aerotransport	
YU–AKJ	Boeing 727-2H9	Jugoslovenski Aerotransport	

Notes	Reg.	Type	Owner or Operator
	YU–AMA	Douglas DC-10-30	Jugoslovenski Aerotransport *Nikola Tesla*
	YU–AMB	Douglas DC-10-30	Jugoslovenski Aerotransport *Edvard Rusijan*
	YU–ANE	Tupolev Tu-134A	Aviogenex *Novi Sad*
	YU–	Douglas DC-9-81	Inex Adria Airways
	YU–	Douglas DC-9-81	Inex Adria Airways
	YU–	Douglas DC-9-81	Inex Adria Airways

YV (Venezuela)

Notes	Reg.	Type	Owner or Operator
	YV–133C	Douglas DC-10-30	Viasa
	YV–134C	Douglas DC-10-30	Viasa
	YV–135C	Douglas DC-10-30	Viasa
	YV–136C	Douglas DC-10-30	Viasa
	YV–137C	Douglas DC-10-30	Viasa
	YV–138C	Douglas DC-10-30	Viasa

ZS (South Africa)

Notes	Reg.	Type	Owner or Operator
	ZS–SAA	Boeing 747-244B	South African Airways
	ZS–SAB	Boeing 747-244B	South African Airways
	ZS–SAL	Boeing 747-244B	South African Airways *Tafelberg*
	ZS–SAM	Boeing 747-244B	South African Airways *Drakensberg*
	ZS–SAN	Boeing 747-244B	South African Airways *Lebombo*
	ZS–SAO	Boeing 747-244B	South African Airways *Magaliesberg*
	ZS–SAP	Boeing 747-244B	South African Airways *Swartberg*
	ZS–SPA	Boeing 747SP-44	South African Airways *Matroosberg*
	ZS–SPB	Boeing 747SP-44	South African Airways *Outeniqua*
	ZS–SPC	Boeing 747SP-44	South African Airways *Maluti*
	ZS–SPD	Boeing 747SP-44	South African Airways *Majuba*
	ZS–SPE	Boeing 747SP-44	South African Airways *Hantam*
	ZS–SPF	Boeing 747SP-44	South African Airways *Soutpansberg*

4R (Sri Lanka)

Notes	Reg.	Type	Owner or Operator
	4R–ALA	Boeing 707-321B	Air Lanka
	4R–ALB	Boeing 707-321B	Air Lanka
	4R–ALE	L.1011-385 TriStar 1	Air Lanka
	4R–	L.1011-385 TriStar 1	Air Lanka

4X (Israel)

Notes	Reg.	Type	Owner or Operator
	4X–ATA	Boeing 707-458	El Al
	4X–ATB	Boeing 707-458	El Al
	4X–ATR	Boeing 707-358B	El Al
	4X–ATS	Boeing 707-358B	El Al
	4X–ATT	Boeing 707-358B	El Al
	4X–ATX	Boeing 707-358C	El Al

Reg.	Type	Owner or Operator	Notes
4X–ATY	Boeing 707-358C	El Al	
4X–AXA	Boeing 747-258B	El Al	
4X–AXB	Boeing 747-258B	El Al	
4X–AXC	Boeing 747-258B	El Al	
4X–AXD	Boeing 747-258C	El Al	
4X–AXF	Boeing 747-258C	El Al	
4X–AXG	Boeing 747-258F	El Al	
4X–AXH	Boeing 747-258B	El Al	
4X–AXZ	Boeing 747-124F	El Al	

5A (Libya)

5A–DAI	Boeing 727-224	Libyan Arab Airlines	
5A–DAK	Boeing 707-3L5C	Libyan Arab Airlines	
5A–DIA	Boeing 727-2L5	Libyan Arab Airlines	
5A–DIB	Boeing 727-2L5	Libyan Arab Airlines	
5A–DIC	Boeing 727-2L5	Libyan Arab Airlines	
5A–DID	Boeing 727-2L5	Libyan Arab Airlines	
5A–DIE	Boeing 727-2L5	Libyan Arab Airlines	
5A–DIF	Boeing 727-2L5	Libyan Arab Airlines	
5A–DIG	Boeing 727-2L5	Libyan Arab Airlines	
5A–DIH	Boeing 727-2L5	Libyan Arab Airlines	
5A–DII	Boeing 727-2L5	Libyan Arab Airlines	
5A–DGE	Canadair CL-44D-4	United African Airlines	
5A–DGJ	Canadair CL-44J	United African Airlines	

5B (Cyprus)

5B–DAG	BAC One-Eleven 537GF	Cyprus Airways	
5B–DAH	BAC One-Eleven 537GF	Cyprus Airways	
5B–DAJ	BAC One-Eleven 537GF	Cyprus Airways	
5B–DAK	Boeing 707-123B	Cyprus Airways	
5B–DAL	Boeing 707-123B	Cyprus Airways	
5B–DAM	Boeing 707-123B	Cyprus Airways	
5B–DAN	Canadair CL-44D-4	Cyprus Airways	
5B–DAO	Boeing 707-123B	Cyprus Airways	
5B–DAP	Boeing 707-123B	Cyprus Airways	

5H (Tanzania)

NOTE: Air Tanzania operates a Boeing 707-331 which retains its US registration N762TW.

5N (Nigeria)

5N–ABJ	Boeing 707-3F9C	Nigeria Airways	
5N–ABK	Boeing 707-3F9C	Nigeria Airways	
5N–ANN	Douglas DC-10-30	Nigeria Airways	
5N–ANO	Boeing 707-3F9C	Nigeria Airways	
5N–ANR	Douglas DC-10-30	Nigeria Airways	

5X (Uganda)

Notes	Reg.	Type	Owner or Operator
	5X–UAC	Boeing 707-351C	Uganda Airlines
	5X–UCF	Lockheed L382G Hercules	Uganda Airlines

5Y (Kenya)

Kenya Airways

Notes	Reg.	Type	Owner or Operator
	5Y–BBI	Boeing 707-351B	Kenya Airways
	5Y–BBJ	Boeing 707-351B	Kenya Airways
	5Y–BBK	Boeing 707-351B	Kenya Airways
	5Y–BBX	Boeing 720-047B	Kenya Airways

6Y (Jamaica)

Notes	Reg.	Type	Owner or Operator
	6Y–JGG	Douglas DC-8-61	Air Jamaica
	6Y–JGH	Douglas DC-8-61	Air Jamaica
	6Y–JII	Douglas DC-8-62	Air Jamaica

7T (Algeria)

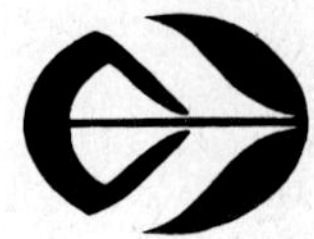

Notes	Reg.	Type	Owner or Operator
	7T–VEA	Boeing 727-2D6	Air Algerie *Tassili*
	7T–VEB	Boeing 727-2D6	Air Algerie *Hoggar*
	7T–VEC	Boeing 727-2D6	Air Algerie *Aures Nemencha*
	7T–VED	Boeing 727-2D6C	Air Algerie *Atlas Saharien*
	7T–VEE	Boeing 727-2D6	Air Algerie *Oasis*
	7T–VEF	Boeing 727-2D6	Air Algerie *Saoura*
	7T–VEG	Boeing 727-2D6	Air Algerie *Monts des Ouleds Neils*
	7T–VEH	Boeing 727-2D6	Air Algerie *Lalla Khadidja*
	7T–VEI	Boeing 727-2D6	Air Algerie *Djebel Amour*
	7T–VEJ	Boeing 727-2D6	Air Algerie *Chrea*
	7T–VEK	Boeing 727-2D6	Air Algerie *Edough*
	7T–VEL	Boeing 727-2D6	Air Algerie *Akfadou*
	7T–VEM	Boeing 727-2D6	Air Algerie *Mont du Ksall*
	7T–VEN	Boeing 727-2D6	Air Algerie *La Soummam*
	7T–VEO	Boeing 737-2D6	Air Algerie *Le Titteri*
	7T–VEP	Boeing 727-2D6	Air Algerie *Mont du Tessala*
	7T–VEQ	Boeing 737-2D6	Air Algerie *Le Zaccar*
	7T–VER	Boeing 737-2D6	Air Algerie *Le Souf*
	7T–VES	Boeing 737-2D6C	Air Algerie *Le Tadmaït*
	7T–VET	Boeing 727-2D6	Air Algerie
	7T–VEU	Boeing 727-2D6	Air Algerie
	7T–VEV	Boeing 727-2D6	Air Algerie
	7T–VEW	Boeing 727-2D6	Air Algerie

9G (Ghana)

Reg.	Type	Owner or Operator	Notes
9G–ABO	V.1102 VC10	Ghana Airways	
9G–ACE	Britannia 253F	Geminair	

9H (Malta)

9H–AAK	Boeing 720-047B	Air Malta	
9H–AAL	Boeing 720-047B	Air Malta	
9H–AAM	Boeing 720-040B	Air Malta	
9H–AAN	Boeing 720-040B	Air Malta	
9H–AAO	Boeing 720-047B	Air Malta	

NOTE: Air Malta also operates Boeing 737s on lease from Transavia Holland.

9J (Zambia)

9J–ADY	Boeing 707–349C	Zambia Airways	
9J–AEB	Boeing 707–351C	Zambia Airways	
9J–AEL	Boeing 707–338C	Zambia Airways	
9J–AEQ	Boeing 707-321C (Cargo)	Zambia Airways	

9K (Kuwait)

9K–ACJ	Boeing 707-369C	Kuwait Airways *Wara*	
9K–ACK	Boeing 707-369C	Kuwait Airways *Kadhma*	
9K–ACL	Boeing 707-369C	Kuwait Airways *Al-Jahra*	
9K–ACM	Boeing 707-369C	Kuwait Airways *Failaka*	
9K–ACN	Boeing 707-369C	Kuwait Airways *Burghan*	
9K–ACS	Boeing 707-321C	Kuwait Airways *Gharnada*	
9K–ACU	Boeing 707-321C	Kuwait Airways *Qurtoba*	
9K–ACX	Boeing 707-311C	Kuwait Airways	
9K–ADA	Boeing 747-269B	Kuwait Airways *Al Sabahiya*	
9K–ADB	Boeing 747-269B	Kuwait Airways *Al Jaberyia*	
9K–ADC	Boeing 747-269B	Kuwait Airways *Al Murbarakiya*	

9L (Sierra Leone)

Sierra Leone Airways' services between Freetown and London are operated with British-registered Boeing 707s, in conjunction with British Caledonian.

9M (Malaysia)

Notes	Reg.	Type	Owner or Operator
	9M–MAS	Douglas DC-10-30	Malaysian Airline System
	9M–MAT	Douglas DC-10-30	Malaysian Airline System
	9M–	Douglas DC-10-30	Malaysian Airline System

9Q (Zaïre)

	9Q–CLG	Douglas DC-8-63CF	Air Zaïre *Domaine de la Nsélé*
	9Q–CLH	Douglas DC-8-63AF	Air Zaïre *Ville de Kinshasa*
	9Q–CLI	Douglas DC-10-30	Air Zaïre *Mont Ngaliema*
	9Q–CLT	Douglas DC-10-30	Air Zaïre *Mont Ngafula*

9V (Singapore)

	9V–SQD	Boeing 747-212B	Singapore Airlines
	9V–SQE	Boeing 747-212B	Singapore Airlines
	9V–SQF	Boeing 747-212B	Singapore Airlines
	9V–SQG	Boeing 747-212B	Singapore Airlines
	9V–SQH	Boeing 747-212B	Singapore Airlines
	9V–SQI	Boeing 747-212B	Singapore Airlines
	9V–SQJ	Boeing 747-212B	Singapore Airlines
	9V–SQK	Boeing 747-212B	Singapore Airlines
	9V–SQL	Boeing 747-212B	Singapore Airlines
	9V–SQM	Boeinf 747-212B	Singapore Airlines
	9V–SQN	Boeing 747-212B	Singapore Airlines
	9V–SQO	Boeing 747-212B	Singapore Airlines
	9V–SQP	Boeing 747-212B	Singapore Airlines
	9V–SQQ	Boeing 747-212B	Singapore Airlines
	9V–SQR	Boeing 747-212B	Singapore Airlines
	9V–SQS	Boeing 747-212B	Singapore Airlines

9Y (Trinidad and Tobago)

	9Y–TED	Boeing 707-351C	B.W.I.A. *Scarlet Ibis*
	9Y–TEE	Boeing 707-351C	B.W.I.A. *Humming Bird*
	9Y–TEJ	Boeing 707-351C	B.W.I.A. *Bird of Paradise*
	9Y–TEK	Boeing 707-351C	B.W.I.A. *Toucan*
	9Y–TEX	Boeing 707-321B	B.W.I.A.
	9Y–TEZ	Boeing 707-321B	B.W.I.A.
	9Y–TGN	L.1011 TriStar 500	B.W.I.A.
	9Y–	L.1011 TriStar 500	B.W.I.A.

BRITISH AIRCRAFT PRESERVATION COUNCIL REGISTER

The British Aircraft Preservation Council was formed in 1967 to co-ordinate the works of all bodies involved in the preservation, restoration and display of historical aircraft. Membership covers the whole spectrum of national, Service, commercial and voluntary groups, and meetings are held regularly at the bases of member organisations. The Council is able to provide a means of communication, helping to resolve any misunderstandings or duplication of effort. Every effort is taken to encourage the raising of standards of both organisation and technical capacity amongst the member groups to the benefit of everyone interested in aviation. To assist historians, the B.A.P.C. register has been set up and provides an identity for those aircraft which do not qualify for a Service serial or inclusion in the UK Civil Register.

Aircraft on the current BAPC Register are as follows:

Reg.	*Type*	*Owner or Operator*	*Notes*
1	Roe Triplane Type IV (replica)	The Shuttleworth Trust, Old Warden, Beds.	
2	Bristol Boxkite (replica)	The Shuttleworth Trust	
3	Blériot XI	The Shuttleworth Trust	
4	Deperdussin monoplane	The Shuttleworth Trust	
5	Blackburn monoplane	The Shuttleworth Trust	
6	Roe Triplane Type IV (replica)	Historic Aircraft Museum, Southend	
7	Southampton University MPA	The Shuttleworth Trust	
8	Dixon ornithopter	The Shuttleworth Trust	
9	Humber Monoplane (replica)	Midland Air Museum, Coventry	
10	Hafner R.II Revoplane	British Rotorcraft Museum	
11	English Electric Wren	The Shuttleworth Trust	
12	Mignet HM.14 Pou-du-Ciel	Museum of Flight, E. Fortune	
13	Mignet HM.14 Pou-du-Ciel	The Aeroplane Collection Ltd.	
14	Addyman standard training glider	N. H. Ponsford	
15	Addyman standard training glider	The Aeroplane Collection Ltd.	
16	Addyman ultra-light aircraft	N. H. Ponsford	
17	Woodhams Sprite	The Aeroplane Collection Ltd.	
18	Killick MP Gyroplane	The Aeroplane Collection Ltd.	
19	Bristol F.2b	Anne Lindsay	
20	Lee-Richards annular biplane (replica)	Newark Air Museum	
22	Mignet HM.14 Pou-du-Ciel (G–AEOF)	Newark Air Museum	
25	Nyborg TGN-111 glider	Midland Air Museum	
26	Auster AOP.9	S. Wales Aircraft Preservation Soc.	
27	Mignet HM.14 Pou-du-Ciel	M. J. Abbey	
28	Wright Flyer (replica)	RAF Museum, Cardington	
29	Mignet HM.14 Pou-du-Ciel (G–ADRY)	J. J. Penney	
31	Slingsby T.7 Tutor	S. Wales Aircraft Preservation Soc.	
32	Crossley Tom Thumb	Midland Air Museum	
33	DFS.108-49 Grunau Baby 116	Russavia Collection, Duxford	
34	DFS.108-49 Grunau Baby 116	D. Elsdon	
35	EoN primary glider	The Shuttleworth Trust	
36	FZG-76 (V.I) (replica)	The Shuttleworth Trust	
37	Blake Bluetit	The Shuttleworth Trust	
38	Bristol Scout replica (A1742)	RAF St. Athan	
40	Bristol Boxkite (replica)	Bristol City Museum	
41	B.E.2C (replica) (6232)	RAF St. Athan	
42	Avro 504 (replica)	RAF St. Athan	
43	Mignet HM.14 Pou-du-Ciel	Lincolnshire Aviation Museum	
44	Miles Magister (L6906)	G. H. R. Johnson	
45	Pilcher Hawk (replica)	Lord Braye	
46	Mignet HM.14 Pou-du-Ciel	Alan McKechnie Racing Ltd.	
47	Watkins monoplane	C. M. Watkins	
48	Pilcher Hawk (replica)	Glasgow Museum of Transport	
49	Pilcher Hawk	Royal Scottish Museum	
50	Roe Triplane Type I	Science Museum	
51	Vickers Vimy IV	Science Museum	
52	Lilienthal glider	Science Museum	
53	Wright Flyer (replica)	Science Museum	
54	JAP-Harding monoplane	Science Museum	
55	Levavasseur Antoinette VII	Science Museum	
56	Fokker E.III	Science Museum	
57	Pilcher Hawk (replica)	Science Museum	
58	Yokosuka MXY–7 Ohka II	Science Museum	

Notes	Reg.	Type	Owner or Operator
	59	Sopwith Camel (replica) (D3419)	RAF St. Athan
	60	Murray M.1 helicopter	The Aeroplane Collection Ltd.
	61	Stewart man-powered ornithopter	Lincolnshire Aviation Museum
	62	Cody Biplane	Science Museum
	63	Hurricane (replica) (L1592)	Torbay Aircraft Museum
	64	Hurricane (replica)	Historic Aircraft Museum, Southend
	65	Spitfire (replica)	Torbay Aircraft Museum
	66	Bf 109 (replica)	Historic Aircraft Museum, Southend
	67	Bf 109 (replica)	Midland Air Museum
	68	Hurricane (replica)	Midland Air Museum
	69	Spitfire (replica)	Midland Air Museum
	70	Auster AOP.5 (TJ472)	Aircraft Preservation Soc. of Scotland
	71	Spitfire (replica)	Norfolk & Suffolk Aviation Museum
	72	Hurricane (replica)	Midland Air Museum
	73	Hurricane (replica)	J. P. Berkeley
	74	Bf 109 (replica)	Torbay Aircraft Museum
	75	Mignet HM.14 Pou-du-Ciel	Nigel Ponsford
	76	Mignet HM.14 Pou-du-Ciel (G–AFFI)	Nostell Aviation Museum
	77	Mignet HM.14 Pou-du-Ciel	Skyfame Collection
	78	Hawker Hind (Afghan) (K5457)	The Shuttleworth Trust
	79	Fiat G.46–4	Historic Aircraft Museum, Southend
	80	Airspeed Horsa	Museum of Army Flying
	81	Hawkridge Dagling	Russavia Collection, Duxford
	82	Hawker Hind (Afghan)	RAF Museum
	83	Kawasaki Ki-100-IB	Aerospace Museum, Cosford
	84	Nakajima Ki-46 (Dinah III)	RAF St. Athan
	85	Weir W-2 autogyro	Museum of Flight, E. Fortune
	86	de Havilland Tiger Moth (replica)	Yorkshire Aircraft Preservation Soc.
	87	Bristol Babe (replica)	Yorkshire Aircraft Preservation Soc.
	88	Fokker Dr I (replica)	Fleet Air Arm Museum
	89	Cayley glider (replica)	RAF Museum
	90	Colditz Cock (replica)	Torbay Aircraft Museum
	91	Fieseler Fi 103/FZG.76 (V.1)	Lashenden Air Warfare Museum
	92	Fieseler Fi 103/FZG.76 (V.1)	RAF Museum
	93	Fieseler Fi 103/FZG.76 (V.1)	RAF St. Athan
	94	Fieseler Fi 103/FZG.76 (V.1)	Cosford Aerospace Museum
	95	Gizmer autogyro	N.E. Aircraft Museum
	96	Brown helicopter	N.E. Aircraft Museum
	97	Luton L.A.4a Minor	N.E. Aircraft Museum
	98	Yokosuka MXY-7 Ohka II	RAF Museum, Henlow
	99	Yokosuka MXY-7 Ohka II	Aerospace Museum, Cosford
	100	Clarke glider	Science Museum
	101	Mignet HM.14 Pou-du-Ciel	Lincolnshire Aviation Museum
	102	Mignet HM.14 Pou-du-Ciel	W. Sneesby
	103	Pilcher glider (replica)	Personal Plane Services Ltd.
	104	Blériot XI (replica)	RAF St. Athan
	105	Blériot XI (replica)	Mosquito Aircraft Museum
	106	Blériot XI	RAF Museum
	107	Blériot XXVII	RAF Museum, Cardington
	108	Fairey Swordfish	RAF Museum, Henlow
	109	Slingsby Kirby Cadet	RAF Museum, Henlow
	110	Fokker D.VII replica (static) (5125)	Leisure Sport Ltd.
	111	Sopwith Triplane replica (static) (N5492)	Leisure Sport Ltd.
	112	D.H.2 replica (static) (5964)	Leisure Sport Ltd.
	113	S.E.5A replica (static) (B4863)	Leisure Sport Ltd.
	114	Vickers Type 60 Viking (static)	Leisure Sport Ltd.
	115	Mignet HM.14 Pou-du-Ciel	Essex Aviation Group
	116	Santos-Dumont Demoiselle (replica)	Cornwall Air Park
	117	B.E.2C (replica)	British Broadcasting Corp.
	118	Albatross D.V. (replica)	British Broadcasting Corp.
	119	Benson B.7	N.E. Aircraft Museum
	120	Mignet HM.14 Pou-du-Ciel	1324 (Brough) Squadron ATC
	121	Mignet HM.14 Pou-du-Ciel (G–AEKR)	Nostell Aviation Museum
	122	Avro 504 (replica)	British Broadcasting Corp.
	123	Vickers FB.5 Gunbus (replica)	British Broadcasting Corp.
	124	Lilienthal Glider Type XI (replica)	Science Museum

Reg.	Type	Owner or Operator	Notes
125	Clay Cherub	Midland Air Museum	
126	D.31 Turbulent (static)	Midland Air Museum	
127	Halton Jupiter	Shuttleworth Trust	
128	Watkinson Cyclogyroplane Mk. IV	British Rotocraft Museum	
129	Blackburn 1911 Monoplane (replica)	Cornwall Aero Park	
130	Blackburn 1912 Monoplane (replica)	Cornwall Aero Park	
131	Pilcher Hawk (replica)	C. Paton	
132	Blériot XI	Aerospace Museum, Cosford	
133	Fokker Dr I (replica)	Torbay Aircraft Museum	
134	Pitts S-2A static (G–RKSF)	Rothmans of Pall Mall	
135	Bristol M.IC (replica) (C4912)	Leisure Sport Ltd.	
136	Deperdussin Seaplane (replica)	Leisure Sport Ltd.	
137	Sopwith Baby Floatplane (replica)	Leisure Sport Ltd.	
138	Hansa Brandenburg W.29 Floatplane (replica) (2292)	Leisure Sport Ltd.	
139	Fokker Dr I (replica) (DR I/17)	Leisure Sport Ltd.	
140	Curtiss R3C-2 Floatplane (replica)	Leisure Sport Ltd.	
141	Macchi M.39 Floatplane (replica)	Leisure Sport Ltd.	
142	SE-5A (replica)	Cornwall Air Park	
143	Paxton MPA	R. A. Paxton	
144	Weybridge Mercury	Weybridge MPAG	
145	Oliver MPA	D. Oliver	
146	Pedal Aeronauts Toucan MPA	Shuttleworth Trust	
147	Benson B.7	Norfolk & Suffolk Aviation Museum	
148	Hawker Fury II (replica) (K7271)	Aerospace Museum, Cosford	
149	Short S.27 (replica)	Fleet Air Arm Museum	
150	SEPECAT Jaguar GR.2 (replica) (XX732)	RAF Exhibition Flight	
151	SEPECAT Jaguar GR.1 (replica) (XX824)	RAF Exhibition Flight	
152	BAe Hawk T.1 (replica) (XX163)	RAF Exhibition Flight	

NOTE: Registrations/Serials carried are mostly false identities. MPA = Man Powered Aircraft.

TRANSPORT TRACKERS

We regret that, due to lack of sufficient interest, this scheme has been cancelled.

ADDENDA

New registrations

Notes	Reg.	Type	Owner or Operator
	G-BILM	—	—
	G-BILN	—	—
	G-BILO	—	—
	G-BILP	—	—
	G-BILR	—	—
	G-BILS	—	—
	G-BILT	Cessna F.172P	Denham Flying Training School Ltd.
	G-BILU	—	—
	G-BILV	Cessna FA.152	Denham Flying Training School Ltd.
	G-BILW	Colt 60 balloon	Hot Air Balloon Co. Ltd.
	G-BILX	Colt 31A balloon	Hot Air Balloon Co. Ltd.
	G-BILY	Beech 200 Super King Air	Eagle Aircraft Services Ltd.
	G-BILZ	Taylor JT.1 Monoplane	G. Beaumont
	G-BIMA	A.300B4-203 Airbus	Laker Airways
	G-BIMB	A.300B4-203 Airbus	Laker Airways
	G-BIMC	A.300B4-203 Airbus	Laker Airways
	G-BIMD	A.300B4-203 Airbus	Laker Airways
	G-BIME	A.300B4-203 Airbus	Laker Airways
	G-BIMF	A.300B4-203 Airbus	Laker Airways
	G-BIMG	—	Laker Airways
	G-BIMH	—	Laker Airways
	G-BIMI	—	Laker Airways
	G-BIMJ	—	Laker Airways
	G-BIMK	Tiger T.2000 Srs. 1 balloon	M. K. Baron
	G-BIML	Turner Super T.40A	R. T. Callow
	G-BIMM	PA-18 Super Cub 135	D. S. & I. M. Morgan
	G-BIMN	Steen Skybolt	C. R. Williamson
	G-BIMO	—	—
	G-BIMP	—	—
	G-BIMR	—	—
	G-BIMS	—	—
	G-BIMT	Cessna FA.152	Denham Flying Training School Ltd.
	G-BIMU	Sikorsky S-61N	British Caledonian Helicopters Ltd.
	G-BIMV	—	—
	G-BIMW	—	—
	G-BIMX	Rutan Vari-Eze	A. S. Knowles
	G-BIMY	—	—
	G-BIMZ	—	—
	G-BINA	Saffery S.9 balloon	A. P. Bashford
	G-BINB	WMB.2A Windtracker balloon	S. R. Woolfries
	G-BINC	Tour de Calais balloon	Cupro Sapphire Ltd.
	G-BIND	—	—
	G-BINE	Scruggs BL.2A Wunda balloon	M. Gilbey
	G-BINF	—	—
	G-BING	—	
	G-BINH	—	—
	G-BINI	—	—
	G-BINJ	—	—
	G-BINK	—	—
	G-BINL	—	—
	G-BINM	—	—
	G-BINN	—	—
	G-BINO	Evans VP-1	J. I. Visser
	G-BINP	—	—
	G-BINR	—	—
	G-BINS	Unicorn UE.2A balloon	I. Chadwick
	G-BINT	Unicorn UE.1A balloon	I. Chadwick
	G-BINU	—	—
	G-BINV	—	—
	G-BINW	—	—
	G-BINX	—	—
	G-BINY	—	—
	G-BINZ	—	—
	G-BIOU	Jodel D.117A	M.S. Printing & Graphics Machinery Ltd.
	G-BIRD	Pitts S-1C Special	R. N. York
	G-BITC	Aerostar 601PE	Bell Industrial Trust Co. Ltd.

Out-of-sequence registrations

Reg.	Type	Owner or Operator	Notes
G–BLCG	SOCATA TB.10 Tobago	A. M. & H. T. Vaughan-Read (G–BHES)	
G–BMAL	Sikorsky S-76A	Management Aviation Ltd.	
G–BNSH	Sikorsky S-76A	Management Aviation Ltd.	
G–BSHL	H.S.125 Srs. 600B	S.H. Services Ltd. (G–BBMD)	
G–BSPE	Cessna F.172P	A & G Aviation Ltd.	
G–CEGA	PA-34-200T-II Seneca	Spooner Aviation Ltd.	
G–CETA	Cessna E.310Q	Ceta Video Ltd. (G–BBIM)	
G–CICI	Cameron R-15 balloon	Ballooning Endeavours Ltd.	
G–CJBC	PA-28 Cherokee 180	I. McDonald	
G–COMM	PA-23 Aztec 250	Commair Aviation Ltd. (G–AZMG)	
G–CULT	Cameron D-38 airship	The Colt Car Co. Ltd.	
G–FLIP	Cessna FA.152	Citation Flying Services Ltd.	
G–FUZZ	PA-19 Super Cub 95	G. W. Chine	
G–GOSS	Jodel DR.221	M. I. Goss	
G–HBUS	Bell 206L LongRanger	Willowbrook International Ltd.	
G–HFCI	Cessna F.150L	Citation Flying Services Ltd.	
G–HWBK	Agusta A.109A	Willowbrook International Ltd.	
G–IMBE	PA-31-300 Navajo	T. Beattie Edwards & Co. Ltd. (G–BXYB/G–AXYB)	
G–JONE	Cessna 172M	A. G. Jones	
G–JOSI	CP.1310-C3 Super Emeraude	J. G. O'Donnell (G–BCHP)	
G–JSGM	Cessna P210N	Air Service Training Ltd.	
G–JTCA	PA-23 Aztec 250	J. D. Tighe & Co. Ltd. (G–BBCU)	
G–JWWT	SOCATA TB.10 Tobago	McNultys World Travel Ltd.	
G–LADE	PA-32-300 Cherokee Six	Appleby Glade Ltd.	
G–MIST	Cessna T.210	Rogers Aviation Sales Ltd. (G–AYGM)	
G–NELL	R. Commander 112A	Arcdeal Ltd.	
G–OODI	Pitts S-1D Special	R. N. Goode (G–BBBU)	
G–PORR	AS.350B Ecureuil	The Colt Car Co. Ltd.	
G–RHFI	Alexander Todd Skybolt	R.H.F. (Estates) Ltd.	
G–SAHI	SAH-1	Trago Mills Ltd.	
G–SCEA	CAARP CAP-10B	C. J. Else & Co. Ltd.	
G–SNIP	Cessna F.172H	H. V. Jones (G–AXSI)	
G–STYR	Beech F90 King Air	The Colt Car Co. Ltd. (G–BHUT)	
G–SVHA	Partenavia P.68B	N.P.D. Aviation Ltd.	
G–TEAM	Cessna 414A	Northair Aviation Ltd. (G–BHJT)	
G–UBHL	Beech 200 Super King Air	United Biscuits (UK) Ltd.	
G–UPVC	Agusta A.109A	Anglian Double Glazing Co.	
G–VAGA	PA-15 Vagabond	R. Harris & I. S. Hodge	
G–WARD	Taylor JT.1 Monoplane	G. & G. D. Ward	

NOTE: Due to the re-registrations included in these late additions, the following entries should be cancelled— G–AXSI, G–AYGM, G–AZMG, G–BBBU, G–BBCU, G–BBIM, G–BBMD, G–BCHP, G–BHES, G–BHJT, G–BHUT, G–BVET, G–BXYB and G–KERK.

Overseas airline addenda

Reg.	Type	Owner or Operator	Notes
B–2442	Boeing 747SP-J6	CAAC	
B–2444	Boeing 747SP-J6	CAAC	
CU–T1208	Ilyushin IL-62	Cubana *Capt. Wifredo Perez*	
CU–T1209	Ilyushin IL-62	Cubana	
CU–T1215	Ilyushin IL-62	Cubana	
CU–T1216	Ilyushin IL-62	Cubana	
CU–T1217	Ilyushin IL-62	Cubana	
F–BNOH	S.E.210 Caravelle 12	Air Inter	
F–BOJF	Boeing 727-228	Air France	
F–BVFF	Concorde 101	Air France	
F–GCIM	FH.227B Friendship	T.A.T.	
F–GCIN	FH.227B Friendship	T.A.T.	
F–GCPT	FH.227B Friendship	T.A.T.	
N744PR	Boeing 747-2F6B	Philippine Airlines	
N1304E	Boeing 747SP-J6	CAAC	
OD–AGZ	Boeing 707–327C	Trans Mediterranean Airways	
VP–WKW	Boeing 707-330B	Air Zimbabwe	
4W–ACF	Boeing 727-2N8	Yemen Airways	
4W–ACG	Boeing 727-2N8	Yemen Airways	
4W–ACH	Boeing 727-2N8	Yemen Airways	
4W–ACI	Boeing 727-2N8	Yemen Airways	
4W–ACJ	Boeing 727-2N8	Yemen Airways	

NOTE: The following Flying Tiger Boeing 747s, N702SW, N703SW, N705SW and N706SW have been re-registered N812FT, N813FT, N815FT and N816FT respectively.

Future Allocations Log

The grid provides the facility to record future in-sequence registrations as they are issued or seen. To trace a particular code, refer to the left hand column which contains the three letters following the G prefix. The final letter can be found by reading across the columns headed A to Z. For example, the box for G–BITD is located five rows down (BIT) and then four across to the D column.

G-	A	B	C	D	E	F	G	H	I	J	K	L	M	N	O	P	R	S	T	U	V	W	X	Y	Z
BIO																									
BIP																									
BIR																									
BIS																									
BIT																									
BIU																									
BIV																									
BIW																									
BIX																									
BIY																									
BIZ																									
BJA																									
BJB																									
BJC																									
BJD																									
BJE																									
BJF																									
BJG																									
BJH																									
BJI																									
BJJ																									
BJK																									
BJL																									
BJM																									
BJN																									
BJO																									
BJP																									
BJR																									
BJS																									
BJT																									
BJU																									
BJV																									
BJW																									
	A	B	C	D	E	F	G	H	I	J	K	L	M	N	O	P	R	S	T	U	V	W	X	Y	Z

Credit: *Wal Gandy*